Study Guide

Quantitative Methods for Business
TENTH EDITION

David R. Anderson
University of Cincinnati

Dennis J. Sweeney
University of Cincinnati

Thomas A. Williams
Rochester Institute of Technology

Prepared by

John S. Loucks
St. Edward's University

SOUTH-WESTERN
CENGAGE Learning

Australia • Brazil • Japan • Korea • Mexico • Singapore • Spain • United Kingdom • United States

Study Guide: Quantitative Methods for Business, Tenth Edition
David R. Anderson, Dennis J. Sweeney, and Thomas A. Williams

VP/Editorial Director: Jack W. Calhoun

Editor-in-Chief: Alex von Rosenberg

Senior Acquisitions Editor: Charles McCormick

Senior Developmental Editor: Alice C. Denny

Senior Marketing Manager: Larry Quails

Senior Production Project Manager: Deanna Quinn

Manager of Technology, Editorial: Vicky True

Technology Project Editor: Kelly Reid

Web Coordinator: Scott Cook

Senior Manufacturing Coordinator: Diane Lohman

Art Director: Stacy Jenkins Shirley

Internal Designer: Michael Stratton/Chris Miller

Cover Designer: Paul Neff

Cover Image(s): © Getty Images, Inc.

© 2006 South-Western, Cengage Learning

ALL RIGHTS RESERVED. No part of this work covered by the copyright herein may be reproduced, transmitted, stored or used in any form or by any means graphic, electronic, or mechanical, including but not limited to photocopying, recording, scanning, digitizing, taping, Web distribution, information networks, or information storage and retrieval systems, except as permitted under Section 107 or 108 of the 1976 United States Copyright Act, without the prior written permission of the publisher.

> For product information and technology assistance, contact us at
> **Cengage Learning Customer & Sales Support, 1-800-354-9706**
> For permission to use material from this text or product,
> submit all requests online at **www.cengage.com/permissions**
> Further permissions questions can be emailed to
> **permissionrequest@cengage.com**

ISBN-13: 978-0-324-31266-9

ISBN-10: 0-324-31266-0

South-Western
5191 Natorp Boulevard
Mason, OH 45040
USA

Cengage Learning is a leading provider of customized learning solutions with office locations around the globe, including Singapore, the United Kingdom, Australia, Mexico, Brazil, and Japan. Locate your local office at **international.cengage.com/region**

Cengage Learning products are represented in Canada by Nelson Education, Ltd.

To learn more about South-Western, visit **www.cengage.com/southwestern**

Purchase any of our products at your local college store or at our preferred online store **www.ichapters.com**

Printed in the United States of America
2 3 4 5 6   15 14 13 12 11

FD332

Contents

Preface	An Introduction to the Use of the Study Guide	1
Chapter 1	Introduction	3
Chapter 2	Introduction to Probability	17
Chapter 3	Probability Distributions	35
Chapter 4	Decision Analysis	51
Chapter 5	Utility and Game Theory	83
Chapter 6	Forecasting	107
Chapter 7	Introduction to Linear Programming	129
Chapter 8	Linear Programming: Sensitivity Analysis and Interpretation of Solution	153
Chapter 9	Linear Programming Applications	179
Chapter 10	Transportation, Assignment, and Transshipment Problems	213
Chapter 11	Integer Linear Programming	237
Chapter 12	Project Scheduling: PERT/CPM	257
Chapter 13	Inventory Models	283
Chapter 14	Waiting Line Models	307
Chapter 15	Simulation	331
Chapter 16	Markov Processes	359
Chapter 17	Multicriteria Decisions	377
Answers		411
Appendices		443

CHAPTER 1

Introduction

4 CHAPTER 1

KEY CONCEPTS

CONCEPT	ILLUSTRATED PROBLEMS	ANSWERED PROBLEMS
Quantitative Approach to Decision Making	1,2	6,7
Mathematical Models Development	3,4	8,9,10,11,12
Cost, Volume, Profit Analysis	5	13
Marginal Analysis	5	13
Breakeven Analysis	5	13

REVIEW

1. The <u>quantitative approach to decision making</u> based on the <u>scientific method</u> of problem solving. This body of knowledge is referred to as <u>management science</u>, <u>operations research</u> or <u>decision science</u>. It had its early roots in World War II and is flourishing in business and industry with the aid of computers in general and the microcomputer in particular.

2. <u>Problem solving</u> is a process designed to better a current state of affairs. It consists of two phases: (1) decision making; and (2) implementation and evaluation.

3. The <u>decision making process</u> involves: (1) structuring the problem; and (2) analyzing the problem.

4. <u>Structuring the problem</u> includes: (1) defining the problem; (2) identifying the alternatives; and (3) choosing the criteria (single or multiple) to be used to evaluate the alternatives.

5. <u>Analyzing the problem</u> consists of: (1) qualitatively and quantitatively evaluating the alternatives; and (2) making a decision recommendation (choosing an alternative).

6. Evaluating alternatives is often accomplished by experimenting with a <u>model</u>. A model is a representation of a real object or situation. Generally, experimenting with a model is <u>less costly</u>, requires <u>less time</u>, and involves <u>less risk</u>.

7. Three forms of models are iconic, analog, and mathematical. <u>Iconic models</u> are physical replicas (scalar representations) of real objects. <u>Analog models</u> are physical in form, but do not physically resemble the object being modeled.

8. <u>Mathematical models</u> (also called symbolic models), represent real world problems through a system of mathematical formulas and expressions. They are idealizations of real-life problems based on key assumptions, estimates, and/or statistical analyses.

9. After the problem has been structured, the steps in the <u>quantitative analysis</u> are: (1) mathematical modeling; (2) data preparation; (3) model solution (and refinement); and (4) report generation.

10. Mathematical models relate <u>decision variables</u> (or <u>controllable inputs</u>) with fixed or variable parameters (or <u>uncontrollable inputs</u>). Frequently mathematical models seek to maximize or minimize some <u>objective function</u> subject to <u>constraints</u>.

11. If any of the uncontrollable inputs is subject to variation the model is said to be <u>stochastic</u>. Otherwise the model is said to be <u>deterministic</u>. Generally, stochastic models are more realistic than deterministic models, but they are more difficult to analyze.

12. The values of the decision variables that provide the mathematically-best output are referred to as the <u>optimal solution</u> for the model.

13. If a particular decision alternative does not satisfy one or more of the model constraints, the decision alternative is rejected as being <u>infeasible</u>. If all constraints are satisfied, the decision alternative is <u>feasible</u> and a candidate for the "best" or recommended decision.

14. <u>Cost/benefit considerations</u> must be made in selecting an appropriate mathematical model. Frequently a less complicated (and perhaps less precise) model is more appropriate than a more complex and accurate one due to cost and ease of solution considerations.

15. Quantitative analysis should be used as <u>one of many input factors</u> for managerial decision making. It is not a replacement for human decision making.

16. Primary reasons for the use of quantitative analysis are: (1) the <u>problem is complex</u>; (2) the <u>problem is important</u> and/or the decision must be thoroughly justified; (3) the <u>problem is new</u> and there is little or no previous experience to rely on; (4) the <u>problem is repetitive and time-consuming</u> (with varying uncontrollable input values); and (5) <u>what-if questions</u> need to be answered in an efficient manner.

17. <u>Break-even analysis</u> uses a basic mathematical model of the relationship between a volume variable and cost, revenue, or profit. The <u>break-even point</u> is the volume that results in total revenue equaling total cost.

18. <u>Fixed cost</u> is the portion of the total cost that does not depend on the volume variable; this cost remains regardless of the value of the volume variable. <u>Variable cost</u> is the portion of the total cost that is dependent on and varies with the volume variable.

19. <u>Marginal cost</u> is the rate of change of the total cost with respect to the volume variable. <u>Marginal revenue</u> is the rate of change of the total revenue with respect to the volume variable.

20. Linear programming, integer programming, network models, and simulation are among the <u>most frequently used quantitative techniques</u> in business.

ILLUSTRATED PROBLEMS

PROBLEM 1

Consider a construction company building a 250-unit apartment complex. The project consists of hundreds of activities involving excavating, framing, wiring, plastering, painting, landscaping, and more. Some of the activities must be done sequentially and others can be done simultaneously. Also, some of the activities can be completed faster than normal by purchasing additional resources (workers, equipment, etc.).

a) How could a quantitative approach to decision making be used to solve this problem?

b) What would be the uncontrollable inputs?

c) What would be the decision variables of the mathematical model? the objective function? the constraints?

d) Is the model deterministic or stochastic?

e) Suggest assumptions that could be made to simplify the model.

SOLUTION 1

a) A quantitative approach to decision making can provide a structured way to determine the minimum project completion time based on the activities' normal times and then based on the activities' expedited (reduced) times.

b) Normal and expedited activity completion times; activity expediting costs; funds available for expediting; precedence relationships of the activities.

c) Decision variables--which activities to expedite and by how much, and when to start each activity; objective function--minimize project completion time; constraints--do not violate any activity precedence relationships and do not expedite in excess of the funds available.

d) Stochastic--activity completion times, both normal and expedited, are uncertain and subject to variation; activity expediting costs are uncertain; the number of activities and their precedence relationships might change before the project is completed due to a project design change.

e) Make the model deterministic by assuming normal and expedited activity times are known with certainty and are constant. The same assumption might be made about the other stochastic, uncontrollable inputs.

PROBLEM 2

Consider a department store that must make weekly shipments of a certain product from two different warehouses to four different stores.

a) How could a quantitative approach to decision making be used to solve this problem?

b) What would be the uncontrollable inputs for which data must be gathered?

c) What would be the decision variables of the mathematical model? the objective function? the constraints?

d) Is the model deterministic or stochastic?

e) Suggest assumptions that could be made to simplify the model.

SOLUTION 2

a) A quantitative approach to decision making can provide a systematic way to determine a minimum shipping cost from the warehouses to the stores.

b) Fixed costs and variable shipping costs; the demand each week at each store; the supplies each week at each warehouse.

c) Decision variables--how much to ship from each warehouse to each store; objective function--minimize total shipping costs; constraints--meet the demand at the stores without exceeding the supplies at the warehouses.

d) Stochastic--weekly demands fluctuate as do weekly supplies; transportation costs could vary depending upon the amount shipped, other goods sent with a shipment, etc.

e) Make the model deterministic by assuming fixed shipping costs per item, demand is constant at each store each week, and weekly supplies in the warehouses are constant.

PROBLEM 3

An auctioneer has developed a simple mathematical model for deciding the starting bid he will require when auctioning a used automobile. Essentially, he sets the starting bid at seventy percent of what he predicts the final winning bid will (or should) be. He predicts the winning bid by starting with the car's original selling price and making two deductions, one based on the car's age and the other based on the car's mileage. The age deduction is $800 per year and the mileage deduction is $.025 per mile.

a) Develop the mathematical model that will give the starting bid (B) for a car in terms of the car's original price (P), current age (A) and mileage (M).

CHAPTER 1

b) Suppose a four-year old car with 60,000 miles on the odometer is up for auction. If its original price was $12,500, what starting bid should the auctioneer require?

c) The model is based on what assumptions?

SOLUTION 3

a) The expected winning bid can be expressed as:
$$P - 800(A) - .025(M)$$

The entire model is:
$B = .7(\text{expected winning bid})$ or
$B = .7(P - 800(A) - .025(M))$ or
$B = .7(P) - 560(A) - .0175(M)$

b) $B = .7(12,500) - 560(4) - .0175(60,000) = \$5460$.

c) The model assumes that the only factors influencing the value of a used car are the original price, age, and mileage (not condition, rarity, or other factors). Also, it is assumed that age and mileage devalue a car in a linear manner and without limit. (Note, the starting bid for a very old car might be negative.)

PROBLEM 4

A firm manufactures two products made from steel and has just received this month's allocation of b pounds of steel. It takes a_1 pounds of steel to make a unit of product 1 and it takes a_2 pounds of steel to make a unit of product 2. Let x_1 and x_2 denote this month's production level of product 1 and product 2 respectively.

Denote by p_1 and p_2 the unit profits for products 1 and 2, respectively. The manufacturer has a contract calling for at least m units of product 1 this month. The firm's facilities are such that at most u units of product 2 may be produced monthly.

a) Write a mathematical model for this problem.

b) Suppose $b = 2000$, $a_1 = 2$, $a_2 = 3$, $m = 60$, $u = 720$, $p_1 = 100$, $p_2 = 200$. Rewrite the model with these specific values for the uncontrollable inputs.

c) The optimal solution to (b) is $x_1 = 60$ and $x_2 = 626\ 2/3$. If the product were engines, explain why this is not a true optimal solution for the "real-life" problem.

SOLUTION 4

a) The total monthly profit
= (profit per unit of product 1) x (monthly production of product 1)
+ (profit per unit of product 2) x (monthly production of product 2)
= $p_1 x_1 + p_2 x_2$.

The total amount of steel used during monthly production
  = (steel per unit of product 1) x (monthly production of product 1)
  + (steel per unit of product 2) x (monthly production of product 2)
  = $a_1x_1 + a_2x_2$.

This quantity must be less than or equal to the allocated b pounds of steel:
$$a_1x_1 + a_2x_2 \leq b.$$

The monthly production level of product 1 must be greater than or equal to m:
$$x_1 \geq m.$$

The monthly production level of product 2 must be less than or equal to u:
$$x_2 \leq u.$$

The production level for product 2 cannot be negative:
$$x_2 \geq 0.$$

Thus, the model is:

MAXIMIZE $p_1x_1 + p_2x_2$

s.t.
$$a_1x_1 + a_2x_2 \leq b$$
$$x_1 \geq m$$
$$x_2 \leq u$$
$$x_2 \geq 0$$

b) Substituting, the model is:

MAXIMIZE $100x_1 + 200x_2$

s.t.
$$2x_1 + 3x_2 \leq 2000$$
$$x_1 \geq 60$$
$$x_2 \leq 720$$
$$x_2 \geq 0$$

c) One cannot produce and sell 2/3 of an engine. Thus the problem is further restricted by the fact that both x_1 and x_2 must be integers. They could remain fractions if it is assumed these fractions are work in progress to be completed the next month.

PROBLEM 5

Ponderosa Development Corporation (PDC) is a small real estate developer operating in the Rivertree Valley. It has seven permanent employees whose monthly salaries are given in the table below:

CHAPTER 1

Employee	Monthly Salary
President	$10,000
VP, Development	6,000
VP, Marketing	4,500
Project Manager	5,500
Controller	4,000
Office Manager	3,000
Receptionist	2,000

PDC leases a building for $2,000 per month. The cost of supplies, utilities, and leased equipment runs another $3,000 per month. PDC builds only one style house in the valley. Land for each house costs $55,000 and lumber, supplies, etc. run another $28,000 per house. Total labor costs are figured at $20,000 per house. The one sales representative of PDC is paid a commission of $2,000 on the sale of each house. The selling price of the house is $115,000.

a) Identify all costs and denote the marginal cost and marginal revenue for each house.

b) Write the monthly cost function $c(x)$, revenue function $r(x)$, and profit function $p(x)$.

c) What is the monthly profit if 12 houses per month are built and sold? Solve algebraically and using a spreadsheet.

d) What is the breakeven point for monthly sales of the houses? Solve algebraically and using a spreadsheet.

SOLUTION 5

a) The monthly salaries total $35,000 and monthly office lease and supply costs total another $5,000. This $40,000 is a monthly fixed cost. The total cost of land, material, labor, and sales commission per house, $105,000, is the marginal cost for a house. The selling price of $115,000 is the marginal revenue per house.

b) $c(x)$ = variable cost + fixed cost = 105,000x + 40,000
$r(x)$ = 115,000x
$p(x) = r(x) - c(x)$ = 10,000x - 40,000

c) $p(12)$ = 10,000(12) - 40,000 = $80,000 monthly profit.

Spreadsheet with formulas

	A	B
1	PROBLEM DATA	
2	Fixed Cost	$40,000
3	Variable Cost Per Unit	$105,000
4	Selling Price Per Unit	$115,000
5	MODEL	
6	Sales Volume	
7	Total Revenue	=B4*B6
8	Total Cost	=B2+B3*B6
9	Total Profit (Loss)	=B7-B8

Spreadsheet with solution

	A	B
1	PROBLEM DATA	
2	Fixed Cost	$40,000
3	Variable Cost Per Unit	$105,000
4	Selling Price Per Unit	$115,000
5	MODEL	
6	Sales Volume	12
7	Total Revenue	$1,380,000
8	Total Cost	$1,300,000
9	Total Profit (Loss)	$80,000

d) $r(x) = c(x)$ or $115,000x = 105,000x + 40,000$. Solving, $x = 4$.

Determining the break-even point using Excel's Goal Seek tool
Step 1: Select the **Tools** pull-down menu
Step 2: Choose the **Goal Seek** option
Step 3: When the **Goal Seek** dialog box appears:
  Enter B9 in the **Set cell** box
  Enter 0 in the **To value** box
  Enter B6 in the **By changing cell** box
  Click **OK**

	A	B
1	PROBLEM DATA	
2	Fixed Cost	$40,000
3	Variable Cost Per Unit	$105,000
4	Selling Price Per Unit	$115,000
5	MODEL	
6	Sales Volume	4
7	Total Revenue	$460,000
8	Total Cost	$460,000
9	Total Profit (Loss)	$0

CHAPTER 1

ANSWERED PROBLEMS

PROBLEM 6

Zipco Printing operates a shop that has five printing machines. The machines differ in their capacities to perform various printing operations due to differences in the machines' designs and operator skill levels. At the start of the workday there are five printing jobs to schedule. The manager must decide what the job-machine assignments should be.

a) How could a quantitative approach to decision making be used to solve this problem?

b) What would be the uncontrollable inputs for which data must be collected?

c) Define the decision variables, objective function, and constraints to appear in the mathematical model.

d) Is the model deterministic or stochastic?

e) Suggest some simplifying assumptions for this problem.

PROBLEM 7

Zizzle Company is a new small local company that is about to manufacture Zizzle briefcases in three styles. The company wants to determine how to use its resources most efficiently to get the product mix that will maximize its profits.

One manager has advocated hiring an outside consulting firm to analyze sales potentials and customer preferences. He suggests doing a complete time and motion study of the production process and analyzing the potential for acquiring additional manpower and material resources. In short, he is advocating an extremely accurate but complex study.

A second manager suggests a simplified model using "best guess", rough approximations and simplifying assumptions as a starting point. The data for this model can be obtained in a short period of time, and the model can be solved in-house at a fraction of the cost of the more complex model.

a) Which manager's advice would you follow? Explain.

b) Would your answer change if Zizzle Company were a large national conglomerate which plans to market tens of thousands of briefcases?

PROBLEM 8

A client of an investment firm has $10,000 available for investment. He has instructed that his money be invested in three stocks so that no more than $5,000 is invested in any one stock but at least $1,000 is invested in each stock. He has further instructed the firm to use its current data and invest in a manner that maximizes his expected overall gain during a one-year period. The stocks, the current price per share, and the firm's projected stock price a year from now are summarized in the following table.

Stock	Current Price	Projected Price 1 Year Hence
James Industries	$25	$35
QM Inc.	$50	$60
Delicious Candy Co.	$100	$125

a) Let s_j = the number of shares of stock j purchased for j = 1 (James), 2 (QM), 3 (Delicious). Formulate a mathematical model for this problem using these decision variables.

b) Let x_j = the number of dollars invested in stock j. Reformulate this mathematical model in terms of these decision variables instead of those used in part (a).

c) If both models were solved using a quantitative approach to decision making, how would you expect the results of the models to compare?

PROBLEM 9

Bank Guard Company provides security service for banks and savings and loan companies during business hours. The number of guards supplied is a function of the average number of people in the facility. One guard is provided for an average of 25 customers.

Let L = the average number of customers in the facility, C_g = the hourly cost per guard, N = the number of guards required, and C = the daily cost for guard service (based on an 8-hour day).

a) Develop a mathematical model for N in terms of L. Then develop a mathematical model for C in terms of N and C_g. Finally, express C in terms of L and C_g.

b) If night guard service costs a flat $$C_n$, develop a mathematical model for 24 hour guard service. Discuss the results with C_g = $10 per hour, C_n = $100 per night and L = 50. Discuss the results with L = 65.

PROBLEM 10

Comfort Plus Inc. (CPI) manufactures a standard dining chair used in restaurants. The demand forecasts for quarter 1 (January-March) and quarter 2 (April-June) are 3700 chairs and 4200 chairs, respectively. CPI has a policy of satisfying all demand in the quarter in which it occurs.

The chair contains an upholstered seat that can be produced by CPI or a subcontractor. The subcontractor currently charges $12.50 per seat, but has announced a new price of $13.75 effective April 1. CPI can produce the seat at a cost of $10.25.

Seats that are produced or purchased in quarter 1 and used to satisfy demand in quarter 2 cost CPI $1.50 each to hold in inventory.

a) What are the controllable inputs?

b) Develop a mathematical model for the total cost (objective) function.

PROBLEM 11

Continuing with problem 10, consider the following additional information. CPI's seat-producing capacity is 3800 seats per quarter. CPI cannot hold more than 300 seats in inventory from quarter to quarter.

a) Complete the mathematical model started in problem 10 by modeling the constraints.

b) Is the model stochastic or deterministic?

c) How would adding a third quarter to the problem change the model?

PROBLEM 12

A retail furniture store has set aside 800 square feet to display its new 18th Century Collection of sofas and chairs. Considering aisle space, it is estimated that each sofa utilizes 50 sq. ft. and each chair utilizes 30 sq. ft. At least five sofas and at least five chairs are to be displayed.

a) Write a mathematical model representing the store's constraints.

b) Suppose the profit on sofas is $200 and on chairs is $100. On a given day, the probability that a displayed sofa will be sold is .03 and that a displayed chair will be sold is .05. Mathematically model each of the following objectives:

    1) Maximize the total pieces of furniture displayed.
    2) Maximize the total expected number of daily sales.
    3) Maximize the total expected daily profit.

PROBLEM 13

Universal Computer is considering producing the new UC15 computer. Its components cost $630 and it will take 1/2 hour to assemble and 1/4 hour to pack. Skilled workers assemble the computer and make $16 per hour. Unskilled laborers pack the computers and make $4 per hour. Packaging material cost $11 per computer.

The computer will be produced in a plant located on Harbor Blvd. and leased for $3,100 per month. Utilities, supplies, insurance, etc. are expected to run another $1,900 per month regardless of the quantity of computers produced. Based on current market conditions, Universal assumes it can sell at $700 each all units it produces.

a) Identify the variable and fixed costs for Universal. Let x = the monthly production quantity of UC15s. Express the monthly cost function in terms of x.

b) Express the revenue function in terms of x.

c) What is the breakeven point?

d) What is the marginal cost of a computer?

e) What will be the monthly profit with a production schedule for UC15 computers of 500 per month?

CHAPTER 1

TRUE/FALSE

___ 14. The optimal solution to a mathematical model is always the policy that should be implemented by the company.

___ 15. A problem to decide monthly shipping patterns is dependent on the amount of product available at the factory. This amount can be modeled by a normal distribution. Thus, this is a stochastic mathematical model.

___ 16. Microcomputers have made most quantitative approaches to decision making more accessible to even moderate size firms.

___ 17. The three most commonly used quantitative approaches to decision making are statistical analyses, simulation, and linear programming.

___ 18. A company seeks to maximize profit subject to limited availability of man-hours. Man-hours is a controllable input.

___ 19. Past data might be used to test the validity of a mathematical model.

___ 20. A toy train layout designed to represent an actual railyard is an example of an analog model.

___ 21. A feasible solution is one that satisfies at least one of the constraints in the problem.

___ 22. The terms stochastic and deterministic have the same meaning in quantitative approaches to decision making.

___ 23. If you are deciding to buy either machine A, B, or C with the objective of minimizing the sum of labor, material and utility costs, you are dealing with a multicriteria decision.

___ 24. If your decision alternatives for dealing with increased product inventory are: (1) build another warehouse, (2) reduce the production rate, and (3) increase the sales rate, you are dealing with a multicriteria decision.

___ 25. Model development should be left to quantitative analysts; the model user's involvement should begin at the implementation stage.

___ 26. The volume that results in marginal revenue equaling marginal cost is called the break-even point.

___ 27. In quantitative analysis, the optimal solution is the mathematically-best solution.

___ 28. Problem solving is narrower in scope than decision making.

CHAPTER 2

Introduction to Probability

CHAPTER 2

KEY CONCEPTS

CONCEPT	ILLUSTRATED PROBLEMS	ANSWERED PROBLEMS
Experiments, Sample Spaces, Counting Rule	1	7,12
Assignment of Probabilities	1-4	13
Probability Relationships: Complement, Union, Intersection, Mutually Exclusive, Addition Law, Conditional Probability, Joint Probability, Marginal Probability, Multiplication Law	1-4	8,9,14,15,16
Bayes' Theorem	5,6	10,11,17

REVIEW

1. <u>Probability</u> is concerned with the study of uncertain or random events. It is a numerical measure of the chance that a particular event will occur. It provides a mechanism for developing a mathematical model which enables an analysis of the uncertainties of future events. Such an analysis is important for decision making concerning these events.

2. An <u>experiment</u> is any process that generates well-defined outcomes.

3. A <u>sample space</u> consists of all outcomes of interest to the experimenter. For example, if one observes the number of foreign cars in a sample of 20, the sample space consists of 21 sample points: E_1 = (0 foreign cars observed), E_2 = (1 foreign car observed), ..., E_{21} = (20 foreign cars observed).

4. Probabilities are non-negative values between 0 and 1 assigned to each sample point. The sum of the probabilities of all the sample points in the sample space must equal 1.

5. The basis for assigning probabilities to outcomes is an attempt to assign to each outcome a numerical value which reflects its likelihood of occurrence.

INTRO. TO PROBABILITY

6. There are three methods for assigning probabilities to sample points:
   (1) If all outcomes are equally likely (e.g. the flip of a fair coin), the <u>classical method</u> assigns to each possible outcome an identical probability.
   (2) If evidence suggests that all outcomes are not equally likely to occur, the <u>relative frequency method</u> assigns the probability based on evidence.
   (3) When neither the classical nor the relative frequency methods can be applied, the <u>subjective method</u> may be used. This assigns the probabilities based on the experimenter's belief concerning the likelihood of each outcome occurring.

7. <u>Events</u> are sets of sample points which are of interest to the experimenter. For example, using the example above, the event, "fewer than three foreign cars" is a set of the sample points consisting of $\{E_1, E_2, E_3\}$. Since events are sets, set operations such as union, intersection, and complementation may be applied.

8. The probabilities for sample outcomes are used to determine the probability for events. The probability of any event is equal to the sum of the probabilities of the sample points in the event. (For example, if a fair die is rolled, the probability of each of the six sample points is 1/6. Thus the probability of the event {rolling a number higher than a 3} will be the sum of the probabilities of {rolling a 4}, {rolling a 5}, or {rolling a 6}. This equals 1/6 + 1/6 + 1/6 = 1/2.)

9. Given an event, A, the <u>complement</u> of A, denoted A^C is defined as consisting of all sample points not in A.

10. The <u>union</u> of two events, A and B, denoted $A \cup B$, is defined as the set of all sample points in A or in B or in both A and B.

11. The <u>intersection</u> of two events, A and B, denoted $A \cap B$, is defined as the set of all sample points in both A and B.

12. Two events, A and B, are said to be <u>mutually exclusive</u> if the events do not have any sample points in common. (A and A^C are mutually exclusive.)

13. One method frequently used for visualizing relationships between event sets is the <u>Venn diagram</u>. This is a picture with a rectangle representing the entire sample space, S, and event sets represented by circles within the rectangle. If events are not mutually exclusive, then corresponding circles will overlap in the Venn diagram.

14. The <u>addition law</u> of probability states that for two events, A and B, $P(A \cup B) = P(A) + P(B) - P(A \cap B)$. Note that if A and B are mutually exclusive, then $P(A \cap B) = 0$ and $P(A \cup B) = P(A) + P(B)$.

15. The probability of the complement of the event A is $P(A^C) = 1 - P(A)$.

16. In certain instances, one may be interested in the probability of one event, A, given knowledge concerning a second event, B. This <u>conditional probability</u> of event A given event B has occurred is written as $P(A|B)$ and its probability is given by: $P(A \cap B)/P(B)$.

17. For mutually exclusive events P(A|B) = P(B|A) = 0, (e.g. the probability a coin comes up heads given it comes up tails is 0.)

18. The <u>multiplication law</u> is derived from the definition of conditional probability and states:
    P(A∩B) = P(A|B)P(B) = P(B|A)P(A).

19. Two events, A and B, are said to be <u>independent</u> if P(B|A) = P(B) or P(A|B) = P(A). Then P(A∩B) = P(A)P(B).

20. Often the <u>joint probability</u> of events A and B, P(A∩B) will be written in a <u>joint probability table</u>. The row sums and the column sums of this table are referred to as <u>marginal probabilities</u>.

21. <u>Bayes' Theorem</u> is used to update the probability of events to account for new information concerning the events of interest. Consider the sample space being partitioned among n mutually exclusive and collectively exhaustive events, A_1, A_2, ..., A_n. Initially, <u>prior probabilities</u>, $P(A_1)$, $P(A_2)$, ..., $P(A_n)$ are assumed. Additional information (such as market surveys, experiments, etc.) is obtained, the result being event being event B.
    Bayes' Theorem then gives a formula for revising the estimates for the probabilities of events A_1, A_2, etc. These <u>posterior probabilities</u>, $P(A_i|B)$, are given by:

$$P(A_i|B) = \frac{P(A_i)P(B|A_i)}{P(A_1)P(B|A_1) + P(A_2)P(B|A_2) + ... + P(A_n)P(A_n|B)}.$$

22. An easy way to compute the posterior probabilities using Bayes' Theorem is through the tabular approach. In this method five columns are prepared listing:
    (1) the mutually exclusive events possible A_i
    (2) the prior probabilities for the events $P(A_i)$
    (3) the given conditional probabilities $P(B|A_i)$
    (4) the joint probabilities (Column (2) x Column (3)) $P(A_i \cap B)$
    (5) the posterior probabilities (Column (4) divided by the sum of the entries in Column (4)). $P(A_i|B)$

ILLUSTRATED PROBLEMS

PROBLEM 1

A market study taken at a local sporting goods store showed that of 20 people questioned, 6 owned tents, 10 owned sleeping bags, 8 owned camping stoves, 4 owned both tents and camping stoves, and 4 owned both sleeping bags and camping stoves.

Let: Event A = owns a tent
      Event B = owns a sleeping bag
      Event C = owns a camping stove

and let the sample space be the 20 people questioned.

a) Find P(A), P(B), P(C), P(A∩C), P(B∩C).

b) Are the events A and C mutually exclusive? Explain briefly.

c) Are the events B and C independent events? Explain briefly.

d) If a person questioned owns a tent, what is the probability he also owns a camping stove?

e) If two people questioned own a tent, a sleeping bag, and a camping stove, how many own only a camping stove? In this case is it possible for 3 people to own both a tent and a sleeping bag, but not a camping stove?

SOLUTION 1

a) Using the relative frequency method, the probability equals the number of sample points in the event divided by the number of sample points in the sample space. Thus,

$$P(A) = 6/20 = .3$$
$$P(B) = 10/20 = .5$$
$$P(C) = 8/20 = .4$$

P(A∩B) = P(owns a tent and owns a camping stove) = 4/20 = .2
P(B∩C) = P(owns a sleeping bag and owns a camping stove) = 4/20 = .2

b) Events B and C are not mutually exclusive because there are people (4 people) who both own a tent and a camping stove.

c) To see whether events B and C are independent, check to see if P(B∩C) = P(B)P(C). Since P(B∩C) = .2 and P(B)P(C) = (.5)(.4) = .2, then these events are independent.

d) Here the probability that a person owns a camping stove <u>given</u> he owns a tent or P(C|A) must be determined. Using conditional probability,

$$P(C|A) = P(A \cap C)/P(A) = .2/.3 = .667.$$

e) Using a Venn diagram gives the following:

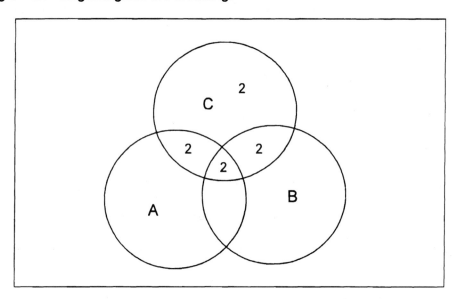

Note that <u>two</u> people own only a camping stove.

To determine whether it is possible for three people to own both a tent and a sleeping bag, but not a camping stove, note that the total size of event A is 6. Therefore the shaded area above could not have a value of three.

PROBLEM 2

The Bidwell Valve Company requires all prospective employees to be interviewed by a three man personnel committee. Each member of the committee votes individually for or against a recommendation of employment. Upon examining the company's records over the past year, the personnel director has noted the following probabilities for eight possible outcomes:

	Vote Of Member			
Outcome	1	2	3	Probability
E_1	for	for	for	.20
E_2	for	for	against	.14
E_3	for	against	for	.12
E_4	for	against	against	.07
E_5	against	for	for	.08
E_6	against	for	against	.11
E_7	against	against	for	.06
E_8	against	against	against	.22

Denote the following events:

A = Member 1 votes for employment
B = Member 2 votes for employment
C = Member 3 votes for employment

a) Find P(A), P(B), and P(C).

b) Which sample outcomes correspond to the event D = A∪B (Member 1 or Member 2 votes for employment)? Find P(D).

c) Which sample outcomes correspond to the event G = A∩B (Member 1 and Member 2 vote for employment)? Use the addition law of probability to find P(G).

d) Let the event F = B^c∩C (Member 2 votes against employment and Member 3 votes for employment). Draw a Venn diagram to represent the event G∪F.

e) If a new employee is hired only if a majority of the committee members vote for employment, what is the probability that a prospective employee is hired? Does a job applicant stand a better or worse chance of being employed going before the committee than one of the individuals?

SOLUTION 2

a) Find the outcomes corresponding to events A, B, and C.

$A = \{E_1, E_2, E_3, E_4\}$. Thus $P(A) = P(E_1) + P(E_2) + P(E_3) + P(E_4)$
$= .20 + .14 + .12 + .07 = .53$.

$B = \{E_1, E_2, E_5, E_6\}$. Thus $P(B) = P(E_1) + P(E_2) + P(E_5) + P(E_6)$
$= .20 + .14 + .08 + .11 = .53$.

$C = \{E_1, E_3, E_5, E_7\}$. Thus $P(C) = P(E_1) + P(E_3) + P(E_5) + P(E_7)$
$= .20 + .12 + .08 + .06 = .46$.

b) The event D corresponds to all sample points which are either in event A or event B or both. Thus $D = \{E_1, E_2, E_3, E_4, E_5, E_6\}$.

$P(D) = P(E_1) + P(E_2) + P(E_3) + P(E_4) + P(E_5) + P(E_6)$
$= .20 + .14 + .12 + .07 + .08 + .11 = .72$.

c) The event G corresponds to sample points in both event A and event B. Thus $G = \{E_1, E_2\}$.

The addition law states: $P(A \cup B) = P(A) + P(B) - P(A \cap B)$
or in this case: $P(G) = P(A) + P(B) - P(D) = .53 + .53 - .72 = .34$.
Note that this checks with $P(G) = P(E_1) + P(E_2) = .20 + .14 = .34$.

d) Event $B^C = \{E_3, E_4, E_7, E_8\}$. $F = B^C \cap C$ = the events common to both B^C and $C = \{E_3, E_7\}$. The Venn diagram for $G \cup F$ is:

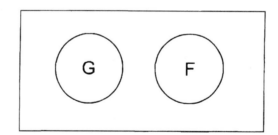

e) Let event L = the event that the majority of the committee members vote for employment, i.e. two or three vote for employment. Thus, $L = \{E_1, E_2, E_3, E_5\}$. $P(L) = P(E_1) + P(E_2) + P(E_3) + P(E_5)$. Thus $P(L) = .20 + .14 + .12 + .08 = .54$. Since this is higher than $P(A)$, $P(B)$, and $P(C)$, the job applicant has a better chance of being employed by going before the committee.

PROBLEM 3

Harry owns shares of stock in both Bidwell Valve Company and Mini Car Motors, Inc. Harry has recorded the performance of these shares on each day for a 200 day period as to whether the price has risen, fallen or remained unchanged. Harry's data is as follows:

Event	Bidwell	Mini Car	# of Days
E_1	Rise	Rise	40
E_2	Rise	Unchanged	14
E_3	Rise	Fall	20
E_4	Unchanged	Rise	18
E_5	Unchanged	Unchanged	12
E_6	Unchanged	Fall	14
E_7	Fall	Rise	28
E_8	Fall	Unchanged	16
E_9	Fall	Fall	38

a) Using this data and the relative frequency approach, find the following probabilities:
   (1) a rise in the price of Bidwell stock
   (2) a fall in the price of Mini Car stock
   (3) a rise in Bidwell <u>and</u> a fall in Mini Car stock
   (4) a rise in Bidwell <u>or</u> a fall in Mini Car stock (Use the addition law.)

b) Suppose Harry is told that Bidwell stock has risen. What is the probability that Mini Car stock has fallen?

c) Suppose Harry is told that Mini Car stock has fallen. What is the probability that Bidwell stock has risen?

d) Are the events "a rise in Bidwell stock" and "a fall in Mini Car stock" independent?

SOLUTION 3

a) Let: event A = "rise in Bidwell stock" = $\{E_1, E_2, E_3\}$
          event B = "fall in Mini Car stock" = $\{E_3, E_6, E_9\}$

Using the relative frequency approach,

$P(E_1) = 40/200 = .20$
$P(E_2) = 14/200 = .07$
$P(E_3) = 20/200 = .10$
$P(E_4) = 18/200 = .09$
$P(E_5) = 12/200 = .06$
$P(E_6) = 14/200 = .07$
$P(E_7) = 28/200 = .14$
$P(E_8) = 16/200 = .08$
$P(E_9) = 38/200 = .19$

(1) $P(A) = P(E_1) + P(E_2) + P(E_3) = .20 + .07 + .10 = .37$
(2) $P(B) = P(E_3) + P(E_6) + P(E_9) = .10 + .07 + .19 = .36$
(3) $P(A \cap B) = P(E_3) = .10$
(4) $P(A \cup B) = P(A) + P(B) - P(A \cap B) = .37 + .36 - .10 = .63$

b) $P(B|A) = P(A \cup B)/P(A) = .10/.37 = .27$.

c) $P(A|B) = P(A \cup B)/P(B) = .10/.36 = .28$.

d) There are several ways to check if events A and B are independent. One method is to see if $P(A|B) = P(A)$. Since $P(A|B) = .28$ and $P(A) = .37$, these events are <u>dependent</u>.

PROBLEM 4

The Board of Directors of Bidwell Valve Company have made the following estimates for the upcoming year's annual earnings:

P(earnings lower than this year) = .30
P(earnings about the same as this year) = .50
P(earnings higher than this year) = .20

After talking with union leaders, the personnel department has drawn the following conclusions:

P(Union will request wage increase|lower earnings next year) = .25
P(Union will request wage increase|same earnings next year) = .40
P(Union will request wage increase|higher earnings next year) = .90

a) Are the probabilities developed by the directors and personnel manager based on the classical, relative frequency, or subjective method?

b) Calculate the following probabilities:
   (1) The company earns the same as this year and the union requests a wage increase
   (2) The company has higher earnings next year and the union does not request a wage increase
   (3) The union requests a wage increase

SOLUTION 4

a) Since the predicted outcomes of earnings and wage increase requests have unequal probabilities, they are not based on the classical method. Similarly, since the outcomes arise from a unique situation, the relative frequency method is not applicable. Instead, the probabilities stated reflect the judgment of the individuals involved, and hence are developed by the <u>subjective method</u>.

b) Define the following events: L = lower earnings next year; S = about the same earnings next year; H = higher earnings next year. Further define R = union requests a pay increase next year, so R^C = union does not request a pay increase next year.

(1) This is $P(S \cap R)$. Now, $P(S) = .50$ and $P(R|S) = .40$.
   Thus, $P(S \cap R) = P(S)P(R|S) = (.50)(.40) = .20$

(2) This is $P(H \cap R^C)$. Now, $P(H) = .20$ and $P(R^C|H) = 1 - P(R|H) = .10$
   Thus, $P(H \cap R^C) = P(H)P(R^C|H) = (.20)(.10) = .02$.

(3) This is $P(R)$. $P(R) = P(R|H)P(H) + P(R|S)P(S) + P(R|L)P(L)$
   $= (.25)(.30) + (.40)(.50) + (.90)(.20) = .455$

PROBLEM 5

An accounting firm has noticed that of the companies it audits, 85% show no inventory shortages, 10% show small inventory shortages and 5% show large inventory shortages. The firm has devised a new accounting test for which it believes the following probabilities hold:

P(company will pass test | no shortage) = .90
P(company will pass test | small shortage) = .50
P(company will pass test | large shortage) = .20

a) If a company being audited fails this test, what is the probability of a large or small inventory shortage?

b) If a company being audited passes this test, what is the probability of no inventory shortage?

SOLUTION 5

a) Let: A_1 = no inventory shortage P = company passes test
        A_2 = small inventory shortage F = company fails test
        A_3 = large inventory shortage

Event	Prior Probabilities $P(A_i)$	Conditional Probabilities $P(F|A_i)$	Joint Probabilities $P(A_i \cap F)$	Posterior Probabilities $P(A_i|F)$
A_1	.85	.10	.085	.486
A_2	.10	.50	.050	.286
A_3	.05	.80	.040	.229
			$P(F) = .175$	

Here, $P(F|A_i) = 1 - P(P|A_i)$; the joint probabilities are the multiples of the prior and conditional probabilities and the posterior probabilities are the joint probabilities divided by P(F). The probability of a large or small shortage = .286 + .229 = .515.

b)

Event	Prior Probabilities $P(A_i)$	Conditional Probabilities $P(P\|A_i)$	Joint Probabilities $P(A_i \cap P)$	Posterior Probabilities $P(A_i\|P)$
A_1	.85	.90	.765	.927
A_2	.10	.50	.050	.061
A_3	.05	.20	.010	.012
			$P(P) = .825$	

The solution is $P(A_1|P) = .927$.

PROBLEM 6

An investment advisor recommends the purchase of shares in Probaballisitics, Inc. He has made the following predictions:

$$P(\text{Stock goes up 20\%} | \text{Rise in GDP}) = .6$$
$$P(\text{Stock goes up 20\%} | \text{Level GDP}) = .5$$
$$P(\text{Stock goes up 20\%} | \text{Fall in GDP}) = .4$$

An economist has predicted that the probability of a rise in the GDP is 30%, whereas the probability of a fall in the GDP is 40%.

a) What is the probability that the stock will go up 20%?

b) We have been informed that the stock has gone up 20%. What is the probability of a rise or fall in the GDP?

SOLUTION 6

a) Given that P(Rise in GDP) = .3 and P(Fall in GDP = .4),
then the P(Level GDP) = 1 - .3 - .4 = .3.

Now, P(Stock goes up by 20%)
= P(Stock goes up by 20% | Rise in GDP) x P(Rise in GDP)
+ P(Stock goes up by 20% | Level GDP) x P(Level GDP)
+ P(Stock goes up by 20% | Fall in GDP) x P(Fall in GDP)
= (.6)(.3) + (.5)(.3) + (.4)(.4) = .49.

b) From Bayes' Law,
P(Rise in GDP|Stock goes up by 20%) =
P(Stock goes up by 20% | Rise in GDP)P(Rise in GDP)/P(Stock goes up 20%)
= (.6)(.3)/.49 = .367

P(Fall in GDP | Stock goes up by 20%) =
P(Stock goes up by 20% | Fall in GDP)P(Fall in GDP)/P(Stock goes up 20%)
= (.4)(.4)/.49 = .327

Thus, the probability of a rise or fall in the GDP is .367 + .327 = .694.

ANSWERED PROBLEMS

PROBLEM 7

Providence Land Development Company has just hired four new salespersons. After six months on the job each salesperson will be rated as either poor, average or excellent and will be compensated accordingly. Assume that Providence is concerned with the number of salespersons in each category.

a) List the outcomes of this experiment.

b) Let the event A = at least two salespersons are rated average and let event B = exactly one salesperson is rated poor. List the outcomes in A∩B.

c) Let the event C = exactly two salespersons are rated excellent. List the outcomes in D = B∩C.

d) Are events D and A mutually exclusive?

e) Are events D and A collectively exhaustive?

f) Let the event E = at most one salesperson is rated average. Are events A and E mutually exclusive and collectively exhaustive?

PROBLEM 8

Global Airlines operates two types of jet planes: jumbo and ordinary. On jumbo jets, 25% of the passengers are on business while on ordinary jets 30% of the passengers are on business. Of Global's airfleet, 40% of its capacity is provided on jumbo jets. (Hint: The 25% and 30% values are <u>conditional</u> probabilities stated as percentages.)

a) What is the probability a randomly chosen business customer flying with Global is on a jumbo jet?

b) What is the probability a randomly chosen nonbusiness customer flying with Global is on an ordinary jet?

PROBLEM 9

The following probability model describes the number of snow storms for Washington, D.C. for a given year:

# of Snowstorms	0	1	2	3	4	5	6
Probability	.25	.33	.24	.11	.04	.02	.01

The probability of 7 or more snowstorms in a year is 0.

a) What is the probability of more than 2 but less than 5 snowstorms?

b) Given this a particularly cold year (in which 2 snowstorms have already been observed), what is the conditional probability that 4 or more snowstorms will be observed?

c) If at the beginning of winter there is a snowfall, what is the probability of at least one more snowstorm before winter is over?

PROBLEM 10

Safety Insurance Company has compiled the following statistics. For any one year period:

$$P(\text{accident} \mid \text{male driver under 25}) = .22$$
$$P(\text{accident} \mid \text{male driver over 25}) = .15$$
$$P(\text{accident} \mid \text{female driver under 25}) = .16$$
$$P(\text{accident} \mid \text{female driver over 25}) = .14$$

The percentage of Safety's policyholders in each category are:

   Male Under 25   20%
   Male Over 25   40%
   Female Under 25   10%
   Female Over 25   30%

a) What is the probability that a randomly selected policyholder will have an accident within the next year?

b) Given that a driver has an accident, what is the probability that the driver is a male over 25?

c) Given that a driver has no accident, what is the probability the driver is a female?

d) Does knowing the fact that a driver has had no accidents give us a great deal of information regarding the driver's sex?

PROBLEM 11

Consider the Bidwell Valve Company of problem 4. The directors believe that the probability of strong first quarter earnings is dependent on the full year's earnings. They estimate:

P(Strong first quarter earnings | lower earnings next year) = .10
P(Strong first quarter earnings | same earnings next year) = .30
P(Strong first quarter earnings | higher earnings next year) = .70

a) Using this data together with data in problem 4, find the probabilities of the company earnings being higher, lower, and the same as this year's given the company has strong first quarter earnings.

b) Using these new posterior probabilities, compute the probability that the union requests a wage increase given the company has a strong first quarter.

c) Compute the probability that the union requests a wage increase given the company does not have a strong first quarter.

PROBLEM 12

Mini Car Motors offers its luxury car in three colors: gold, silver and blue. The vice president of advertising is interested in the order of popularity of the color choices by customers during the first month of sales.

a) How many sample points are there in this experiment?

b) If the event A = gold is the most popular color, list the outcome(s) in event A.

c) If the event B = blue is the least popular color, list the outcome(s) in A∩B.

d) List the outcome(s) in A∩B^C.

PROBLEM 13

Sales of the first 500 luxury Mini Cars were as follows: 250 gold, 150 silver, and 100 blue. Assume the relative frequency method is used to assign probabilities for color choice and the color of each car sold is independent of that of any other car sold.

a) What is the probability that the next two cars sold will be gold?

b) What is the probability that neither of the next two cars sold will be silver?

c) What is the probability that of the next two cars sold, one will be silver and the other will be blue?

PROBLEM 14

The sales manager for Widco Distributing Company has estimated demand for a new combination microwave oven and color television will be between 0 and 2 units per day. He believes the probability of selling no units is .65, of one unit is .25, and two units is .10. The company is interested in sales over a two day period.

a) What is the probability of selling no units during the two days?

b) What is the probability of selling one unit during the two days?

c) What is the probability of selling three or more units during the two days?

d) What is the probability of selling two units during the two days?

PROBLEM 15

Super Cola sales breakdown as 80% regular soda and 20% diet soda. While 60% of the regular soda is purchased by men, only 30% of the diet soda is. If a woman purchases Super Cola, what is the probability that it is a diet soda? (Hint: The 60% and 30% values are conditional probabilities stated as percentages.)

PROBLEM 16

Stanton Marketing conducted a taste preference test among married and single persons for the Super Cola Company. Among single people, 11% of the population questioned preferred Super Cola over all other brands. For married people the data was:

	Wife Prefers	Wife Does Not Prefer
Husband Prefers	.08	.06
Does Not Prefer	.07	.79

a) Using this study as a basis for a probability measure, find the probability that if two single people are questioned: (1) they both prefer Super Cola; (2) they both do not prefer Super Cola; and (3) one prefers Super Cola and the other does not

b) What is the probability that a married female prefers Super Cola?

c) Given that a husband prefers Super Cola, what is the probability his wife also prefers it?

d) Is the event "husband prefers Super Cola" independent of the event "wife prefers Super Cola"?

e) Do married people prefer Super Cola more than single people?

PROBLEM 17

Higbee Manufacturing Corp. has recently received 5 cases of a certain part from one of its suppliers. The defect rate for the parts is normally 5%, but the supplier has just notified Higbee that one of the cases shipped to them has been made on a misaligned machine that has a defect rate of 97%. So the plant manager selects a case at random and tests a part.

a) What is the probability that the part is defective?

b) Suppose the part is defective, what is the probability that this is from the case made on the misaligned machine?

c) After finding that the first part was defective, suppose a second part from the case is tested. However, this part is found to be good. Using the revised probabilities from part (b) compute the new probability of these parts being from the defective case.

d) Do you think you would obtain the same posterior probabilities as in part (c) if the first part was not found to be defective but the second part was?

e) Suppose, because of other evidence, the plant manager was 80% certain this case was the one made on the misaligned machine. How would your answer to part (b) change?

TRUE/FALSE

___ 18. Two events that are mutually exclusive cannot be independent.

___ 19. $P(A|B) = P(B|A)$ for all events A and B.

___ 20. $P(A|B) = 1 - P(B|A)$ for all events A and B.

___ 21. $P(A|B) = 1 - P(A^c|B)$ for events A and B.

___ 22. One would use the classical method to assign probabilities for customers' preference of automobile models.

___ 23. The intersection of A and A^c is the entire sample space.

___ 24. If A and B are mutually exclusive, $P(A) + P(B)$ must equal 1.

___ 25. If A and B are mutually exclusive, $P(A \cap B) = 0$.

___ 26. When using Bayes' Theorem, if A_1 and A_2 are prior events, then $P(A_1 \cap A_2) = 0$.

___ 27. $P(A|B) + P(A|B^c) = 1$ for all A and B

___ 28. If A and B are independent, $P(A \cap B) = 0$.

___ 29. A joint probability can have a value greater than 1.

___ 30. The probability of at least one head in two flips of a coin is 0.75.

___ 31. Two events that are independent cannot be mutually exclusive.

___ 32. Posterior probabilities are conditional probabilities.

CHAPTER 3

Probability Distributions

KEY CONCEPTS

CONCEPT	ILLUSTRATED PROBLEMS	ANSWERED PROBLEMS
Continuous and Discrete Random Variables	1-3	9,10
Expected Value and Variance	3	9,10,15
Spreadsheet Example	3	
Discrete Distributions		
Relative Frequency	3	15
Binomial	4	11,16
Poisson	6	17,21
Continuous Distributions		
Uniform	5	13,18
Normal	7	12,14,19,20
Exponential	6,8	21

REVIEW

1. The result of an experiment can be defined so that each possible outcome generates one numerical value. The result of such an experiment is called a <u>random variable</u>.

2. A random variable which may only take on a countable number of values is called a <u>discrete random variable</u>.

3. Associated with each discrete random variable is a <u>probability distribution</u>, $f(x)$, defining the probability of the random variable being equal to the specific value x.

4. The <u>expected value</u> or <u>mean</u> of a random variable (designated μ) is a weighted average of all possible values of the variable, with the weights being the probabilities. Hence, for discrete random variables, $\mu = \Sigma x f(x)$. The expected value can be interpreted as the long run average value for the experiment.

5. The <u>variance</u>, (denoted σ^2) of a random variable is a measure of dispersion of the variable about its mean. The formula used to calculate the variance for discrete random variables is:

$$\sigma^2 = \Sigma(x - \mu)^2 f(x).$$

6. The <u>standard deviation</u> of a random variable is the square root of its variance.

7. A <u>Bernoulli Process</u> is a sequence of independent trials with two possible outcomes (success, $x = 1$ and failure, $x = 0$) and the probability of success (denoted by p) remains constant.

8. The <u>binomial distribution</u> is the probability distribution of the number of successes, x, in n independent Bernoulli trials. For a binomial distribution, its mean, $\mu = np$, and its variance, $\sigma^2 = np(1-p)$. The probability distribution for the binomial distribution of x successes in n trials is:

$$f(x) = \frac{n!}{(x!)(n-x)!} p^x (1-p)^{n-x}$$

Appendix A provides a table of binomial probabilities for various values of n and p for $p \leq .50$. If p is greater than .50, then one can use the table by focusing on the number of failures as follows: the probability of a single failure is $1-p$ (a number less than .50), and if x denotes the number of successes, then the number of failures is $(n-x)$.

9. The <u>Poisson probability function</u> is often used for describing the number of occurrences of an event over a specified interval of time. If the average number of events in the specified time interval is denoted by λ, then its mean, $\mu = \lambda$, and variance $\sigma^2 = \lambda$. The Poisson distribution for x events occurring in the time interval is given by:

$$f(x) = \frac{\lambda^x e^{-\lambda}}{x!}$$

Selected Poisson probabilities are given in Appendix B while values for $e^{-\lambda}$ are tabulated in Appendix D.

10. A process can be defined by the Poisson distribution if the following three assumptions hold:
    1) In any short interval there can be at most one occurrence of the event;
    2) The probability of an occurrence of the event is the same for any two intervals of equal length;
    3) The occurrence or nonoccurrence of the event in any interval is independent of the occurrence or nonoccurrence of the event in any other interval.

11. A random variable which, at least theoretically, may take on any possible value within an interval (e.g. time, distance, weight, etc.) is called a <u>continuous random variable</u>.

12. Probabilities for continuous random variables are defined over intervals by a <u>probability density function</u>, $f(x)$ with the following properties:
    (1) $f(x) \geq 0$ for all x;
    (2) The total area under the curve $f(x)$ is equal to 1;
    (3) The probability that the random variable takes on a value between two values a and b, is equal to the area under the curve, $f(x)$, between the points a and b.

13. For continuous random variables, (3) above implies the probability of any specific value is equal to 0. This is not to say that such events are impossible, however, the probability is defined this way so as to be mathematically consistent.

14. If a random variable is restricted to be within some interval and the probability density function is constant over the interval, (a,b) the continuous random variable is said to have a <u>uniform distribution</u> between a and b. Its density function is given by:

$$f(x) = 1/(b-a) \text{ for } a \leq x \leq b, \text{ and } = 0 \text{ outside this interval.}$$

15. The <u>normal distribution</u> is perhaps the most widely used distribution for describing a continuous random variable. Its probability density function is a bell shaped curve which is symmetric about the mean and defined over all values of x.

16. A continuous random variable which has a normal distribution with a mean of 0 and a standard deviation of 1 is said to have a <u>standard normal distribution</u>. Because of the complexity of the density function for the normal distribution, random variables with a normal distribution are transformed to a standard normal distribution, for which tables are readily available (Appendix C). For a normal random variable with mean = μ and standard deviation = σ, a value x is transformed to its standard normal value, z, by: $z = (x - \mu)/\sigma$.

17. A continuous probability distribution frequently used for computing the probability of the time to complete a task is the <u>exponential distribution</u>. If the average time to complete a task is denote by μ, then the probability density function for the amount of time, x, to complete the task is given by:

$$f(x) = (1/\mu)e^{-(x/\mu)} \text{ for } x \geq 0 \text{ and } \mu > 0.$$

From this distribution, the probability a task is completed within a specified time, x_0, is:

$$P(x < x_0) = 1 - e^{-(x_0/\mu)}$$

18. If customers arrive according to a Poisson distribution with a mean of λ customers, then the interarrival times of customers follows an exponential distribution with $\mu = 1/\lambda$.

PROBABILITY DISTRIBUTIONS 39

ILLUSTRATED PROBLEMS

PROBLEM 1

Dollar Department Stores is planning to open a new store on the corner of Main and Vine Streets. It has asked the Stanton Marketing Company to do a market study of randomly selected families within a five mile radius of the store. Among the questions it wishes Stanton to ask each homeowner are: (a) family income; (b) family size; (c) distance from home to the store site; and, (d) whether or not the family owns a dog or a cat.

For each of the four questions, develop a random variable of interest to Dollar Department Stores. Denote which of these are discrete and which are continuous random variables.

SOLUTION 1

Question	Random Variable	Discrete or Continuous
(a) Family income	x = Annual dollar gross income the family reported on their tax return	Discrete
(b) Family size	x = Number of dependents in the family reported on their tax return	Discrete
(c) Distance from home to store	x = The distance in miles from home to the store site	Continuous

(d) Dog/Cat	x = 1 if own no pet;
	   = 2 if own dog(s) only;
	   = 3 if own cat(s) only;
	   = 4 if own dog(s) and cat(s)
		Discrete

PROBLEM 2

Stanton Marketing reported back to Dollar Department Stores the following information. Out of 400 families surveyed, 260 owned no pet, 120 owned dogs and 50 owned cats.

a) On the basis of this information, find the probability distribution for the random variable x, defined in (d) in problem 1.

b) Dollar Department Stores is considering opening a pet department if the expected number of families owning pets shopping at its store exceeds 4,000. If Dollar expects to serve 12,000 families, should it open a pet department?

SOLUTION 2

Since 260 owned no pets, 140 owned pets. Since 120 owned dogs and 50 owned cats (total = 170), then 30 must own both dogs and cats.

a) Since 120 owned dogs and 30 owned dogs and cats, 120 - 30 = 90 own dogs only. Similarly, 50 - 30 = 20 own cats only. Using the relative frequency method to calculate f(x):
$f(1) = 260/400 = .65$ $f(3) = 20/400 = .05$
$f(2) = 90/400 = .225$ $f(4) = 30/400 = .075$

b) The probability of owning a pet = 1 - the probability of not owning a pet = $1 - f(1) = 1 - .65 = .35$. Multiply this probability by the total number of families Dollar expects to serve to obtain the expected number of families owning pets = $(.35)(12,000) = 4,200$. Since this is greater than 4,000, Dollar should open a pet department.

PROBLEM 3

The salespeople at Gold Key Realty sell up to 9 houses per month. the probability distribution of a salesperson selling x houses in a month is as follows:

Sales (x)	0	1	2	3	4	5	6	7	8	9
Probability f(x)	.05	.10	.15	.20	.15	.10	.10	.05	.05	.05

a) What are the mean and standard deviation for the number of houses sold by a salesperson per month?

b) Use a spreadsheet to verify your calculations in part a).

c) Any salesperson selling more houses than the amount equal to the mean plus two standard deviations receives a bonus. How many houses per month must a salesperson sell to receive a bonus?

SOLUTION 3

a) The mean, $\mu = \Sigma x f(x) = (0)(.05) + (1)(.10) + (2)(.15) + (3)(.20) + (4)(.15)$
$+ (5)(.10) + (6)(.10) + (7)(.05) + (8)(.05) + (9)(.05)$
$= 3.9$

The variance, $\sigma^2 = \Sigma(x - \mu)^2 f(x) = (0 - 3.9)^2(.05) + (1 - 3.9)^2(.10) + (2 - 3.9)^2(.15)$
$+ (3 - 3.9)^2(.20) + (4 - 3.9)^2(.15) + (5 - 3.9)^2(.10)$
$+ (6 - 3.9)^2(.10) + (7 - 3.9)^2(.05) + (8 - 3.9)^2(.05)$
$+ (9 - 3.9)^2(.05)$
$= 5.49$

The standard deviation σ is the square root of σ^2, so $\sigma = 2.34$.

b)   Spreadsheet showing formulas: Spreadsheet showing results:

	A	B	C	D
1	x	f(x)	xf(x)	$(x-\mu)^2 f(x)$
2	0	0.05	=A2*B2	=(A2-C$12)^2*B2
3	1	0.10	=A3*B3	=(A3-C$12)^2*B3
4	2	0.15	=A4*B4	=(A4-C$12)^2*B4
5	3	0.20	=A5*B5	=(A5-C$12)^2*B5
6	4	0.15	=A6*B6	=(A6-C$12)^2*B6
7	5	0.10	=A7*B7	=(A7-C$12)^2*B7
8	6	0.10	=A8*B8	=(A8-C$12)^2*B8
9	7	0.05	=A9*B9	=(A9-C$12)^2*B9
10	8	0.05	=A10*B10	=(A10-C$12)^2*B10
11	9	0.05	=A11*B11	=(A11-C$12)^2*B11
12			=SUM(C2:C11)	=SUM(D2:D11)
13			Exp. Value (μ)	Variance (σ^2)

	A	B	C	D
1	x	f(x)	xf(x)	$(x-\mu)^2 f(x)$
2	0	0.05	0.00	0.7605
3	1	0.10	0.10	0.8410
4	2	0.15	0.30	0.5415
5	3	0.20	0.60	0.1620
6	4	0.15	0.60	0.0015
7	5	0.10	0.50	0.1210
8	6	0.10	0.60	0.4410
9	7	0.05	0.35	0.4805
10	8	0.05	0.40	0.8405
11	9	0.05	0.45	1.3005
12			3.90	5.49
13			Exp. Value (μ)	Variance (σ^2)

c) The number of houses a salesperson must sell to be two standard deviations from the mean is $\mu + 2\sigma = 3.9 + (2)(2.34) = 8.58$ or 9 houses.

PROBLEM 4

Ralph's Gas Station is running a giveaway promotion. With every fill-up of gasoline, Ralph gives out a lottery ticket which has a 25% chance of being a winning ticket. Customer who collect four winning lottery tickets are eligible for the "BIG SPIN" for large payoffs.
What is the probability of qualifying for the big spin if a customer fills up: (a) 3 times; (b) 4 times; (c) 7 times?

SOLUTION 4

a) If a customer fills up only three times there is no possibility of 4 winning lottery tickets. Hence the probability is 0.

b) If a customer fills up 4 times, he must obtain 4 winning tickets in 4 tries. From the binomial table, $n = 4$, $x = 4$, $p = .25$, this probability is .0039.

c) If a customer fills up $n = 7$ times, he will qualify for the big spin if he has $x = 4$, $x = 5$, $x = 6$, or $x = 7$ winning tickets.
From the tables with $p = .25$ this is $= (.0577) + (.0115) + (.0013) + (.0000) = .0705$.

PROBLEM 5

Suppose a random variable x has a continuous uniform distribution with values ranging from 5 to 15.

a) What is the probability that x has a value between 8 and 10?

b) What is the probability that the value for x is less than 7 or greater than 12?

c) What is the probability that x has a value less than 20?

d) What is the probability that x equals 11?

SOLUTION 5

Since x has a uniform distribution between 5 and 15, $f(x) = 1/10$ for $5 \leq x \leq 15$
$ = 0$ elsewhere.

a) To find the probability x has a value between 8 and 10, multiply the interval width (2) by $f(x)$. This equals $(2)(.1) = .2$.

b) The event "7 or less" is the interval between 5 and 7 = 2, and "greater than 12" is between 12 and 15 = 3. The total interval width of the event is 3+2 = 5. Hence its probability is $5(.1) = .50$

c) Since x can only be 15 or less, the probability $x \leq 20$ is 1.

d) The probability that x exactly equals 11 is 0 by definition.

PROBLEM 6

Telephone calls arrive at the Global Airline reservation office in Lemonville according to a Poisson distribution with a mean of 1.2 calls per minute.

a) What is the probability of receiving exactly one call during a one minute interval?

b) What is the probability of receiving at most 2 calls during a one minute interval?

c) What is the probability of receiving at least two calls during a one minute interval?

d) What is the probability of receiving exactly 4 calls during a <u>five</u> minute interval?

e) What is the probability that at most 2 minutes elapse between one call and the next?

SOLUTION 6

For this problem $\lambda = 1.2$ per minute.

a)
$$f(1) = \frac{1.2^1 e^{-1.2}}{1!} = (1.2)(.3012)/1 = .36.$$

b) The probability of receiving at most two calls is $f(0) + f(1) + f(2)$.

$$f(0) = \frac{(1.2)^0 e^{-1.2}}{0!} = .30 \qquad f(2) = \frac{(1.2)^2 e^{-1.2}}{2!} = \frac{(1.44)(.3012)}{2} = .22$$

Hence the probability is $.30 + .36 + .22 = .88$.

c) The probability of receiving at least two calls is 1 - the probability of receiving at most one call $= 1 - f(0) - f(1) = 1 - .30 - .36 = .34$.

d) To find the probability of receiving 4 calls in a five minute interval, one must first calculate g for this interval. This is simply $5(1.2) = 6$. Hence,

$$f(4) = \frac{6^4 e^{-6}}{4!} = \frac{(1296)(.0025)}{24} = .135.$$

e) To find the probability that at most two minutes elapse between two calls, use the exponential distribution with $\mu = \lambda = 1.2$.

$$P(x \leq 2) = 1 - e^{-(1.2)(2)} = 1 - e^{-2.4} = 1 - .0907 = .9093.$$

PROBLEM 7

The time at which the mailman delivers the mail to Ace Bike Shop follows a normal distribution with mean 2:00 PM and standard deviation of 15 minutes.

a) What is the probability the mail will arrive after 2:30 PM?

b) What is the probability the mail will arrive before 1:36 PM?

c) What is the probability the mail will arrive between 1:48 PM and 2:09 PM?

SOLUTION 7

The mail delivery time follows a normal distribution with mean μ = 2:00 PM and σ = 15 minutes. To transform any time x to a standard normal random variable z, $z = (x - \mu)/\sigma$.

a) The probability that the mail arrives after 2:30 is the probability that z is greater than (2:30 - 2:00)/15 or the $P(z > 2)$. See figure below. From Appendix C, $P(0 < z < 2) = .4772$. Hence, $P(z > 2) = .5000 - .4772 = .0228$.

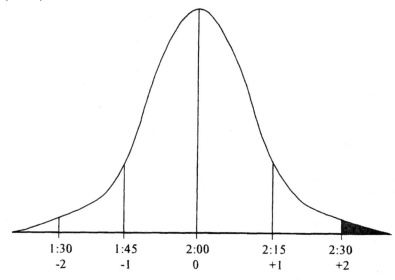

b) The probability the mail will arrive before 1:36 PM is equivalent to the $P(z < (1:36 - 2:00)/15) = P(z < -1.6) = .5000 - P(z$ is between -1.6 and $0) = .5000 - .4452 = .0548$.

c) The probability that the mail arrives between 1:48 PM and 2:09 PM is the same as the sum of the probabilities between 1:48 PM and 2:00 PM and between 2:00 PM and 2:09 PM. 1:48 has a z-value of (1:48 - 2:00)/15 = -0.80 and 2:09 has a z-value of (2:09 - 2:00)/15 = 0.60. From the table these two areas sum to .2257 + .2881 = .5138.

PROBLEM 8

A light bulb manufacturer claims his light bulbs will last 500 hours on the average. The lifetime of a light bulb is assumed to follow an exponential distribution.

a) What is the probability that the light bulb will have to replaced within 500 hours?

b) What is the probability that the light bulb will last more than 1000 hours?

c) What is the probability that the light bulb will last between 200 and 800 hours.

SOLUTION 8

a) As the average life of a light bulb is 500, using the exponential distribution with $\mu = 500$, gives

$$P(x \leq 500) = 1 - e^{-(500/500)} = 1 - e^{-1} = 1 - .368 = .632$$

b) $P(x > 1000) = e^{-(1000/500)} = e^{-2} = .135$

c) $P(200 \leq x \leq 800) = P(x > 200) - P(x > 800)$

$$= e^{-(200/500)} - e^{-(800/500)}$$

$$= e^{-.4} - e^{-1.6} = .670 - .202 = .468$$

ANSWERED PROBLEMS

PROBLEM 9

Ace Mopeds sells mopeds which on which it gives a one-month warranty. Over a one-year period the following data was compiled.

Month	Number Sold in Month	Number Returned For Warranty Service	Month	Number Sold in Month	Number Returned For Warranty Service
Jan	4	1	Jul	20	4
Feb	6	2	Aug	20	4
Mar	3	0	Sep	20	4
Apr	20	1	Oct	16	3
May	10	2	Nov	16	3
Jun	16	4	Dec	10	3

a) Based on this data determine the probability distribution for: (1) the number of mopeds sold in a month; (2) the number of mopeds returned for warranty in a given month; (3) the percentage of mopeds returned for warranty in a given month.

b) Which of these distributions are for discrete and which are for continuous random variables.

c) Which distribution would be of concern to (1) the quality control engineer at the moped factory; (2) the sales department at Ace; and, (3) the service department at Ace.

d) What is the probability that the percentage of Mopeds returned for service in a randomly selected month is between 10% and 21%?

e) What is the probability that the number of mopeds returned for service in a randomly selected month is greater than 1 and less than 4?

f) What is the expected number of mopeds Ace sells in a month?

g) What is the expected number of mopeds returned for service in a month?

h) What is the expected percentage of mopeds returned for service in a month?

i) Is the answer to part (h) equal to the answer in part (g) divided by the answer to part (f)?

PROBLEM 10

Two headache remedies, <u>Relief</u> and <u>Comfort</u> are waging an advertising campaign. Each claims it eliminates a headache faster. The data compiled by an independent testing agency is as follows:

Time (in minutes) after taking remedy	Percentage of headaches cured at that time using	
	Relief	Comfort
5	.60	0
10	0	.10
15	0	.75
20	0	.15
25	0	0
30	.40	0

a) What are the means and variances of the time until headache cure (1) using Relief; (2) using Comfort?

b) Which remedy has the maximum probability of relieving a headache within 10 minutes?

c) Which remedy has the maximum probability of relieving a headache within 20 minutes?

PROBLEM 11

Sandy's Pet Center grooms large and small dogs. It takes Sandy 40 minutes to groom a small dog and 70 minutes to groom a large dog. Large dogs account for 20% of Sandy's business. Sandy has 5 appointments on August 15.

a) What is the probability that all 5 dogs are small?

b) What is the probability that two of the dogs are large?

c) What is the expected amount of time to finish all five dogs? (Hint: Find the expected number of small and large dogs and multiply by the time required for grooming.)

PROBABILITY DISTRIBUTIONS   47

PROBLEM 12

The township of Middleton sets the speed limit on its roads by conducting a traffic study and determining the speed (to the nearest 5 miles per hour) at which 80% of the drivers travel at or below. A study was done on Brown's Dock Road which indicated that drivers' speeds follow a normal distribution with a mean of 36.25 miles per hour and a variance of 6.25.

a) What should the speed limit be?

b) What percent of the drivers travel below that speed?

PROBLEM 13

The Harbour Island Ferry leaves on the hour and at 15 minute intervals. The time, x, it takes John to drive from his house to the ferry has a uniform distribution with x between 10 and 20 minutes. One morning John leaves his house at precisely 8:00a.m.

a) What is the probability John will wait less than 5 minutes for the ferry?

b) What is the probability John will wait less than 10 minutes for the ferry?

c) What is the probability John will wait less than 15 minutes for the ferry?

d) What is the probability John will not have to wait for the ferry?

e) Suppose John leaves at 8:05a.m. What is the probability John will wait (1) less than 5 minutes for the ferry; (2) less than 10 minutes for the ferry?

f) Suppose John leaves at 8:10a.m. What is the probability John will wait (1) less than 5 minutes for the ferry; (2) less than 10 minutes for the ferry?

g) What appears to be the best time for John to leave home if he wishes to maximize the probability of waiting less than 10 minutes for the ferry?

PROBLEM 14

Dollar Department Stores has compiled the following data concerning its daily sales. The sales for each day of the week are normally distributed with the following parameters:

Day	Mean = μ	Std. Dev. = σ	Day	Mean = μ	Std. Dev. = σ
Monday	$120,000	$20,000	Thursday	$120,000	$40,000
Tuesday	$100,000	$25,000	Friday	$140,000	$20,000
Wednesday	$100,000	$10,000	Saturday	$160,000	$50,000

For each day of the week, find the probability that the store sales are between $110,000 and $150,000.

PROBLEM 15

June's Specialty Shop sells designer original dresses. On 10% of her dresses, June makes a profit of $10, on 20% of her dresses she makes a profit of $20, on 30 of her dresses she makes a profit of $30, and on 40% of her dresses she makes a profit of $40.

a) What is the expected profit June earns on the sale of a dress?

b) On a given day, the probability of June having no customers is .05, one customer is .10, two customers is .20, three customers is .35, four customers is .20, and five customers is .10. June's daily operating cost is $0 per day. Using your answer to (a), find the expected net profit June earns per day. (Hint: To find the expected daily gross profit, multiply the expected profit per dress by the expected number of customers per day.)

c) June is considering moving to a larger store. She estimates that doing so will double the expected number of customers. If the larger store will increase her operating costs to $100 per day, should she make the move?

PROBLEM 16

Chez Paul is an exclusive French restaurant that seats only 10 couples for dinner. Paul is famous for his "truffle salad for two" which must be prepared one day in advance. The probability of any couple ordering the salad is .4 and each couple orders independently of other couples.

a) What is the expected number of "truffle salads for two" that Paul serves per dinner? What is the variance?

b) What is the probability that on a given evening, at most three couples want a "truffle salad for two"?

c) How many salads should Paul prepare if he wants the probability of not having enough salads for all customers who desire one to be no greater than .10?

d) There is a 70% chance a couple will order coffee after dinner. What is the probability that on a given evening, exactly eight of the ten couples will order coffee? (Again assume that couples order independently of each other.)

PROBLEM 17

The number of customers at Winkies Donuts between 8:00a.m. and 9:00a.m. is believed to follow a Poisson distribution with a mean of 2 customers per minute.

a) During a randomly selected one minute interval during this time period, what is the probability of 6 customers arriving to Winkies?

b) What is the probability that at least 2 minutes elapse between customer arrivals?

PROBLEM 18

Delicious Candy markets a two pound box of assorted chocolates. Because of imperfections in the candy making equipment, the actual weight of the chocolate has a continuous uniform distribution ranging from 31.8 and 32.6 ounces.

a) Define a probability density function for the weight of the box of chocolate.

b) What is the probability that a box weighs (1) exactly 32 ounces; (2) more than 32.3 ounces; (3) less than 31.8 ounces?

c) The government requires that at least 60% of all products sold weigh at least as much as the stated weight. Is Delicious violating government regulations?

PROBLEM 19

Mark Investment Service is currently recommending the purchase of shares of Dollar Department Stores selling at 18 per share. Mark estimates that in one year the price of the shares will be at x, where x is a random variable which is approximately normally distributed with mean of 20 and a standard deviation of 2.

a) What is the probability that in a year the shares will be selling for (1) exactly $20; (2) more than $20; (3) less than $20; and (4) less than $18.

b) What is the expected profit per share within a year?

PROBLEM 20

Joe's Record World has two stores. The sales at each store follow a normal distribution. For store 1, $\mu = \$2,000$ and $\sigma = \$200$ per day; for store 2, $\mu = \$1,900$ and $\sigma = \$400$ per day.

a) Which store has the higher average daily sales?

b) What is the probability that daily sales are greater than $2,200 for store 1? for store 2?

c) Is there a contradiction between parts (a) and (b)? Explain.

PROBLEM 21

During lunch time, customers arrive at Bob's Drugs according to a Poisson distribution with $\lambda = 4$ per minute.

a) During a one-minute interval, determine the following probabilities: (1) no arrivals; (2) one arrival; (3) two arrivals; and, (4) three or more arrivals.

b) What is the probability of two arrivals in a two-minute period?

c) What is the probability that no more than 30 seconds elapses between arrivals?

TRUE/FALSE

___ 22. For a probability distribution, the variance will always be greater than its mean.

___ 23. If a random variable can take on a countably infinite number of values, one would use a discrete distribution.

___ 24. If one wanted to find the probability of ten customer arrivals in an hour at a service station, one would generally use the Poisson distribution.

___ 25. To use the standard normal distribution, one subtracts the mean from the variable value and divides this result by the variance.

___ 26. The variance of a random variable will always be greater than its standard deviation.

___ 27. An arrival process follows a Poisson distribution with a mean of 5 per hour. The probability of 0 arrivals in an hour is the same as the probability of the interarrival time being at least 1 hour when computed using an exponential distribution with $\mu = 5$.

___ 28. The probability of exactly 3 heads in 6 tosses of a fair coin is 1/2.

___ 29. Suppose the weight of adult males follows a normal distribution with a mean of 160 lbs. and a standard deviation of 15 lbs. Then the probability that a randomly selected male weighs exactly 160 is .50.

___ 30. Suppose the lifetime of a particular appliance follows an exponential distribution with a mean of 10 years. The probability the appliance will fail within ten years is .50.

___ 31. Suppose that the arrival time for the next bus at a bus stop is uniformly distributed between 1 and 25 minutes. The probability that the bus will arrive within the next 5 minutes is 1/6.

___ 32. The binomial distribution is most symmetric when p equals 0.5.

___ 33. An exponential distribution, like the Poisson distribution, can be described by a single parameter.

___ 34. Each of the five probability distributions studied can be classified as always left-skewed, always right-skewed, or never skewed.

___ 35. A Bernoulli process is a sequence of independent trials with two possible outcomes whose probabilities can change between trials.

___ 36. The probability of a continuous variable having a specific value is 0.

CHAPTER 4
Decision Analysis

52 CHAPTER 4

KEY CONCEPTS

CONCEPT	ILLUSTRATED PROBLEMS	ANSWERED PROBLEMS
Payoff Tables	1,2,3	6-14
Nonprobabilistic Decision Criteria: Optimistic, Conservative, Minimax Regret Approaches	1,2	6,7,8,10,13
Spreadsheet Approach	1	
Probabilistic Decision Criteria: Expected Monetary Value, Expected Opportunity Loss	1-4	6-14
Spreadsheet Approach	3	
Decision Trees	1,3,4	8-14
Expected Value of Perfect Information	1,3	7,8,9,11,12,14
Spreadsheet Approach	3	
Bayes' Rule: Revising Probabilities	3,4	9,11,12,14
Spreadsheet Approach	3	
Expected Value of Sample Information	3	9,11,12,14
Efficiency of Sample Information	3	9,11
Sensitivity Analysis	5	9
Influence Diagrams	3	

REVIEW

1. A <u>decision problem</u> is characterized by decision alternatives, states of nature, and resulting payoffs.

2. The <u>decision alternatives</u> are the different strategies the decision maker can employ.

DECISION ANALYSIS 53

3. The <u>states of nature</u> refer to future events, not under the control of the decision maker, which may occur. States of nature should be defined so that they are mutually exclusive and collectively exhaustive.

4. For each decision alternative and state of nature, there is a resulting <u>payoff</u>. These are often represented in matrix form called a <u>payoff table</u>.

5. A decision is said to <u>dominate</u> another decision if the payoffs for every state of nature for one is at least equal to the corresponding payoffs for the other and is greater for at least one state of nature.

6. One way to solve a complex decision problem is by the use of a <u>decision tree</u>. This is a <u>chronological representation</u> of the decision problem.

7. Each decision tree has two types of nodes. <u>Round nodes</u> correspond to the states of nature while <u>square nodes</u> correspond to the decision alternatives. The branches leaving each round node represent the different states of nature while the branches leaving each square node represent the different decision alternatives.

8. At the end of each limb of a decision tree are the payoffs attained from the series of branches making up that limb. To solve the problem one "<u>folds back the tree</u>", working backwards from the ends of the branches towards the root node of the tree.

9. <u>Decision making under certainty</u> occurs when the decision maker knows with certainty which state of nature will occur.

10. If the decision maker does not know with certainty which state of nature will occur, then he is said to be doing <u>decision making under uncertainty</u>.

11. Three <u>commonly used criteria</u> for decision making under uncertainty when probability information regarding the likelihood of the states of nature is unavailable are: (1) the optimistic, (2) the conservative, and (3) the minimax regret approach.

12. The <u>optimistic approach</u> would be used by an optimistic decision maker. The decision with the largest possible payoff is chosen. (If the payoff table were in terms of costs, the decision that had the lowest cost would be chosen.)

13. The <u>conservative approach</u> would be used by a conservative decision maker. For each decision the minimum payoff is listed. Then the decision corresponding to the maximum of these minimum payoffs is selected. Hence, the minimum possible payoff is maximized. (If the payoffs were in terms of costs, the maximum costs would be determined for each decision. The decision having the minimum of these maximum costs would be selected.)

14. The <u>minimax regret approach</u> requires the construction of a <u>regret table</u> or an <u>opportunity loss table</u>. This is done by calculating for each state of nature the difference between each payoff and the largest payoff for that state of nature. Then, using this regret table, the maximum regret for each possible decision is listed. The decision chosen is the one corresponding to the minimum of the maximum regrets.

54   CHAPTER 4

15. If probabilistic information regarding he states of nature is available, one may use the <u>expected value (EV)</u> approach. Here the expected return for each decision is calculated by summing the products of the payoff under each state of nature and the probability of the respective state of nature occurring. The decision yielding the best expected return is chosen.

16. <u>Risk analysis</u> is the study of the possible payoffs associated with a decision alternative or a decision strategy. A tool used in risk analysis is graphically depicting the <u>risk profile</u> for a decision alternative. The risk profile shows the possible payoffs for the decision alternative along with their associated probabilities. Sometimes a review of the risk profile associated with an optimal decision alternative may cause the decision maker to choose another decision alternative even though the expected value of the other alternative is not as good.

17. <u>Sensitivity analysis</u> can be used to determine the probability range over which a decision will remain optimal. Such an analysis can provide a better perspective on management's original judgment regarding the state of nature probabilities.

18. Frequently information is available which can improve the probability estimates for the states of nature. The <u>expected value of perfect information (EVPI)</u> is the <u>increase</u> in the expected profit that would result if one knew with certainty which state of nature would occur. This quantity provides an upper bound on the expected value of any sample or survey information. EVPI can be calculated as follows:
    (1) determine the optimal return corresponding to each state of nature;
    (2) compute the expected value of these optimal returns;
    (3) subtract the EV of the optimal decision from the amount determined in step (2).

19. Knowledge of sample or survey information can be used to revise the probability estimates for the states of nature. Prior to obtaining this information, the probability estimates for the states of nature are called <u>prior probabilities</u>. With knowledge of <u>conditional probabilities</u> for the outcomes or indicators of the sample or survey information, these prior probabilities can be revised by employing <u>Bayes' Theorem</u>. The outcomes of this analysis are called <u>posterior probabilities</u>.

20. <u>Posterior probabilities</u> are calculated as follows:
    (1) For each state of nature, multiply the prior probability by its conditional probability for the indicator -- this gives the <u>joint probabilities</u> for the states and indicator.
    (2) Sum these joint probabilities over all states -- this gives the <u>marginal probability</u> for the indicator.
    (3) For each state, divide its joint probability by the marginal probability for the indicator -- this gives the posterior probability distribution.

21. The <u>expected value of sample information (EVSI)</u> is the additional expected profit possible through knowledge of the sample or survey information. EVSI is calculated as follows:
    (1) determine the optimal decision and its expected return for the possible outcomes of the sample or survey using the posterior probabilities for the states of nature;
    (2) compute the expected value of these optimal returns;
    (3) subtract the EV of the optimal decision obtained without using the sample information from the amount determined in step (2).

DECISION ANALYSIS 55

22. <u>Efficiency</u> of sample information is the ratio of EVSI to EVPI. As the EVPI provides an upper bound for the EVSI, efficiency is always a number between 0 and 1.

23. An <u>influence diagram</u> is a graphical device that shows the relationship among decisions, chance events, and consequences for a decision problem. Such a device is particularly useful in understanding and analyzing more-complex decision problems.

FLOW CHART FOR DECISION MAKING WITHOUT PROBABILITIES

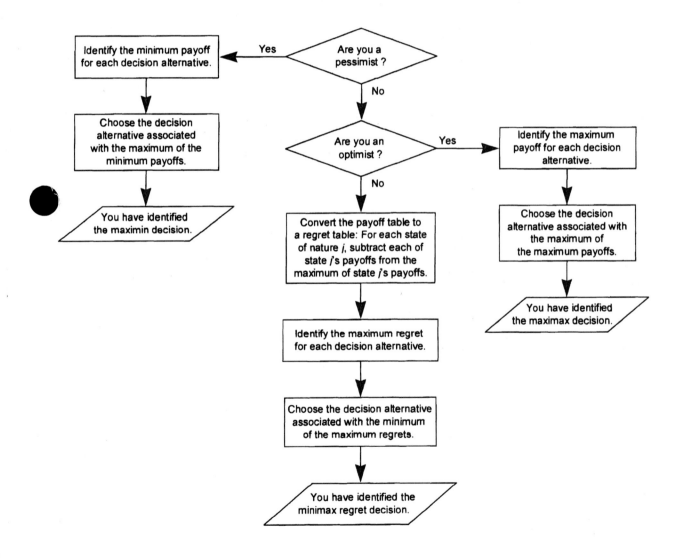

CHAPTER 4

FLOW CHART FOR DECISION MAKING WITH PROBABILITIES (PART A)

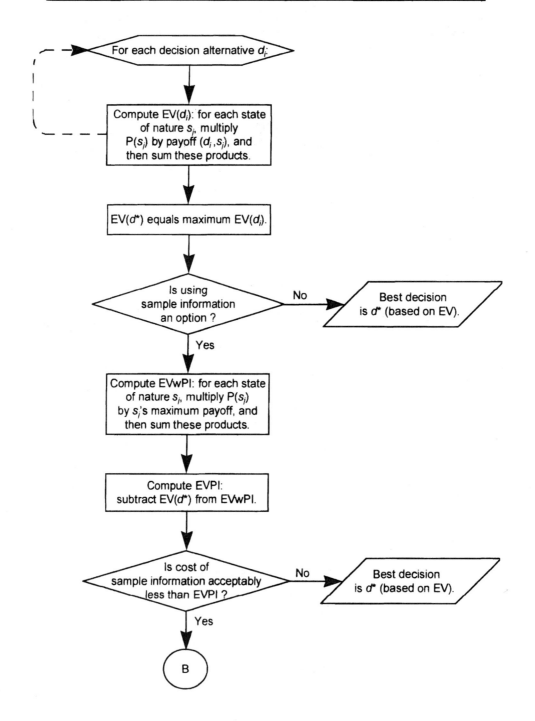

FLOW CHART FOR DECISION MAKING WITH PROBABILITIES (PART B)

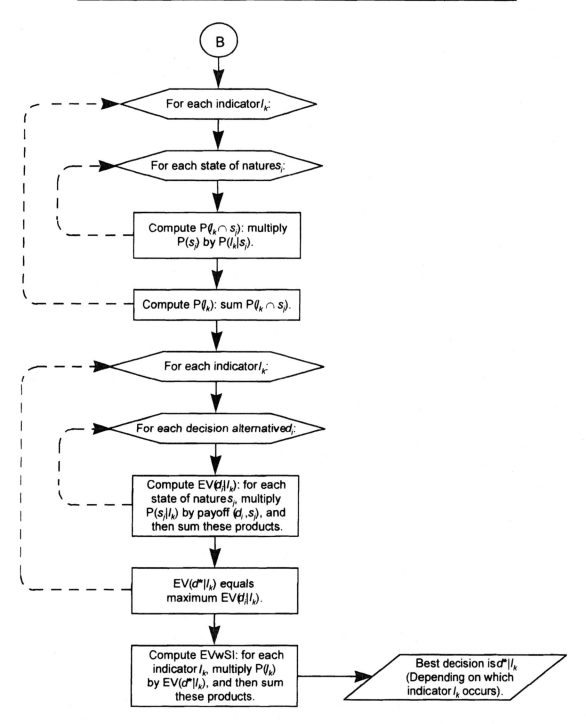

CHAPTER 4

ILLUSTRATED PROBLEMS

PROBLEM 1

Consider the following problem with three decision alternatives and three states of nature with the following payoff table representing profits:

		\multicolumn{3}{c}{States of Nature}		
		s_1	s_2	s_3
Decisions	d_1	4	4	-2
	d_2	0	3	-1
	d_3	1	5	-3

a) What is the optimal decision if the decision maker were conservative?

b) What is the optimal decision if the decision maker were optimistic?

c) What is the optimal decision using the minimax regret approach?

d) Use spreadsheets to find the optimal decisions in parts a), b), and c) above.

e) Use a decision tree to find the optimal decision if $P(s_1) = .20$, $P(s_2) = .50$, $P(s_3) = .30$.

f) Given the probabilities in (d), calculate the EVPI.

SOLUTION 1

a) A conservative decision maker would use the conservative approach. List the minimum payoff for each decision. Choose the decision with the maximum of these minimum payoffs.

Decision	Minimum Payoff
d_1	-2
d_2	-1 (maximum, choose d_2)
d_3	-3

b) An optimistic decision maker would use the optimistic approach. All we really need to do is to choose the decision that has the largest single value in the payoff table. This largest value is 5, and hence the optimal decision is d_3.

c) For the minimax regret approach, first compute a regret table by subtracting each payoff in a column from the largest payoff in that column. In this example, in the first column subtract 4, 0, and 1 from 4; in the second column, subtract 4, 3, and 5 from 5; etc.

DECISION ANALYSIS

The resulting regret table is:

	s_1	s_2	s_3
d_1	0	1	1
d_2	4	2	0
d_3	3	0	2

Then, for each decision list the maximum regret. Choose the decision with the minimum of these values:

Decision	Maximum Regret
d_1	1 (minimum, choose d_1)
d_2	4
d_3	3

d) **Conservative Approach**

Formula Spreadsheet for Conservative Approach

	A	B	C	D	E	F
1	PAYOFF TABLE					
2						
3	Decision	State of Nature			Minimum	Recommended
4	Alternative	s1	s2	s3	Payoff	Decision
5	d1	4	4	-2	=MIN(B5:D5)	=IF(E5=$E$9,A5,"")
6	d2	0	3	-1	=MIN(B6:D6)	=IF(E6=$E$9,A6,"")
7	d3	1	5	-3	=MIN(B7:D7)	=IF(E7=$E$9,A7,"")
8						
9		Best Payoff			=MAX(E5:E7)	

Solution Spreadsheet for Conservative Approach

	A	B	C	D	E	F
1	PAYOFF TABLE					
2						
3	Decision	State of Nature			Minimum	Recommended
4	Alternative	s1	s2	s3	Payoff	Decision
5	d1	4	4	-2	-2	
6	d2	0	3	-1	-1	d2
7	d3	1	5	-3	-3	
8						
9		Best Payoff			-1	

Optimistic Approach

Formula Spreadsheet for Optimistic Approach

	A	B	C	D	E	F
1	PAYOFF TABLE					
2						
3	Decision	State of Nature			Maximum	Recommended
4	Alternative	s1	s2	s3	Payoff	Decision
5	d1	4	4	-2	=MAX(B5:D5)	=IF(E5=$E$9,A5,"")
6	d2	0	3	-1	=MAX(B6:D6)	=IF(E6=$E$9,A6,"")
7	d3	1	5	-3	=MAX(B7:D7)	=IF(E7=$E$9,A7,"")
8						
9			Best Payoff		=MAX(E5:E7)	

Solution Spreadsheet for Optimistic Approach

	A	B	C	D	E	F
1	PAYOFF TABLE					
2						
3	Decision	State of Nature			Maximum	Recommended
4	Alternative	s1	s2	s3	Payoff	Decision
5	d1	4	4	-2	4	
6	d2	0	3	-1	3	
7	d3	1	5	-3	5	d3
8						
9			Best Payoff		5	

Minimax Regret Approach

Formula Spreadsheet for Minimax Regret Approach

	A	B	C	D	E	F
1	PAYOFF TABLE					
2	Decision	State of Nature				
3	Altern.	s1	s2	s3		
4	d1	4	4	-2		
5	d2	0	3	-1		
6	d3	1	5	-3		
7						
8	OPPORTUNITY LOSS TABLE					
9	Decision	State of Nature			Maximum	Recommended
10	Altern.	s1	s2	s3	Regret	Decision
11	d1	=MAX($B$4:$B$6)-B4	=MAX($C$4:$C$6)-C4	=MAX($D$4:$D$6)-D4	=MAX(B11:D11)	=IF(E11=$E$14,A11,"")
12	d2	=MAX($B$4:$B$6)-B5	=MAX($C$4:$C$6)-C5	=MAX($D$4:$D$6)-D5	=MAX(B12:D12)	=IF(E12=$E$14,A12,"")
13	d3	=MAX($B$4:$B$6)-B6	=MAX($C$4:$C$6)-C6	=MAX($D$4:$D$6)-D6	=MAX(B13:D13)	=IF(E13=$E$14,A13,"")
14			Minimax Regret Value		=MIN(E11:E13)	

Solution Spreadsheet for Minimax Regret Approach

1	PAYOFF TABLE					
2	Decision	State of Nature				
3	Alternative	s1	s2	s3		
4	d1	4	4	-2		
5	d2	0	3	-1		
6	d3	1	5	-3		
7						
8	OPPORTUNITY LOSS TABLE					
9	Decision	State of Nature			Maximum	Recommended
10	Alternative	s1	s2	s3	Regret	Decision
11	d1	0	1	1	1	**d1**
12	d2	4	2	0	4	
13	d3	3	0	2	3	
14		Minimax Regret Value			1	

e) The tree diagram looks as follows:

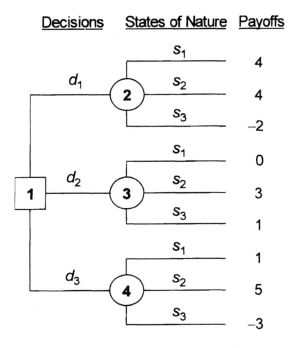

To calculate the expected values at nodes 2, 3, and 4, multiply the payoffs by the corresponding probabilities and then sum. The decision with the maximum expected value of 2.2, d_1, is then chosen.

$$EV(\text{Node 2}) = .20(4) + .50(4) + .30(-2) = 2.2 \text{ (maximum, choose } d_1\text{)}$$
$$EV(\text{Node 3}) = .20(0) + .50(3) + .30(-1) = 1.2$$
$$EV(\text{Node 4}) = .20(1) + .50(5) + .30(-3) = 1.8$$

62 CHAPTER 4

> NOTE: The optimal decision based on minimizing expected regret will always be the same as the optimal decision based on maximizing expected payoff. If you compute the expected regret for each decision, using the regret table in part (c), you will find that d_1 is optimal just as it is in part (d) above.

f) The EVPI is calculated by multiplying the maximum payoff for each state by the corresponding probability, summing these values, and then subtracting the expected value of the optimal decision from this sum.
Thus, the EVPI = $[.20(4) + .50(5) + .30(-1)] - 2.2 = .8$

> NOTE: There is an alternative approach to computing the EVPI. EVPI also equals the expected regret associated with the minimax decision. If you compute the expected regret associated with d_1, using the regret table in part (c), you will see that it agrees with the answer to part (e) above.

PROBLEM 2

Jim has been employed at Gold Key Realty at a salary of $2,000 per month during the past year. Because Jim is considered to be a top salesman, the manager of Gold Key is offering him one of three salary plans for the next year: (1) a 25% raise to $2,500 per month; (2) a base salary of $1,000 plus $600 per house sold; or, (3) a straight commission of $1,000 per house sold. Over the past year, Jim has sold up to 6 homes in a month.

a) Compute the monthly salary payoff table for Jim.

b) For this payoff table find Jim's optimal decision using: (1) the conservative approach, (2) minimax regret approach.

c) Suppose that during the past year the following is Jim's distribution of home sales. If one assumes that this a typical distribution for Jim's monthly sales, which salary plan should Jim select?

Home Sales	Number of Months
0	1
1	2
2	1
3	2
4	1
5	3
6	2

DECISION ANALYSIS

SOLUTION 2

a) There are three decision alternatives (salary plans) and seven states of nature (the number of houses sold monthly).

	\multicolumn{7}{c}{Number of Homes Sold}
	0	1	2	3	4	5	6
Plan I	2500	2500	2500	2500	2500	2500	2500
Plan II	1000	1600	2200	2800	3400	4000	4600
Plan III	0	1000	2000	3000	4000	5000	6000

b) (1) <u>Conservative Approach</u>

Decision	Minimum Payoff
Plan I	2500 (maximum; choose Plan I)
Plan II	1000
Plan III	0

(2) <u>Minimax Regret Approach</u>

Construct a regret table by subtracting all numbers in a column from the maximum number in the column:

<u>Regret Table</u>

	\multicolumn{7}{c}{Number of Homes Sold}
	0	1	2	3	4	5	6
Plan I	0	0	0	500	1500	2500	3500
Plan II	1500	900	300	200	600	1000	1400
Plan III	2500	1500	500	0	0	0	0

Choose the decision with the minimum of the maximum regrets.

Decision	Maximum Regret
Plan I	3500
Plan II	1500 (minimum; choose Plan II)
Plan III	2500

c) Use the relative frequency method for determining the probabilities:

Homes Sold	Frequency	Probability
0	1	1/12
1	2	2/12
2	1	1/12
3	2	2/12
4	1	1/12
5	3	3/12
6	2	2/12

Use the expected value (EV) approach:

EV(Plan I) = 1/12(2500) + 2/12(2500) + 1/12(2500) + 2/12(2500)
             + 1/12(2500) + 3/12(2500) + 2/12(2500) = 2500

EV(Plan II) = 1/12(1000) + 2/12(1600) + 1/12(2200) + 2/12(2800)
              + 1/12(3400) + 3/12(4000) + 2/12(4600) = 3050

EV(Plan III) = 1/12(0) + 2/12(1000) + 1/12(2000) + 2/12(3000)
               + 1/12(4000) + 3/12(5000) + 2/12(6000) = 3417

Choose Plan III, the plan with the highest EV.

PROBLEM 3

Burger Prince Restaurant is contemplating opening a new restaurant on Main Street. It has three different models, each with a different seating capacity. Burger Prince estimates that the average number of customers per hour will be 80, 100, or 120. The payoff table for the three models is as follows:

	Average Number of Customers Per Hour		
	$s_1 = 80$	$s_2 = 100$	$s_3 = 120$
Model A	$10,000	$15,000	$14,000
Model B	$ 8,000	$18,000	$12,000
Model C	$ 6,000	$16,000	$21,000

Burger Prince estimates the probability of 80 customers per hour is the same as the probability of 120 customers per hour and twice as much as the probability of 100 customers per hour.

a) What is the optimal decision using the expected value approach?

b) What is the expected value of perfect information?

c) Answer parts a) and b) above using a spreadsheet.

d) Contrast the risk profiles of the Model A and Model C decision alternatives.

e) Burger Prince must decide whether or not to purchase a marketing survey from Stanton Marketing for $1,000. The results of the survey are "favorable" or "unfavorable". The conditional probabilities are:

P(favorable | 80 customers per hour) = .2
P(favorable | 100 customers per hour) = .5
P(favorable | 120 customers per hour) = .9

Should Burger Prince have the survey performed by Stanton Marketing?

DECISION ANALYSIS   65

f) Answer part e) above using a spreadsheet.

g) What is the efficiency of the survey?

h) Construct an influence diagram for this problem.

SOLUTION 3

a) (1) Determine the probabilities for 80, 100, and 120 customers:
   Given: $P(80) = P(120)$
           $P(80) = 2P(100)$.
           $P(80) + P(100) + P(120) = 1$
   Thus: $P(80) + .5P(80) + P(80) = 1$
           $2.5P(80) = 1 \longrightarrow P(80) = .4 \longrightarrow P(100) = .2, P(120) = .4$.

(2) Calculate the expected value for each decision. The following decision tree can assist in this calculation. Here d_1, d_2, d_3 represent the decision alternatives of models A, B, C, and s_1, s_2, s_3 represent the states of nature of 80, 100, and 120.

Building Model	Customers Per Hour	Payoff
Model A (d_1) — 2	$s_1 = 80$ $P(s_1) = .4$	$10,000
	$s_2 = 100$ $P(s_2) = .2$	$15,000
	$s_3 = 120$ $P(s_3) = .4$	$14,000
Model B (d_2) — 3	$s_1 = 80$ $P(s_1) = .4$	$ 8,000
	$s_2 = 100$ $P(s_2) = .2$	$18,000
	$s_3 = 120$ $P(s_3) = .4$	$12,000
Model C (d_3) — 4	$s_1 = 80$ $P(s_1) = .4$	$ 6,000
	$s_2 = 100$ $P(s_2) = .2$	$16,000
	$s_3 = 120$ $P(s_3) = .4$	$21,000

Calculating the expected value for each decision gives:

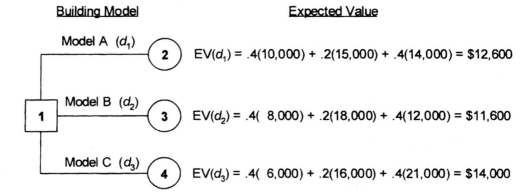

(3) Choose the model with largest EV -- Model C.

b) Calculate the expected value for the optimum payoff for each state of nature and subtract the EV of the optimal decision: EVPI= .4(10,000) + .2(18,000) + .4(21,000) - 14,000 = $2,000.

c) Using a spreadsheet:

Formula Spreadsheet for Expected Value Approach

	A	B	C	D	E	F
1	PAYOFF TABLE					
2						
3	Decision	State of Nature			Expected	Recommended
4	Alternative	s1 = 80	s2 = 100	s3 = 120	Value	Decision
5	Model A	10,000	15,000	14,000	=$B$8*B5+$C$8*C5+$D$8*D5	=IF(E5=$E$9,A5,"")
6	Model B	8,000	18,000	12,000	=$B$8*B6+$C$8*C6+$D$8*D6	=IF(E6=$E$9,A6,"")
7	Model C	6,000	16,000	21,000	=$B$8*B7+$C$8*C7+$D$8*D7	=IF(E7=$E$9,A7,"")
8	Probability	0.4	0.2	0.4		
9		Maximum Expected Value			=MAX(E5:E7)	

Spreadsheet for Expected-Value Decision and EVPI

	A	B	C	D	E	F
1	PAYOFF TABLE					
2						
3	Decision	State of Nature			Expected	Recommended
4	Alternative	s1 = 80	s2 = 100	s3 = 120	Value	Decision
5	d1 = Model A	10,000	15,000	14,000	12600	
6	d2 = Model B	8,000	18,000	12,000	11600	
7	d3 = Model C	6,000	16,000	21,000	14000	d3 = Model C
8	Probability	0.4	0.2	0.4		
9		Maximum Expected Value			14000	
10						
11		Maximum Payoff			EVwPI	EVPI
12		10,000	18,000	21,000	16000	2000

d) Risk profile for Model A alternative: Risk profile for Model C alternative:

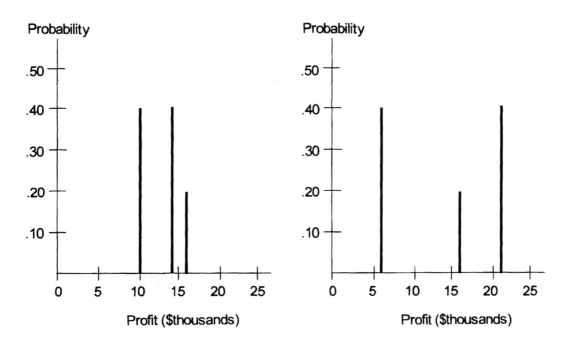

A decision maker <u>might</u> prefer the more equal payoffs associated the Model A decision alternative, even thought its expected value is less than that of the Model C alternative.

e) Find the posterior probabilities:

Favorable

State	Prior	Conditional	Joint	Posterior*
80	.4	.2	.08	.148
100	.2	.5	.10	.185
120	.4	.9	.36	.667
		Total	.54	1.000

Unfavorable

State	Prior	Conditional	Joint	Posterior*
80	.4	.8	.32	.696
100	.2	.5	.10	.217
120	.4	.1	.04	.087
		Total	.46	1.000

Note: * Posterior probability = (Joint Probability)/(Sum of Joint Probabilities)

68 CHAPTER 4

The decision tree for this problem:

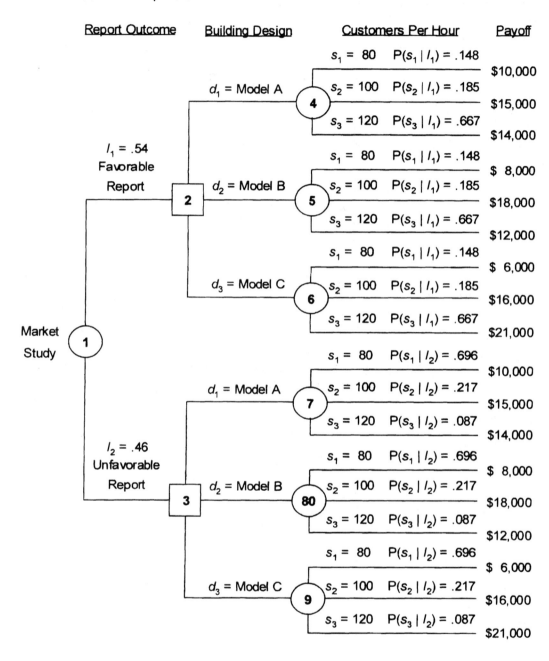

DECISION ANALYSIS

Calculate expected values:

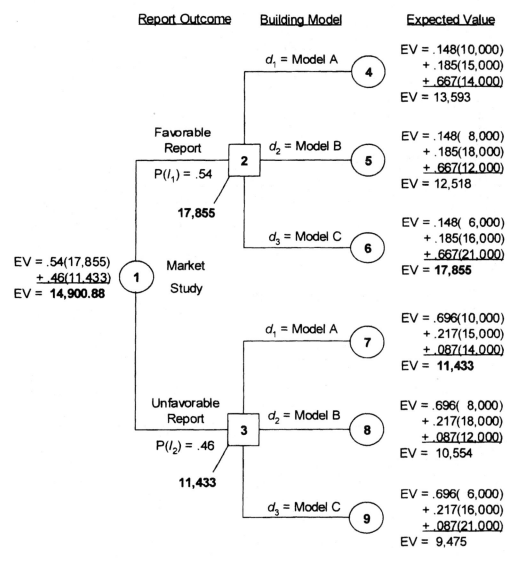

Hence, if the outcome of the survey is "favorable" choose Model C. If it is unfavorable, choose model A.

EVSI = .54($17,855) + .46($11,433) - $14,000 = $900.88

Since this is less than the cost of the survey, the survey should not be purchased.

f) Formula Spreadsheet for Posterior Probabilities

	A	B	C	D	E
1	Market Research Favorable				
2		Prior	Conditional	Joint	Posterior
3	State of Nature	Probabilities	Probabilities	Probabilities	Probabilities
4	s1 = 80	0.4	0.2	=B4*C4	=D4/$D$7
5	s2 = 100	0.2	0.5	=B5*C5	=D5/$D$7
6	s3 = 120	0.4	0.9	=B6*C6	=D6/$D$7
7			P(Favorable) =	=SUM(D4:D6)	
8					
9	Market Research Unfavorable				
10		Prior	Conditional	Joint	Posterior
11	State of Nature	Probabilities	Probabilities	Probabilities	Probabilities
12	s1 = 80	0.4	0.8	=B12*C12	=D12/$D$15
13	s2 = 100	0.2	0.5	=B13*C13	=D13/$D$15
14	s3 = 120	0.4	0.1	=B14*C14	=D14/$D$15
15			P(Unfavorable) =	=SUM(D12:D14)	

Solution Spreadsheet for Posterior Probabilities

	A	B	C	D	E
1	Market Research Favorable				
2		Prior	Conditional	Joint	Posterior
3	State of Nature	Probabilities	Probabilities	Probabilities	Probabilities
4	s1 = 80	0.4	0.2	0.08	0.148
5	s2 = 100	0.2	0.5	0.10	0.185
6	s3 = 120	0.4	0.9	0.36	0.667
7			P(Favorable) =	0.54	
8					
9	Market Research Unfavorable				
10		Prior	Conditional	Joint	Posterior
11	State of Nature	Probabilities	Probabilities	Probabilities	Probabilities
12	s1 = 80	0.4	0.8	0.32	0.696
13	s2 = 100	0.2	0.5	0.10	0.217
14	s3 = 120	0.4	0.1	0.04	0.087
15			P(Favorable) =	0.46	

g) The efficiency = EVSI/EVPI = ($900.88)/($2000) = .4504.

h) Influence diagram:

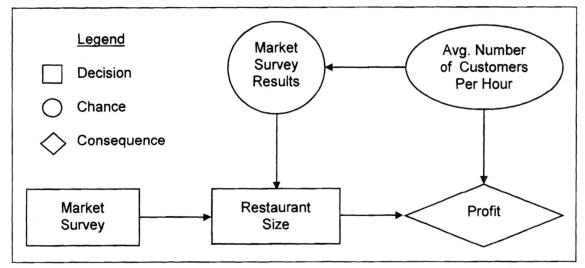

PROBLEM 4

The past few years have seen a general decline in the economic conditions of the resort community of Pacific City. Certain town officials, as well as other interested parties from Las Vegas, believe that the legalization of casino gambling will greatly improve the city's future. They have succeeded in having a gambling referendum placed on the ballot in next month's special election.

One effect of the declining economic condition of Pacific City has been the bankruptcy last year of the St. Carlton Inn, the city's largest hotel. The St. Carlton is scheduled to be sold at a sealed bid auction next week. The terms of the auction call for an immediate 10% down payment by the highest bidder, with the remaining balance due in two months.

One party interested in bidding on the St. Carlton is Justin Thyme, a real estate promoter and publisher of Fantasy Magazine. Justin has learned from associates that there are at least three other parties interested in the property, including a syndicate from Chicago.

Justin has decided that if he submits a bid it will be for $5 million for the property. If he bids $5 million, he estimates that he has a 40% chance of winning the auction. He also estimates that if the gambling referendum passes, he can sell the property for $7.5 million. If the initiative fails, he will sacrifice his 10% deposit.

Justin believes, based on polls taken to date, that the gambling referendum has a 30% chance of passing. He is considering hiring the noted pollster, Harris Gallup, before the auction to give his opinion on the outcome of the referendum. Based on Gallup's past record, Justin estimates that the probability Gallup will correctly predict the outcome is .8. Gallup's fee is $100,000.

What should Justin do?

SOLUTION 4

To solve this problem, first construct a tree diagram with possible courses of actions and outcomes properly sequenced.

Justin's first decision is whether or not to hire Gallup. If he does, then he will either obtain a prediction from him that the referendum will pass or fail. At that point, or if he does not hire Gallup, he must decide whether or not to bid.

Following this, he will learn if his bid is a winning one. If it is, then he will next be concerned with whether the referendum passes or not. If it does, he sells his property for $7.5 million. If it does not pass, he sacrifices his $.5 million deposit.

The tree diagram for this problem:

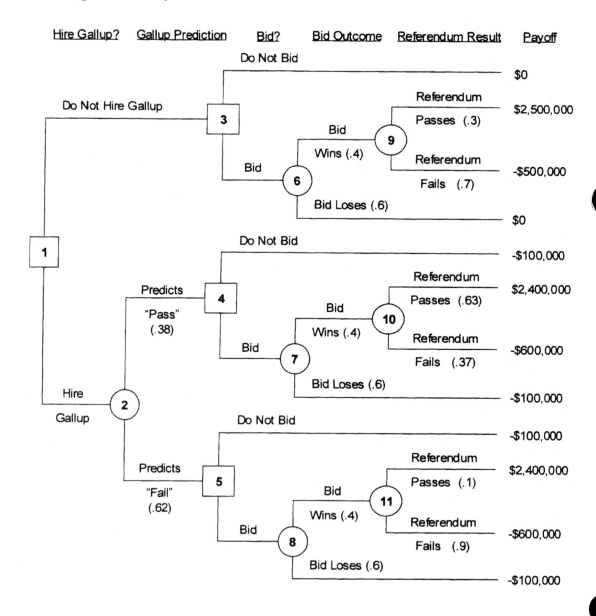

DECISION ANALYSIS   73

Beginning at the ends of the tree and working towards the root gives the expected returns on the following page:

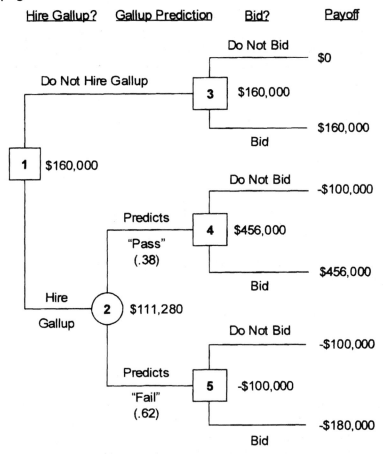

Therefore, if Justin does not hire Gallup, he should bid. If he does hire Gallup, he should bid if Gallup predicts the referendum will pass, and he should not bid if Gallup predicts the referendum will fail.

The expected return if Justin hires Gallup is: .38(456,000) + .62(-100,000) = $111,280
The expected return if Justin does not hire Gallup = $160,000. Thus, his optimal strategy is not to hire Gallup and submit a $5 million bid.

The probabilities for the Pass/Fail branches were obtained as follows:

Indicator Information I_1 -- Gallup Predicts Pass

State	Prior	Conditional	Joint	Posterior
Referendum Passes	.30	.80	.24	.63
Referendum Fails	.70	.20	.14	.37
		Total	.38	1.00

Indicator Information I_2 -- Gallup Predicts Fail

State	Prior	Conditional	Joint	Posterior
Referendum Passes	.30	.20	.06	.10
Referendum Fails	.70	.80	.56	.90
		Total	.62	1.00

PROBLEM 5

East West Distributing is in the process of trying to determine where they should schedule next year's production of a popular line of kitchen utensils that they distribute. Manufacturers in four different countries have submitted bids to East West. However, a pending trade bill in Congress will greatly affect the cost to East West due to proposed tariffs, favorable trading status, etc.

After careful analysis, East West has determined the following cost breakdown for the four manufacturers (in $1,000's) based on whether or not the trade bill passes:

	Bill Passes	Bill Fails
Country A	260	210
Country B	320	160
Country C	240	240
Country D	275	210

a) If East West estimates that there is a 40% chance of the bill passing, which country should they choose for manufacturing?

b) Over what range of values for the "bill passing" will the solution in part (a) remain optimal?

SOLUTION 5

a) Using the expected value approach, calculate the expected value for each country. Note that country D's costs are dominated by country A's costs (whether the bill passes or not, production is at least as expensive in country D as in country A) and therefore country D need not be considered. (Note that the probability that the bill will fail is $1 - .4 = .6$.)

$$EV(A) = .4(260) + .6(210) = 230$$
$$EV(B) = .4(320) + .6(160) = 224$$
$$EV(C) = .4(240) + .6(240) = 240$$

East West should choose the country with the lowest expected cost: country B.

b) To determine the range for the probability of the bill passing over which country B will be optimal, compare choosing country B versus country A, and then compare choosing country B versus country C. Now let,

p = the probability of the trade bill passing.

Country B would be preferred to country A as long as:

$$EV(B) \leq EV(A)$$

or

$$p(320) + (1-p)160 \leq p(260) + (1-p)(210)$$
$$160p + 160 \leq 50p + 210$$
$$110p \leq 50$$
$$p \leq .455$$

Similarly, Country B would be preferred to C as long as EV(B) ≤ EV(C). Using the same approach as above, this is equivalent to $p(320) + (1-p)160 \leq 240$. Solving, $p \leq .50$.

Thus as long as the probability of the bill passing is less than .455, then East West should choose Country B. The following graph illustrates the expected value of the decisions as a function of p:

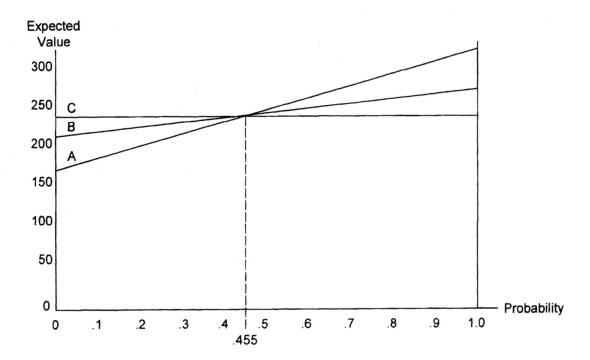

ANSWERED PROBLEMS

PROBLEM 6

Mark Investment Advisors has just completed an analysis on the returns of five utility stocks for next year. Mark hypothesizes that next year the economy will either be in a depression (s_1), a recession (s_2), an upward period (s_3), or a major expansionary period (s_4). The per dollar growth for the five stocks corresponding to each state of the economy are:

	s_1	s_2	s_3	s_4
Stock A	-.40	-.20	+.10	+.60
Stock B	-.30	-.10	0	+.30
Stock C	-.10	0	+.05	+.30
Stock D	0	+.05	+.10	+.15
Stock E	+.05	+.15	-.10	-.20

a) If you could purchase shares of Stock B or Stock C, which would you purchase? Why?

b) Find the best stock to purchase under each approach: (1) optimistic; (2) minimax regret; (3) conservative.

c) Suppose Mark's financial outlook for next year is $P(s_1) = .2$; $P(s_2) = .4$; $P(s_3) = .2$; and $P(s_4) = .2$. Using the expected value approach, which stock would Mark recommend?

PROBLEM 7

Transrail is bidding on a project that it figures will cost $400,000 to perform. Using a 25% markup, it will charge $500,000, netting a profit of $100,000. However, it has been learned that another company, Rail Freight, is also considering bidding on the project. If Rail Freight does submit a bid, it figures to be a bid of about $470,000.

Transrail really wants this project and is considering a bid with only a 15% markup to $460,000 to ensure winning regardless of whether or not Rail Freight submits a bid.

a) Prepare a profit payoff table from Transrail's point of view.

b) What decision would be made if Transrail were conservative?

c) If Rail Freight is known to submit bids on only 25% of the projects it considers, what decision should Transrail make?

d) Given the information in (c), how much would a corporate spy be worth to Transrail to find out if Rail Freight will bid?

DECISION ANALYSIS

PROBLEM 8

The Super Cola Company must decide whether or not to introduce a new diet soft drink. Management feels that if it does introduce the diet soda it will yield a profit of $1 million if sales are around 100 million, a profit of $200,000 if sales are around 50 million, or it will lose $2 million if sales are only around 1 million bottles. If Super Cola does not market the new diet soda, it will suffer a loss of $400,000.

a) Construct a payoff table for this problem.

b) Construct a regret table for this problem.

c) Should Super Cola introduce the soda if the company: (1) is conservative; (2) is optimistic; (3) wants to minimize its maximum disappointment?

d) An internal marketing research study has found P(100 million in sales) = 1/3; P(50 million in sales) = 1/2; P(1 million in sales) = 1/6. Should Super Cola introduce the new diet soda?

e) A consulting firm can perform a more thorough study for $275,000. Should management have this study performed?

PROBLEM 9

Super Cola is also considering the introduction of a root beer drink. The company feels that the probability that the product will be a success is .6. The payoff table is as follows:

	Success (s_1)	Failure (s_2)
Produce (d_1)	$250,000	-$300,000
Do Not Produce (d_2)	-$50,000	-$20,000

The company has a choice of two research firms to obtain information for this product. Stanton Marketing has market indicators, I_1 and I_2 for which $P(I_1|s_1) = .7$ and $P(I_1|s_2) = .4$. New World Marketing has indicators J_1 and J_2 for which $P(J_1|s_1) = .6$ and $P(J_1|s_2) = .3$.

a) What is the optimal decision if neither firm is used? Over what probability of success range is this decision optimal?

b) What is the EVPI?

c) Find the EVSIs and efficiencies for Stanton and New World.

d) If both firms charge $5,000, which firm should be hired?

e) If Stanton charges $10,000 and New World charges $4,000, which firm should Super Cola hire? Why?

PROBLEM 10

Metropolitan Cablevision is investigating the installation of a cable TV system in town. The engineering department estimates the cost of the system (in present worth dollars) to be $700,000. The sales department has investigated four pricing plans. For each pricing plan the marketing division has estimated the revenue per household in present worth dollars to be:

Plan	Revenue Per Household
I	$15
II	$18
III	$20
IV	$24

The sales department estimates the number of household subscribers would be approximately either 10,000, 20,000, 30,000, 40,000, 50,000 or 60,000.

a) Construct a payoff table for this problem.

b) What is the company's optimal decision under (1) the optimistic approach; (2) the conservative approach; and (3) the minimax regret approach?

c) Suppose the sales department has determined that the number of subscribers will be a function of the pricing plan. The probability distributions for the pricing plans are given below. Use decision tree analysis to determine which pricing plan is optimal under the expected value approach.

Number of Subscribers	Probability under Pricing Plan			
	I	II	III	IV
10,000	0	.05	.10	.20
20,000	.05	.10	.20	.25
30,000	.05	.20	.20	.25
40,000	.40	.30	.20	.15
50,000	.30	.20	.20	.10
60,000	.20	.15	.10	.05

PROBLEM 11

Dicom Corporation has developed a new high speed computer that it intends to sell for $150,000. Dicom's salesmen have scheduled demonstrations with four clients next month. For each client, Dicom estimates there is a 40% chance of his purchasing the computer.

Dicom plans to make at least one computer next month but could make as many as four. Production costs are $100,000 for producing one computer, $190,000 for two, $260,000 for three, and $310,000 for four.

Any unsold computers produced will be sold at $60,000. Also, if a client wants to purchase a computer, but all computers produced have been sold, the company estimates it loses $20,000.

DECISION ANALYSIS   79

a) Determine a payoff table for this problem.

b) What is the optimal strategy using the expected monetary value criterion? (HINT: Use the binomial distribution to determine the probabilities for the states of nature.)

c) Dicom can have a survey performed to indicate the market impression of the new computer -- favorable or unfavorable. The following probabilities are believed to hold:

P(favorable | 0 sold) = .1 P(favorable | 3 sold) = .9
P(favorable | 1 sold) = .2 P(favorable | 4 sold) = 1.0
P(favorable | 2 sold) = .6

How much should Dicom pay for this survey and what is its efficiency?

PROBLEM 12

Dollar Department Stores has just acquired the chain of Wenthrope and Sons Custom Jewelers. Dollar has received an offer from Harris Diamonds to purchase the Wenthrope store on Grove Street for $120,000.

Dollar has determined probability estimates of the store's future profitability, based on economic outcomes, as: P($80,000) = .2, P($100,000) = .3, P($120,000) = .1, and P($140,000) = .4.

a) Should Dollar sell the store on Grove Street?

b) What is the EVPI?

c) Dollar can have an economic forecast performed, costing $10,000, that produces indicators I_1 and I_2, for which $P(I_1 | 80,000)$ = .1; $P(I_1 | 100,000)$ = .2; $P(I_1 | 120,000)$ = .6; $P(I_1 | 140,000)$ = .3. Should Dollar purchase the forecast?

PROBLEM 13

Cashman Co. will be leasing a new copier and is considering four plans.

Plan	Monthly Lease	Unit Copy Cost
I	$100	$.020 for the first 10,000; $.016 thereafter
II	$200	$.012 for all copies
III	$150	first 5,000 free; $.022 thereafter
IV	$300	$.005 for all copies

The company has determined it will make either 12,600, 14,400, 16,200, 18,000, 19,800, 21,600 copies per month with probabilities of .05, .10, .15, .25, .25, and .20 respectively.

a) Construct a monthly payoff table for Cashman in terms of costs.

b) What is the optimal plan if the company uses the (1) optimistic approach or (2) the conservative approach?

c) What is the optimal plan using the expected value approach?

PROBLEM 14

An appliance dealer must decide how many (if any) new microwave ovens to order for next month. The ovens cost $220 and sell for $300. Because the oven company is coming out with a new product line in two months, any ovens not sold next month will have to be sold at the dealer's half price clearance sale. Additionally, the appliance dealer feels he suffers a loss of $25 for every oven demanded when he is out of stock. On the basis of past months' sales data, the dealer estimates the probabilities of monthly demand (D) for 0, 1, 2, or 3 ovens to be .3, .4, .2, and .1, respectively.

The dealer is considering conducting a telephone survey on the customers' attitudes towards microwave ovens. The results of the survey will either be favorable (F), unfavorable (U) or no opinion (N). The dealer's probability estimates for the survey results based on the number of units demanded are:

$P(F|D=0) = .1$ $P(F|D=2) .3$ $P(U|D=0) = .8$ $P(U|D=2) = .1$
$P(F|D=1) = .2$ $P(F|D=3) .9$ $P(U|D=1) = .3$ $P(U|D=3) = .1$

a) What is the dealer's optimal decision without conducting the survey?

b) What is the EVPI?

c) Based on the survey results what is the optimal decision strategy for the dealer?

d) What is the maximum amount he should pay for this survey?

DECISION ANALYSIS

TRUE/FALSE

15. The expected value of sample information can never be less than the expected value of perfect information.

16. The expected value of perfect information must always be nonnegative.

17. The expected value of sample information is the difference between the expected value with perfect information and the expected value without perfect information.

18. Using the optimistic and conservative approaches will never give the same optimal decision.

19. The $P(I_k \mid s_j)$ must equal $P(s_j \mid I_k)$.

20. For each indicator, I_k, the joint probabilities must sum to 1.

21. Posterior probabilities are calculated by dividing each joint probability by the sum of the joint probabilities for that indicator.

22. The sum of $P(I_k \mid s_j)$ for all s_j is 1.0.

23. Sample information with an efficiency rating of 100% is perfect information.

24. Sensitivity analysis on state-of-nature probability estimates cannot be performed on problems with more than two states-of-nature.

25. If the expected value of sample information exceeds the cost of purchasing the sample information, one should not attempt to obtain the sample information.

26. The states of nature in a decision problem must be mutually exclusive and collectively exhaustive.

27. Maximizing the expected payoff and minimizing the expected opportunity loss result in the same recommended decision.

28. The expected value approach is more appropriate for a one-time decision than a repetitive decision.

29. $P(I_k \mid s_j)$ is a posterior probability.

CHAPTER 5

Utility and Game Theory

CHAPTER 5

KEY CONCEPTS

CONCEPT	ILLUSTRATED PROBLEMS	ANSWERED PROBLEMS
Determining Utilities	3,4	10,11,13,14,15
Decision Making Using Utility	1,2,3	8,9,10,11,12,13
Attitudes Towards Risk	1,2	8,9,10,11,13,14
Two-Person Zero-Sum Games		
Pure Strategy	5	16
Mixed Strategy	6	17
Dominated Strategies	7	17

REVIEW

1. <u>Utility</u> is a measure of the total worth of a particular outcome, reflecting the decision maker's attitude towards a collection of factors. Some of these factors may be profit, loss, and risk. Utilities are used in decision making when the decision criteria must be based on more than just expected monetary values.

2. One common way to define a utility measure is the concept of a <u>lottery</u>. To use the lottery concept, define the utility of the best payoff to be 1 and the utility of the worst payoff to be 0. The decision maker is then asked to assign a probability, p, that would make him indifferent between choosing the payoff, or playing a lottery in which he will get the best payoff with probability p and the worst payoff with probability $1-p$. These <u>indifference probabilities</u> define the utility function for the decision maker.

3. Once a utility function has been determined, the optimal decision can be chosen using the <u>expected utility approach</u>. Here, for each decision alternative, the utility corresponding to each state of nature is multiplied by the probability for that state of nature. The sum of these products for each decision alternative represents the expected utility for that alternative. The decision alternative with the highest expected utility is chosen.

4. Utility values are influenced by the attitude of the decision maker towards risk. A <u>risk avoider</u> will have a concave utility function when utility is measured on the vertical axis and monetary value is measured on the horizontal axis. Individuals purchasing insurance exhibit risk avoidance behavior.

UTILITY AND GAME THEORY

5. <u>Risk takers</u>, such as gamblers, pay a premium to obtain risk. Their utility function is convex. This reflects the decision maker's increasing marginal value for money.

6. A <u>risk neutral</u> decision maker has a linear utility function. In this case, the expected value approach can be used.

7. Most individuals' risk avoidance or risk taking depends on the amount of money involved. For some relatively large amounts of money they are risk avoiders; for relatively small amounts of money they are risk takers. This explains why the same individual will both purchase insurance and also a lottery ticket.

8. While determining utility values is not an easy task, such an analysis should be performed in cases where payoffs can assume extremely high or extremely low values.

9. In decision analysis, a single decision maker seeks to select an optimal decision alternative. In <u>game theory</u>, there are two or more decision makers, called <u>players</u>, who compete as adversaries against each other.

10. In game theory, each player selects a strategy independently without knowing in advance the strategy of the other player(s).

11. In game theory, the combination of strategies chosen by the players determines the <u>value of the game</u> to each player.

12. <u>Game theory applications</u> have been developed for situations where the competing players are teams, companies, political candidates, armies, and contract bidders.

13. <u>Two-person zero-sum games</u> involve two players and the gain (or loss) for one player is equal to the corresponding loss (or gain) for the other player.

14. A <u>pure strategy</u> is a game solution in which there is a single best strategy for each player.

15. <u>Saddle point</u> is a condition which exists when pure strategies are optimal for both players in a two-person zero-sum game.

16. The saddle point is at the intersection of the optimal strategies for the players and the value of the saddle point is the <u>value of the game</u>.

17. A <u>mixed strategy</u> is a game solution in which the player randomly selects the strategy to play from among several strategies with positive probabilities.

18. The solution to the mixed-strategy game identifies the <u>probabilities (relative frequencies)</u> that each player should use to randomly select the strategy to play.

19. A <u>dominated strategy</u> exists if another strategy is at least as good regardless of what the opponent does. A dominated strategy will never be selected by the player and can be eliminated in order to reduce the size of the game.

CHAPTER 5

FLOW CHART FOR DECISION MAKING WITH UTILITY

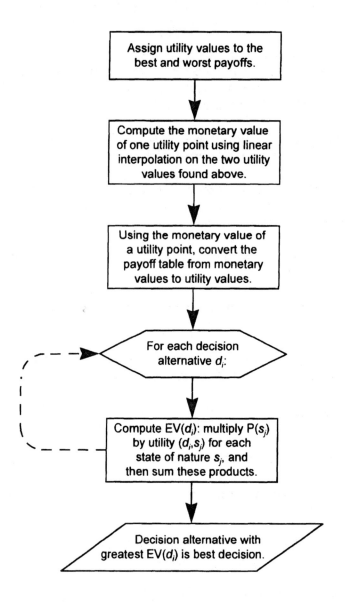

UTILITY AND GAME THEORY 87

FLOW CHART OF TWO-PERSON ZERO-SUM GAME SOLUTION PROCEDURE

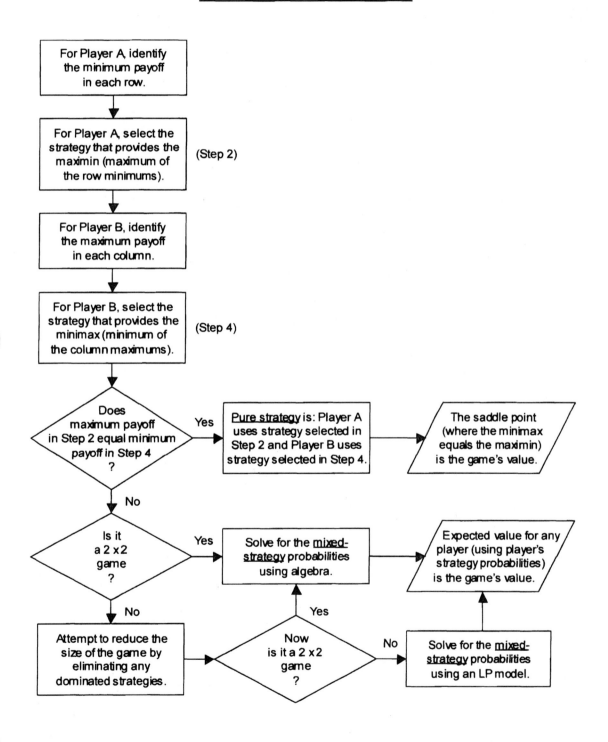

CHAPTER 5

ILLUSTRATED PROBLEMS

PROBLEM 1

Consider the following three state, four decision problem with the following payoff table (in $'s):

	s_1	s_2	s_3
d_1	+100,000	+40,000	-60,000
d_2	+ 50,000	+20,000	-30,000
d_3	+ 20,000	+20,000	-10,000
d_4	+ 40,000	+20,000	-60,000

The probabilities for the three states of nature are: $P(s_1) = .1$, $P(s_2) = .3$, and $P(s_3) = .6$.

a) If the decision maker is risk neutral, what is the optimal decision?

b) Suppose two decision makers have the following utility values:

Amount	Utility Decision Maker I	Utility Decision Maker II
$100,000	100	100
$ 50,000	94	58
$ 40,000	90	50
$ 20,000	80	35
-$ 10,000	60	18
-$ 30,000	40	10
-$ 60,000	0	0

Graph the utility curves for the two decision makers.

c) Classify each of the two decision makers as either a risk avoider, a risk taker, or risk neutral.

d) Find the optimal decision for each of the decision makers.

SOLUTION 1

a) If the decision maker is risk neutral, the expected value approach is applicable.
$$EV(d_1) = .1(100{,}000) + .3(40{,}000) + .6(-60{,}000) = -\$14{,}000$$
$$EV(d_2) = .1(50{,}000) + .3(20{,}000) + .6(-30{,}000) = -\$7{,}000$$
$$EV(d_3) = .1(20{,}000) + .3(20{,}000) + .6(-10{,}000) = +\$2{,}000$$

Note the EV for d_4 need not be calculated as decision d_4 is dominated by decision d_2. The optimal decision is d_3.

b)

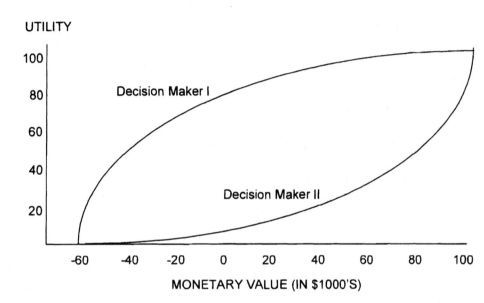

MONETARY VALUE (IN $1000'S)

c) Decision Maker I has a concave utility function -- he is a risk avoider. Decision Maker II has a convex utility function -- he is a risk taker.

d) Note that d_4 is dominated by d_2 and hence is not considered. For the other decisions, compute the expected utility for each decision maker using the probabilities given.

Utility Table I

	s_1	s_2	s_3	Expected Utility
d_1	100	90	0	37.0
d_2	94	80	40	57.4
d_3	80	80	60	**68.0**
Probability	.1	.3	.6	

The largest expected utility is 68.0. Decision Maker I should make decision d_3.

Utility Table II

	s_1	s_2	s_3	Expected Utility
d_1	100	50	0	**25.0**
d_2	58	35	10	22.3
d_3	35	35	18	24.8
Probability	.1	.3	.6	

The largest expected utility is 25.0. Decision Maker II should make decision d_1.

PROBLEM 2

In problem 1 suppose the probabilities for the three states of nature are changed to: $P(s_1) = .5$, $P(s_2) = .3$, and $P(s_3) = .2$.

a) Find the optimal decision for a risk neutral decision maker.

b) Find the optimal decision for Decision Makers I and II.

c) Approximately what is the value of this decision problem to Decision Maker I and Decision Maker II? What conclusion can you draw from this?

SOLUTION 2

a) Use the EV approach for a risk neutral decision maker:

$EV(d_1) = .5(100,000) + .3(40,000) + .2(-60,000) =$ **50,000**
$EV(d_2) = .5(50,000) + .3(20,000) + .2(-30,000) = 25,000$
$EV(d_3) = .5(20,000) + .3(20,000) + .2(-10,000) = 14,000$

Hence, the risk neutral optimal decision is d_1.

b) Use the expected utility criterion for each decision maker:

Decision Maker I

$EU(d_1) = .5(100) + .3(90) + .2(0) = 77.0$
$EU(d_2) = .5(94) + .3(80) + .2(40) =$ **79.0**
$EU(d_3) = .5(80) + .3(80) + .2(60) = 76.0$

Hence, Decision Maker I's optimal decision is d_2.

Decision Maker II

$EU(d_1) = .5(100) + .3(50) + .2(0) =$ **65.0**
$EU(d_2) = .5(58) + .3(35) + .2(10) = 41.5$
$EU(d_3) = .5(35) + .3(35) + .2(18) = 31.6$

Hence, Decision Maker II's optimal decision is d_1.

c) To determine the value of the decision problem to a decision maker, find the amount that corresponds to the expected utility of the optimal decision.

For Decision Maker I, the optimal expected utility is 79. He assigned a utility of 80 to +$20,000, and a utility of 60 to -$10,000. Linearly interpolating in this range 1 point is worth $30,000/20 = $1,500. Thus a utility of 79 is worth about $20,000 - $1,500 = $18,500.

For Decision Maker II, the optimal expected utility is 65. He assigned a utility of 100 to 100,000, and a utility of 58 to $50,000. In this range, 1 point is worth $50,000/42 = $1190.

Thus a utility of 65 is worth about $50,000 + 7($1190) = $58,330. Thus, the decision problem is worth more to Decision Maker II.

PROBLEM 3

Each project manager at Rockford Research Corp. is given three development projects per year to supervise. The manager's salary is determined by the number of successfully completed projects.

Managers have a choice of three salary payment plans:

	Number Of Successfully Completed Projects			
	0	1	2	3
Plan 1	25,000	25,000	25,000	25,000
Plan 2	15,000	20,000	30,000	50,000
Plan 3	20,000	25,000	25,000	30,000

For a lottery having a payoff of $50,000 with probability p and $15,000 with probability $(1-p)$, three managers expressed the following indifference probabilities:

	Indifference Probability (p)		
Amount	Manager 1	Manager 2	Manager 3
$30,000	.5	.6	.8
$25,000	.4	.4	.5
$20,000	.3	.2	.1

If the probability of successfully completing a single project is .5, determine the best salary plan for each of the three managers.

SOLUTION 3

Using the binomial distribution with the probability of success = .5, determine the probabilities of the number of projects successfully completed to be P(0) = .125, P(1) = .375, P(2) = .375, P(3) = .125. For each manager construct the utility table corresponding to his indifference probabilities:

Manager 1

	Number of Projects Completed				Expected Utility
	0	1	2	3	
Plan 1	.4	.4	.4	.4	.4000
Plan 2	0	.3	.5	1.0	**.4250**
Plan 3	.3	.4	.4	.5	.4000
Probability	.125	.375	.375	.125	

Manager 1 should choose plan 2.

Manager 2

	Number of Projects Completed				Expected Utility
	0	1	2	3	
Plan 1	.4	.4	.4	.4	.4000
Plan 2	0	.2	.6	1.0	**.4125**
Plan 3	.2	.4	.4	.6	.4000
Probability	.125	.375	.375	.125	

Manager 2 should choose plan 2.

Manager 3

	Number of Projects Completed				Expected Utility
	0	1	2	3	
Plan 1	.5	.5	.5	.5	**.5000**
Plan 2	0	.1	.8	1.0	.4623
Plan 3	.1	.5	.5	.8	.4875
Probability	.125	.375	.375	.125	

Manager 3 should choose plan 1.

PROBLEM 4

Mark Investment Advisors have analyzed the profit potential of five different investments. The probabilities of the gains on $1000 are as follows:

Investment	Gain			
	$0	$200	$500	$1000
A	.9	0	0	.1
B	0	.8	.2	0
C	.05	.9	0	.05
D	0	.8	.1	.1
E	.6	0	.3	.1

One of Mark's investors informs Mark that he is indifferent between investments A, B, and C. Based on this information, would he prefer investment D to investment E? Why?

SOLUTION 4

Assign a utility of 10 to a $1000 gain and a utility of 0 to a gain of $0. Let x = the utility of a $200 gain and y = the utility of a $500 gain. The expected utility on investment A is then $.9(0) + .1(10) = 1$.

Since the investor is indifferent between investments A and C, this must mean the expected utility of investment C = the expected utility of investment A = 1. But the expected utility of investment C = $.05(0) + .90x + .05(10)$. Since this must equal 1, solving for x, gives $x = 5/9$.

Also since the investor is indifferent between A, B, and C, the expected utility of investment B must be 1. Thus, $0(0) + .8(5/9) + .2y + 0(10) = 1$. Solving for y, gives $y = 25/9$. Thus the utility values for gains of 0, 200, 500, and 1000 are 0, 5/9, 25/9, and 10, respectively.

For investment D, $EU(D) = 0(0) + .8(5/9) + .1(25/9) + .1(10) = 1.72$.
For investment E, $EU(E) = .6(0) + 0(5/9) + .3(25/9) + .1(10) = 1.83$.

Thus the investor should prefer investment E.

PROBLEM 5

Suppose that there are only two vehicle dealerships (A and B) in a small city. Each dealership is considering three strategies that are designed to take sales of new vehicles from the other dealership over a period of four months. The strategies, assumed to be the same for both dealerships, are:

Strategy 1: Offer a cash rebate on a new vehicle.
Strategy 2: Offer free optional equipment on a new vehicle.
Strategy 3: Offer a 0% loan on a new vehicle.

The payoff table (in number of new vehicle sales gained per week by Dealership A (or lost by Dealership B) is shown below.

Dealership A		Dealership B Cash Rebate b_1	Dealership B Free Options b_2	Dealership B 0% Loan b_3
Cash Rebate	a_1	2	2	1
Free Options	a_2	-3	3	-1
0% Loan	a_3	3	-2	0

Identify the pure strategy for this two-person zero-sum game. What is the value of the game?

SOLUTION 5

Step 1: Compute the minimum payoff for each row (Dealership A).
Step 2: For Dealership A, select the strategy that provides the maximum of the row minimums (the maximin payoff).

Dealership B

Dealership A		Cash Rebate b_1	Free Options b_2	0% Loan b_3	Minimum Payoff	
Cash Rebate	a_1	2	2	1	1	Maximin
Free Options	a_2	-3	3	-1	-3	
0% Loan	a_3	3	-2	0	-2	

Strategy a_1 (Cash Rebate) provides the <u>maximin</u> (1 new vehicle sale).

Step 3: Compute the maximum payoff for each column (Dealership B).
Step 4: For Dealership B, select the strategy that provides the minimum of the column maximums (the minimax payoff).

Dealership B

Dealership A		Cash Rebate b_1	Free Options b_2	0% Loan b_3	Minimum Payoff
Cash Rebate	a_1	2	2	1	1
Free Options	a_2	-3	3	-1	-3
0% Loan	a_3	3	-2	0	-2
Maximum Payoff		3	3	1	

Minimax

Strategy b_3 (0% Loan) provides the <u>minimax</u> (1 new vehicle sale).

Step 5: If the maximin value equals the minimax value, an optimal pure strategy exists for both players. The value of the game is given by the value at the saddle point (where the optimal strategies for both dealerships intersect in the payoff table).

		Dealership B		
		Cash Rebate	Free Options	0% Loan
Dealership A		b_1	b_2	b_3
Cash Rebate	a_1	2	2	**1**
Free Options	a_2	-3	3	-1
0% Loan	a_3	3	-2	0

The minimax (1) and maximin (1) are equal. An optimal pure strategy exists for this game. The value of the game (the value at the saddle point) is 1 new vehicle.

Summary of the solution to the game:

Dealership A should offer a cash rebate on new vehicles.
Dealership A can expect to gain a minimum of 1 new vehicle sale per week.
Dealership B should offer a 0% loan on new vehicles.
Dealership B can expect to lose a maximum of 1 new vehicle sale per week.

PROBLEM 6

Consider the following two-person zero-sum game. Assume the two players have the same two strategy options. The payoff table shows the gains for Player A.

	Player B	
Player A	Strategy b_1	Strategy b_2
Strategy a_1	4	8
Strategy a_2	11	5

Determine the optimal strategy for each player. What is the value of the game?

SOLUTION 6

Step 1: Compute the minimum payoff for each row (Player A).
Step 2: For Player A, select the strategy that provides the maximum of the row minimums (the maximin payoff).
Step 3: Compute the maximum payoff for each column (Player B).
Step 4: For Player B, select the strategy that provides the minimum of the column maximums (the minimax payoff).

	Player B		Minimum Payoff	
Player A	Strategy b_1	Strategy b_2		
Strategy a_1	4	8	4	
Strategy a_2	11	5	5	Maximin
Max. Payoff	11	8		

Minimax

Strategy a_2 provides the <u>maximin</u> for Player A. Strategy b_2 provides the <u>minimax</u> for Player B.

Step 5: If the maximin value does <u>not</u> equal the minimax value, a mixed strategy is best. In this case, solve for the optimal mixed-strategy probabilities for each player.

<u>Player A</u>:

p = the probability Player A selects strategy a_1
$(1 - p)$ = the probability Player A selects strategy a_2

If Player B selects b_1: EV = $4p + 11(1 - p)$
If Player B selects b_2: EV = $8p + 5(1 - p)$

To solve for the optimal probabilities for Player A we set the two expected values equal and solve for the value of p.

$$4p + 11(1 - p) = 8p + 5(1 - p)$$
$$4p + 11 - 11p = 8p + 5 - 5p$$
$$11 - 7p = 5 + 3p$$
$$-10p = -6$$
$$p = .6$$
$$\text{and } (1 - p) = .4$$

Player A should select Strategy a_1 with a .6 probability and Strategy a_2 with a .4 probability.

<u>Player B</u>:

q = the probability Player B selects strategy b_1
$(1 - q)$ = the probability Player B selects strategy b_2

If Player A selects a_1: EV = $4q + 8(1 - q)$
If Player A selects a_2: EV = $11q + 5(1 - q)$

To solve for the optimal probabilities for Player B we set the two expected values equal and solve for the value of q.

$$4q + 8(1 - q) = 11q + 5(1 - q)$$
$$4q + 8 - 8q = 11q + 5 - 5q$$
$$8 - 4q = 5 + 6q$$
$$-10q = -3$$
$$q = .3$$
$$\text{and } (1 - q) = .7$$

Player B should select Strategy b_1 with a .3 probability and Strategy b_2 with a .7 probability.

Value of the Game:

For Player A: EV = $4p + 11(1 - p) = 4(.6) + 11(.4) = 6.8$ = expected gain
For Player B: EV = $4q + 8(1 - q) = 4(.3) + 8(.7) = 6.8$ = expected loss

PROBLEM 7

Consider the following two-person zero-sum game. Assume the two players have the same three strategy options. The payoff table shows the gains for Player A.

	Player B		
Player A	Strategy b_1	Strategy b_2	Strategy b_3
Strategy a_1	6	5	-2
Strategy a_2	1	0	3
Strategy a_3	3	4	-3

Is there an optimal pure strategy for this game? If so, what is it? If not, can the mixed-strategy probabilities be found algebraically?

SOLUTION 7

Step 1: Compute the minimum payoff for each row (Player A).
Step 2: For Player A, select the strategy that provides the maximum of the row minimums (the maximin payoff).
Step 3: Compute the maximum payoff for each column (Player B).
Step 4: For Player B, select the strategy that provides the minimum of the column maximums (the minimax payoff).

	Player B			Minimum Payoff	
Player A	Strategy b_1	Strategy b_2	Strategy b_3		
Strategy a_1	6	5	-2	-2	
Strategy a_2	1	0	3	0	Maximin
Strategy a_3	3	4	-3	-3	
Max. Payoff	6	5	3		
			Minimax		

Strategy a_2 provides the <u>maximin</u> for Player A. Strategy b_3 provides the <u>minimax</u> for Player B.

Step 5: If the maximin value does <u>not</u> equal the minimax value, a mixed strategy is best. In this case, solve for the optimal mixed-strategy probabilities for each player.

This is a 3 x 3 game. In order to solve algebraically for the probabilities, the game must be a 2 x 2. If there are dominated strategies, we might be able to reduce the game to a 2 x 2. (Otherwise, linear programming – which is beyond the scope of the textbook – must be used to solve for the probabilities.)

First, we look to see if any of Player A's strategies are dominated (meaning another strategy is at least as good for every strategy that the opponent may employ).

	Player B		
Player A	Strategy b_1	Strategy b_2	Strategy b_3
Strategy a_1	6	5	-2
Strategy a_2	1	0	3
Strategy a_3	3	4	-3

Player A's Strategy a_3 is dominated by Strategy a_1, so it can be eliminated.

Next, we look to see if any of Player B's strategies are dominated.

	Player B		
Player A	Strategy b_1	Strategy b_2	Strategy b_3
Strategy a_1	6	5	-2
Strategy a_2	1	0	3

Player B's Strategy b_2 is dominated by Strategy b_1, so it can be eliminated.

We now have the following 2 x 2 game:

	Player B	
Player A	Strategy b_1	Strategy b_3
Strategy a_1	6	-2
Strategy a_2	1	3

We can now solve algebraically for the optimal mixed-strategy probabilities.

<u>Player A</u>:

p = the probability Player A selects strategy a_1
$(1 - p)$ = the probability Player A selects strategy a_2

If Player B selects b_1:   EV = $6p + 1(1 - p)$
If Player B selects b_3:   EV = $-2p + 3(1 - p)$

To solve for the optimal probabilities for Player A we set the two expected values equal and solve for the value of p.

$$6p + 1(1 - p) = -2p + 3(1 - p)$$
$$1 + 5p = 3 + -5p$$
$$10p = 2$$
$$p = .2$$
$$\text{and } (1 - p) = .8$$

Player A should select Strategy a_1 with a .2 probability and Strategy a_2 with a .8 probability.

<u>Player B</u>:

q = the probability Player B selects strategy b_1
$(1 - q)$ = the probability Player B selects strategy b_3

If Player A selects a_1:   EV = $6q - 2(1 - q)$
If Player A selects a_2:   EV = $1q + 3(1 - q)$

To solve for the optimal probabilities for Player B we set the two expected values equal and solve for the value of q.

$$6q - 2(1 - q) = 1q + 3(1 - q)$$
$$-2 + 8q = 3 - 2q$$
$$10q = 5$$
$$q = .5$$
$$\text{and } (1 - q) = .5$$

Player B should select Strategy b_1 with a .5 probability and Strategy b_3 with a .5 probability.

<u>Value of the Game</u>:

For Player A: EV = $6p + 1(1 - p) = 6(.2) + 1(.8) = 2$ = expected gain
For Player B: EV = $6q - 2(1 - q) = 6(.5) - 2(.5) = 2$ = expected loss

ANSWERED PROBLEMS

PROBLEM 8

Burger Prince Restaurant is considering the purchase of a $100,000 fire insurance policy. The fire statistics indicate that in a given year the probability of property damage in a fire is as follows:

Fire Damage	$100,000	$75,000	$50,000	$25,000	$10,000	$0
Probability	.006	.002	.004	.003	.005	.980

a) If Burger Prince was risk neutral, how much would they be willing to pay for fire insurance?

b) If Burger Prince has the utility values given below, approximately how much would they be willing to pay for fire insurance?

Amount Of Loss	$100,000	$75,000	$50,000	$25,000	$10,000	$5,000	$0
Utility	0	30	60	85	95	99	100

PROBLEM 9

Super Cola is considering the introduction of a new 8 oz. root beer. The probability that the root beer will be a success is believed to equal .6. The payoff table is as follows:

	Success (s_1)	Failure (s_2)
Produce	$250,000	-$300,000
Do Not Produce	-$50,000	-$20,000

Company management has determined the following utility values:

Amount	$250,000	-$20,000	-$50,000	-$300,000
Utility	100	60	55	0

a) Is the company a risk taker, risk averse, or risk neutral?

b) What is Super Cola's optimal decision?

c) Stanton Marketing has market indicators, I_1 and I_2 for which $P(I_1|s_1) = .7$ and $P(I_1|s_2) = .4$. New World Marketing has market indicators J_1 and J_2 for which $P(J_1|s_1) = .6$ and $P(J_1|s_2) = .3$. If both indicators are available for free, which one should be selected?

PROBLEM 10

The president of Metropolitan Cablevision has asked two of his vice presidents for a recommendation concerning the offering of pay TV. Metropolitan has the choice of using one of three pay TV systems. Profits are believed to be a function of customer acceptance. The payoff to Metropolitan for the three systems is:

	System		
Acceptance Level	I	II	III
High	$150,000	$200,000	$200,000
Medium	$ 80,000	$ 20,000	$ 80,000
Low	$ 20,000	-$ 50,000	-$100,000

The probabilities of customer acceptance for each system are:

	System		
Acceptance Level	I	II	III
High	.4	.3	.3
Medium	.3	.4	.5
Low	.3	.3	.2

The first vice president believes that the indifference probabilities for Metropolitan should be:

Amount	Probability
$150,000	.90
$ 80,000	.70
$ 20,000	.50
-$ 50,000	.25

The second vice president believes Metropolitan should assign the following utility values:

Amount	Utility
$200,000	125
$150,000	95
$ 80,000	55
$ 20,000	30
-$ 50,000	10
-$100,000	0

a) Which vice president is a risk taker? Which one is risk averse?

b) Which system will each vice president recommend?

c) What system would a risk neutral vice president recommend?

PROBLEM 11

Chez Paul is contemplating either opening another restaurant or expanding its existing location. The payoff table for these two decisions is:

	s_1	s_2	s_3
New Restaurant	-$80,000	$20,000	$160,000
Expand	-$40,000	$20,000	$100,000

Paul has calculated the indifference probability for the lottery having a payoff of $160,000 with probability p and -$80,000 with probability $(1-p)$ as follows:

Amount	Indifference Probability (p)
-$ 40,000	.4
$ 20,000	.7
$100,000	.9

a) Is Paul a risk avoider, risk taker, or risk neutral?

b) Suppose Paul has defined the utility of -$80,000 to be 0 and the utility of $160,000 to be 80. What would be the utility values for -$40,000, $20,000, and $100,000 based on the indifference probabilities?

c) Suppose the utility of -$80,000 is defined to be -80 and the utility of $160,000 is defined to be +160. Now what are the utility values for -$40,000, $20,000, and $100,000?

d) Suppose $P(s_1) = .4$, $P(s_2) = .3$, and $P(s_3) = .3$. Which decision should Paul make? Compare with the decision using the Expected Value approach.

e) A competitor is willing to pay Paul $20,000 not to open another restaurant or expand his existing restaurant, i.e. maintain the status quo. Should Paul accept the offer? Why?

PROBLEM 12

Dollar Department Stores has the opportunity of acquiring either 3, 5, or 10 leases from the bankrupt Granite Variety Store chain. Dollar estimates the profit potential of the leases depends on the state of the economy over the next five years. The payoff table is given on the next page (payoffs are in $1,000,000's).

There are four possible states of the economy as modeled by Dollar Department Stores and its president estimates $P(s_1) = .4$, $P(s_2) = .3$, $P(s_3) = .1$, and $P(s_4) = .2$. The utility has also been estimated. Given the payoff and utility tables below, which decision should Dollar make?

Payoff Table

	State Of The Economy Over The Next 5 Years			
Decision	s_1	s_2	s_3	s_4
d_1 -- buy 10 leases	10	5	0	-20
d_2 -- buy 5 leases	5	0	-1	-10
d_3 -- buy 3 leases	2	1	0	-1
d_4 -- do not buy	0	0	0	0

Utility Table

Payoff (in $1,000,000's)	+10	+5	+2	0	-1	-10	-20
Utility	+10	+5	+2	0	-1	-20	-50

PROBLEM 13

Consider the following problem with four states of nature, three decision alternatives, and the following payoff table (in $'s):

	s_1	s_2	s_3	s_4
d_1	200	2600	-1400	200
d_2	0	200	-200	200
d_3	-200	400	0	200

The indifference probabilities for three individuals are:

	Indifference Probabilities		
Payoff	Person 1	Person 2	Person 3
$2600	1.00	1.00	1.00
$ 400	.40	.45	.55
$ 200	.35	.40	.50
$ 0	.30	.35	.45
-$ 200	.25	.30	.40
-$1400	0	0	0

a) Plot the utility function for these three people.

b) Classify each person as a risk avoider, risk taker, or risk neutral.

c) For the payoff of $400, what is the premium the risk avoider will pay to avoid risk? What is the premium the risk taker will pay to have the opportunity of the high payoff?

d) Suppose each state is equally likely. What are the optimal decisions for each of these three people?

PROBLEMS 14 & 15 DEPEND ON
YOUR PERSONAL ASSIGNMENT OF PROBABILITIES.

PROBLEM 14

You can purchase a lottery ticket whose prize is $10,000.

a) What would be the minimum probability of winning that you would accept to buy the ticket if it cost $1; $5; $10; $100; $500; $1000.

b) Now you are given the option of receiving an amount, $x, or a ticket in a lottery which wins $10,000 with probability p and wins $0 with probability (1-p). Find the indifference probability for x equal to $1; $5; $10; $100; $500; $1000.

c) Using your answers to (a) and (b), plot your utility function.

d) Does this indicate you are risk neutral or a risk avoider or risk taker?

PROBLEM 15

Consider the following gambling situations:

a) You are offered an amount of money, $x, or a lottery ticket. The lottery ticket pays $10,000 with probability .5 and pays $0 with probability .5. What amount, $x, would make you indifferent to this lottery?

b) You are offered an amount of money, $y, or a lottery ticket. The lottery ticket pays $10,000 with probability .5 and pays $x (the amount in part (a)) with probability .5. What amount, $y, would make you indifferent to this lottery?

c) You are offered an amount of money, $z, or a lottery ticket. The lottery ticket pays $x (the amount in part (a)) with probability .5 and pays $0 with probability .5. What amount, $z, would make you indifferent to this lottery.

d) You are offered an amount of money, $d, or a lottery ticket. The lottery ticket pays $y (your answer to part (b)) with probability .5 and pays $z (your answer to part (c)) with probability .5 What amount, $d, would make you indifferent to this lottery?

e) If the utility of $10,000 is defined to be 100 and the utility of $0 is defined to be 0, find the utility of $x, $y, $z, and $d. Is the utility of $x greater than, less than, or equal to the utility of $d? Are your values for $d and $x consistent with their utilities?

UTILITY AND GAME THEORY

PROBLEM 16

Two banks (Franklin and Lincoln) compete for customers in the growing city of Logantown. Both banks are considering opening a branch office in one of three new neighborhoods: Hillsboro, Fremont, or Oakdale. The strategies, assumed to be the same for both banks, are:

    Strategy 1: Open a branch office in the Hillsboro neighborhood.
    Strategy 2: Open a branch office in the Fremont neighborhood.
    Strategy 3: Open a branch office in the Oakdale neighborhood.

Values in the payoff table below indicate the gain (or loss) of customers (in thousands) for Franklin Bank based on the strategies selected by the two banks..

		Lincoln Bank		
		Hillsboro b_1	Fremont b_2	Oakdale b_3
Franklin Bank				
Hillsboro	a_1	4	2	3
Fremont	a_2	6	-2	-3
Oakdale	a_3	-1	0	5

Identify the neighborhood in which each bank should locate a new branch office. What is the value of the game?

PROBLEM 17

Consider the following two-person zero-sum game. Assume the two players have the same three strategy options. The payoff table below shows the gains for Player A.

	Player B		
Player A	Strategy b_1	Strategy b_2	Strategy b_3
Strategy a_1	3	2	-4
Strategy a_2	-1	0	2
Strategy a_3	4	5	-3

Is there an optimal pure strategy for this game? If so, what is it? If not, can the mixed-strategy probabilities be found algebraically? What is the value of the game?

TRUE/FALSE

___ 18. If one outcome is preferred to another, it will have a higher utility value.

___ 19. Given two decision makers, one risk neutral and the other a risk avoider, the risk avoider will always give a lower utility value for a given outcome.

___ 20. A risk avoider will have a concave utility function.

___ 21. A risk neutral decision maker will choose decisions identical to those chosen using the expected value approach.

___ 22. If an outcome is certain, it is given a utility value of 1.

___ 23. If a decision maker is indifferent between receiving $1,000 or playing a lottery in which he wins nothing with probability .8 and $10,000 with probability .2, then the decision maker could be characterized as risk averse.

___ 24. Consider the decision maker in question 6. If he assigns a utility of 0 to the outcome of a $0 return and a utility of 10 to the $10,000 return, then he would assign a utility value of 1 to $1,000.

___ 25. When using the expected utility approach, a risk avoider and a risk taker will never choose the same decision.

___ 26. The utility function for a risk avoider typically shows a diminishing marginal return for money.

___ 27. A risk neutral decision maker will have a linear utility function.

___ 28. The decision to buy state lottery tickets has a negative expected monetary value.

___ 29. Measuring a decision maker's utility is, at least in part, subjective.

___ 30. In most cases, the decision to buy insurance for a house has a positive expected monetary value.

___ 31. The expected monetary value approach and the expected utility approach to decision making usually result in the same decision choice unless extreme payoffs are involved.

___ 32. The logic of game theory assumes that each player has different information and will select a strategy that provides the best possible outcome from his point of view.

CHAPTER 6

Forecasting

KEY CONCEPTS

CONCEPT	ILLUSTRATED PROBLEMS	ANSWERED PROBLEMS
Smoothing Methods		
Moving Average	1	9
Weighted Moving Average	2	6
Exponential Smoothing	1	6,7
Linear Trend Projection	2	10,12
Multiplicative Time Series Model	3,4	8,11,13
Regression Analysis	5	14
Mean Squared Error	1	6,7,9

REVIEW

1. A <u>time series</u> is a set of observations measured at successive points in time or over successive periods of time. A time series is analyzed so that one may determine good forecasts or predictions of future values for the time series.

2. While a time series may consist of numerous components, a usual assumption is that four separate components combine to affect the values of a time series. These <u>four components</u> are: (1) a trend component; (2) a cyclical component; (3) a seasonal component; and, (4) irregular components.

3. The <u>trend component</u> accounts for the gradual shifting of the time series over a long period of time.

4. Any regular pattern of sequences of values above and below the trend line is attributable to the <u>cyclical component</u> of the series.

5. The <u>seasonal component</u> of the series accounts for regular patterns of variability within certain time periods, such as over a year.

6. The <u>irregular component</u> of the series is caused by short-term, unanticipated and non-recurring factors that affect the values of the time series. One cannot attempt to predict its impact on the time series in advance.

FORECASTING   109

7. In cases in which the time series is fairly stable and has no significant trend, seasonal, or cyclical effects, one can use <u>smoothing methods</u> to average out the irregular components of the time series.

8. The <u>moving average</u> smoothing method consists of computing an average of the most recent n data values for the series and using this average for forecasting the value of the time series for the next period.

9. The <u>centered moving average</u> method consists of computing an average of n periods' data and associating it with the midpoint of the periods. For example, the average for periods 5, 6, and 7 is associated with period 6. This methodology is useful in the process of computing season indexes.

10. In the <u>weighted moving average</u> smoothing method for computing the average of the most recent n periods, the more recent observations are typically given more weight than older observations. (For convenience, the weights usually sum to 1.)

11. One difficulty of both the moving average and the weighted moving average methods is that n historical data points must be stored in order to compute the forecast for the next period. In <u>exponential smoothing</u> only two pieces of information are needed to compute the forecast: (1) the forecasted value for the current period, and (2) the actual value for the current period.

12. Using <u>exponential smoothing</u>, the forecast is calculated by:
α[the actual value for the current period]
+ $(1-\alpha)$[the forecasted value for the current period],
where the <u>smoothing constant</u>, α, is a number between 0 and 1.

13. Another way to view <u>exponential smoothing</u> is that the forecast for the next period is equal to the forecast for the current period plus a proportion (α) of the forecast error in the current period.

14. It is essential that forecasts be as accurate as possible. One measure of <u>forecast accuracy</u> is known as the <u>mean squared error</u>. In this measure the average of the squared forecast errors for the historical data is calculated. The forecasting method or parameter(s) which minimize this mean squared error is then selected.

15. An alternative measure for the performance of a forecasting technique is the <u>mean absolute deviation (MAD)</u>. In this measure, the mean of the absolute values of all forecast errors is calculated, and the forecasting method or parameter(s) which minimize this measure is selected. The mean absolute deviation measure is less sensitive to individual large forecast errors than the mean squared error measure.

16. If a time series exhibits a linear trend, the <u>method of least squares</u> may be used to determine a trend line (projection) for future forecasts. This statistical technique, also used in regression analysis, determines the unique trend line forecast which minimizes the mean square error between the trend line forecasts and the actual observed values for the time series.

17. Using the method of least squares, the formula for the trend projection is: $T_t = b_0 + b_1 t$.
    Here,
    - T_t = the trend forecast for time period t
    - b_1 = the slope of the trend line
    - b_0 = the trend line projection for time 0

    Where $$b_1 = \frac{n\sum tY_t - \sum t \sum Y_t}{n\sum t^2 - (\sum t)^2} \qquad b_0 = \bar{Y} - b_1 \bar{t}$$

    Here,
    - Y_t = the observed value of the time series at time period t
    - $\bar{Y}$ = the average of the observed values for Y_t
    - $\bar{t}$ = the average time period for the n observations

18. In the case of <u>nonlinear trend</u>, a more advanced statistical technique might possibly be used to develop the forecasting curve.

19. The <u>multiplicative time series model</u> assumes that the actual time series value, Y_t, is equal to the product of the four time series components: (1) trend (T_t); (2) cyclical (C_t); (3) seasonal (S_t); and (4) irregular (I_t). Thus, $Y_t = T_t C_t S_t I_t$.

20. In situations in which no historical data is available or when historical data will not give an accurate picture of the future, <u>nonquantitative techniques</u> for forecasting may be used.

21. An example of a nonquantitative forecasting technique is the <u>delphi approach</u>. A panel of experts, each of whom is physically separated from the others and is anonymous, is asked to respond to a sequential series of questionnaires. After each questionnaire, the responses are tabulated and the information and opinions of the entire group are made known to each of the other panel members so that they may revise their previous forecast response. The process continues until some degree of consensus is achieved.

22. Another nonquantitative approach, <u>scenario writing</u>, consists of developing a conceptual scenario of the future based on a well defined set of assumptions. After several different scenarios have been developed, the decision maker determines which is most likely to occur in the future and makes decisions accordingly.

23. <u>Subjective</u> or <u>interactive qualitative approaches</u>, commonly known as "brainstorming sessions" are another way to perform a nonquantitative forecast. It is important in such sessions that any ideas or opinions be permitted to be presented without regard to its relevancy and without fear of criticism.

MULTIPLICATIVE TIME SERIES PROCEDURE

1. **Calculate the centered moving averages (CMAs).**
   The centered moving average represents the combined trend and cyclical components of the series. Calculate K-period moving averages (where K is the number of seasons, i.e. quarterly data would have four seasons whereas monthly data would have twelve seasons.)

2. **Center the CMAs on integer-valued periods.**
   Associate each moving average with the middle period of the K data points comprising the average. When K is an even number there is no distinct middle period. In this case, taking the average of two successive moving averages (one centered just above the period and one centered just below the period) gives the moving average associated with that period.

3. **Determine the seasonal and irregular factors $(S_t I_t)$.**
   For each centered moving average found in step 2, divide this value into the observed value, Y_t. This quotient represents the seasonal and irregular factors.

4. **Determine the average seasonal factors.**
   For each season, average the corresponding quotients found in step 3 to smooth out the irregular component and isolate the seasonal factors.

5. **Scale the seasonal factors (S_t).**
   To ensure that the seasonal factors average to 1, adjust the seasonal factors by dividing each by the average seasonal factor value.

6. **Determine the deseasonalized data.**
   Divide each data value, Y_t, by its seasonal factor.

7. **Determine a trend line of the deseasonalized data.**
   Use the method of least squares on this data set to identify the trend line for the data.

8. **Determine the deseasonalized predictions.**
   Determine the trend forecast(s) associated with the future period(s) by using the trend line equation found in step 7.

9. **Take into account the seasonality.**
   Multiply each deseasonalized prediction by the appropriate seasonal factor.

CHAPTER 6

FLOW CHART OF MULTIPLICATIVE TIME SERIES PROCEDURE

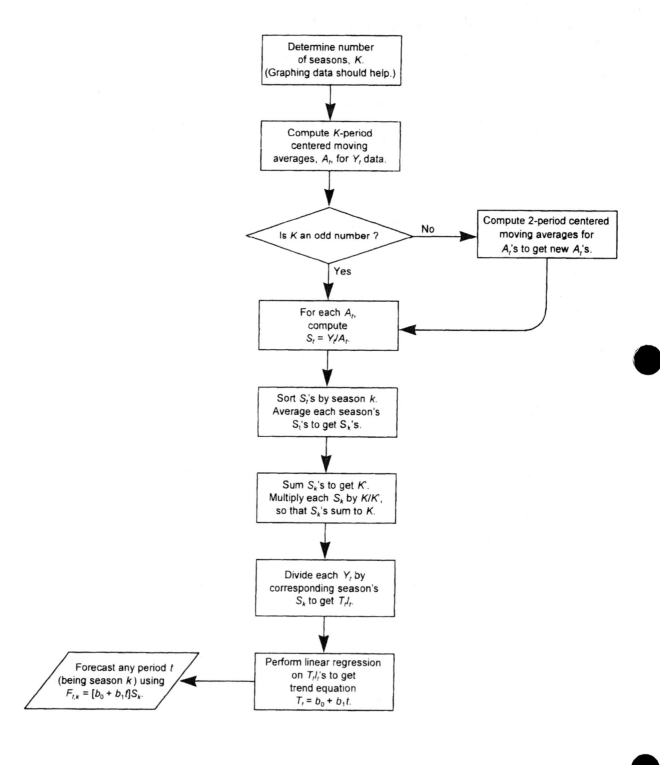

FORECASTING

ILLUSTRATED PROBLEMS

PROBLEM 1

During the past ten weeks, sales of cases of Comfort brand headache medicine at Robert's Drugs have been as follows:

Week	Sales	Week	Sales
1	110	6	120
2	115	7	130
3	125	8	115
4	120	9	110
5	125	10	130

a) If Robert's uses exponential smoothing to forecast sales, which value for the smoothing constant α, $\alpha = .1$ or $\alpha = .8$, gives better forecasts?

b) Using your value for α in part (a) that gave better forecasts, forecast sales for week 11.

c) Forecast sales in week 11 using a three-week moving average.

SOLUTION 1

a) To evaluate the two smoothing constants, determine how the forecasted values would compare with the actual historical values in each case. Let

Y_t = actual sales in week t
F_t = forecasted sales in week t

For $\alpha = .1$, $1 - \alpha = .9$

$F_1 = Y_1 = 110$. For other weeks, $F_{t+1} = .1Y_t + .9F_t$

$F_2 = .1Y_1 + .9F_1 = .1(110) + .9(110) = 110$
$F_3 = .1Y_2 + .9F_2 = .1(115) + .9(110) = 110.5$
$F_4 = .1Y_3 + .9F_3 = .1(125) + .9(110.5) = 111.95$
$F_5 = .1Y_4 + .9F_4 = .1(120) + .9(111.95) = 112.76$
$F_6 = .1Y_5 + .9F_5 = .1(125) + .9(112.76) = 113.98$
$F_7 = .1Y_6 + .9F_6 = .1(120) + .9(113.98) = 114.58$
$F_8 = .1Y_7 + .9F_7 = .1(130) + .9(114.58) = 116.12$
$F_9 = .1Y_8 + .9F_8 = .1(115) + .9(116.12) = 116.01$
$F_{10} = .1Y_9 + .9F_9 = .1(110) + .9(116.01) = 115.41$

For α = .8, 1 - α = .2
F_1 = 110
F_2 = .8(110) + .2(110) = 110
F_3 = .8(115) + .2(110) = 114
F_4 = .8(125) + .2(114) = 122.80
F_5 = .8(120) + .2(122.80) = 120.56
F_6 = .8(125) + .2(120.56) = 124.11
F_7 = .8(120) + .2(124.11) = 120.82
F_8 = .8(130) + .2(120.82) = 128.16
F_9 = .8(115) + .2(128.16) = 117.63
F_{10} = .8(110) + .2(117.63) = 111.53.

In order to determine which smoothing constant gives the better performance, calculate, for each, the mean squared error for the nine weeks of forecasts, weeks 2 through 10 by:

$$[(Y_2-F_2)^2 + (Y_3-F_3)^2 + (Y_4-F_4)^2 + ... + (Y_{10}-F_{10})^2] / 9$$

Week	Y_t	α = .1 F_t	$(Y_t - F_t)^2$	α = .8 F_t	$(Y_t - F_t)^2$
1	110				
2	115	110.00	25.00	110.00	25.00
3	125	110.50	210.25	114.00	121.00
4	120	111.95	64.80	122.80	7.84
5	125	112.76	149.94	120.56	19.71
6	120	113.98	36.25	124.11	16.91
7	130	114.58	237.73	120.82	84.23
8	115	116.12	1.26	128.16	173.30
9	110	116.01	36.12	117.63	58.26
10	130	115.41	212.87	111.53	341.27
		Sum	974.22	Sum	847.52
	MSE	Sum/9	108.25	Sum/9	94.17

Hence, based on the mean squared error criterion, using α = .8 gives a slightly better forecast than using α = .1.

b) If α = .8, then the forecast for week 11 will be

$$.8Y_{10} + .2F_{10} = .8(130) + .2(111.53) = 126.31$$

c) Using a three week moving average, the forecast for week 11 will be the average of the preceding three weeks: weeks 8, 9, and 10.

$$F_{11} = (115 + 110 + 130)/3 = 118.33$$

PROBLEM 2

The number of plumbing repair jobs performed by Auger's Plumbing Service in each of the last nine months are listed below.

Month	Jobs	Month	Jobs	Month	Jobs
March	353	June	374	September	399
April	387	July	396	October	412
May	342	August	409	November	408

a) Assuming a linear trend function, forecast the number of repair jobs Auger's will perform in December using the least squares method.

b) What is your forecast for December using a three-period weighted moving average with weights of .6, .3, and .1? How does it compare with your forecast from part (a)?

SOLUTION 2

> NOTE: The method of least squares requires time periods to be numbered. If your periods are labeled with words (e.g. February or Thursday) or the number labels are large (e.g. 1983), simply assign the first period in your data set the value 1, etc. Determine the value of the time period you want to forecast accordingly. For example, if 1983 is period 1, then 1998 is period 16.

a) The trend line is $T_t = b_0 + b_1 t$. The least squares method gives:

(month) t	Y_t	tY_t	t^2
(Mar.) 1	353	353	1
(Apr.) 2	387	774	4
(May) 3	342	1026	9
(June) 4	374	1496	16
(July) 5	396	1980	25
(Aug.) 6	409	2454	36
(Sep.) 7	399	2793	49
(Oct.) 8	412	3296	64
(Nov.) 9	408	3672	81
Sum 45	3480	17844	285

Thus, $\bar{t} = 5$ $\bar{Y} = 386.667$

$$b_1 = \frac{n\Sigma tY_t - \Sigma t \Sigma Y_t}{n\Sigma t^2 - (\Sigma t)^2} = \frac{(9)(17844) - (45)(3480)}{(9)(285) - (45)^2} = 7.4$$

$$b_0 = \bar{Y} - b_1 \bar{t} = 386.667 - 7.4(5) = 349.667$$

$$T_{10} = 349.667 + (7.4)(10) = 423.667$$

b) Using a three-month weighted moving average, the forecast for December will be the weighted average of the preceding three months: September, October, and November.

$$F_{10} = .1 Y_{Sep.} + .3 Y_{Oct.} + .6 Y_{Nov.} = .1(399) + .3(412) + .6(408) = 408.3$$

Due to the positive trend component in the time series, the least squares method produced a forecast that is more in tune with the trend that exists. The weighted moving average, even with heavy (.6) placed on the current period, produced a forecast that is lagging behind the changing data.

PROBLEM 3

Quarterly revenues (in $1,000,000's) for a national restaurant chain for a five year period were as follows:

Quarter	Year 1	2	3	4	5
1	33	42	54	70	85
2	36	40	53	67	82
3	35	42	54	70	87
4	38	47	62	77	99

Forecast the revenues for the next four quarters.

SOLUTION 3

Assume the data values are a multiplicative function of the data's trend, cyclical, seasonal, and irregular factors, i.e.

$$Y_t = T_t * C_t * S_t * I_t$$

Step 1: Calculate the centered moving averages (CMAs).
   First, use a moving average over the four quarters to mask the effects of the seasonal and irregular factors. For each four quarter period, calculate the moving average and associate it with the "middle period". For example, the first moving average is: (33+36+35+38)/4 = 35.5. The second equals = 37.75, etc.

Step 2: Center the CMAs on integer-valued periods.
   When the number of quarters is even, there is no integer valued "middle period" (the middle of the first four quarters would be quarter 2.5). In order to have the moving average "centered" at a particular quarter, average the half-period moving average preceding this quarter and the half-period moving average succeeding this quarter. For example, the moving averages of quarters 2.5 and 3.5 are 35.5 and 37.75, respectively. Thus, the centered moving average for quarter 3 is (35.5 + 37.75)/2 = 36.625.

FORECASTING 117

Year	Quarter	Revenues	Four Quarter Moving Average	Centered Moving Average
1	1	33		
	2	36		
	(2.5)		35.50	
	3	35		36.625
	(3.5)		37.75	
	4	38		38.250
	(4.5)		38.75	
2	1	42		39.625
	(1.5)		40.50	
	2	40		41.625
	(2.5)		42.75	
	3	42		44.250
	(3.5)		45.75	
	4	47		47.375
	(4.5)		49.00	
3	1.	54		50.500
	(1.5)		52.00	
	2	53		53.875
	(2.5)		55.75	
	3	54		57.750
	(3.5)		59.75	
	4	62		61.500
	(4.5)		63.25	
4	1	70		65.250
	(1.5)		67.25	
	2	67		69.125
	(2.5)		71.00	
	3	70		72.875
	(3.5)		74.75	
	4	77		76.625
	(4.5)		78.50	
5	1	85		80.625
	(1.5)		82.75	
	2	82		85.500
	(2.5)		88.25	
	3	87		
	4	99		

Step 3: Determine the seasonal and irregular factors ($S_t I_t$).

The centered moving averages represent the combine effects of the trend and cyclical factors ($T_t C_t$). Since $Y_t = T_t C_t S_t I_t$, $S_t I_t = Y_t/(T_t C_t) = Y_t/$(centered moving average for period t). So, dividing each data point by its centered moving average gives an estimate of $S_t I_t$.

For example for period 3 (year 1, quarter 3), the data point, $Y_3 = 35$, and its centered moving average $T_3 C_3 = 36.625$. Thus, for this period, $S_3 I_3 = 35/36.625 = .956$

Continue this procedure for determining $S_t I_t$ for all periods:

CHAPTER 6

Year	Quarter	Revenues (Y_t)	Average ($T_t C_t$)	$S_t I_t$
1	3	35	36.625	.956
	4	38	38.250	.993
2	1	42	39.625	1.060
	2	40	41.625	.961
	3	42	44.250	.949
	4	47	47.375	.992
3	1	54	50.500	1.069
	2	53	53.875	.984
	3	54	57.750	.935
	4	62	61.500	1.008
4	1	70	65.250	1.073
	2	67	69.125	.969
	3	70	72.875	.961
	4	77	76.625	1.005
5	1	85	80.625	1.054
	2	82	85.500	.959

Step 4: Determine the average seasonal factors.

To eliminate the irregular effects, take the average of the $S_t I_t$ over the four years. That is, to find the average of $S_t I_t$ for quarter 1, average the quarter 1 values for years 2, 3, 4 and 5. Do the same for quarter 2. For quarters 3 and 4, the average would be over years 1, 2, 3, and 4. This gives:

$$S_1 = (1.060 + 1.069 + 1.073 + 1.054) / 4 \quad = 1.064$$
$$S_2 = (.961 + .984 + .969 + .959) / 4 \quad = .968$$
$$S_3 = (.956 + .949 + .935 + .961) / 4 \quad = .950$$
$$S_4 = (.993 + .992 + 1.008 + 1.005) / 4 \quad = 1.000$$

Step 5: Scale the seasonal factors (S_t).

Each seasonal average must be adjusted by the average of the seasonal factors, $(1.064 + .968 + .950 + 1.000) / 4 = .9955$, giving:

$$S_1 = 1.064/.9955 = 1.069$$
$$S_2 = .968/.9955 = .973$$
$$S_3 = .950/.9955 = .954$$
$$S_4 = 1.000/.9955 = 1.004$$

Step 6: Determine the deseasonalized data.

The seasonal factors (S_t) can now be removed from the data by dividing each data point by its seasonal factor. This gives deseasonalized data which will only be a function of trend, cyclical and irregular factors.

Year	Quarter	Y_t	Deseasonalized (Y_t/S_t)
1	1	33	33/1.069 = 30.87
	2	36	36/0.973 = 37.00
	3	35	35/0.954 = 36.69
	4	38	38/1.004 = 37.84
2	1	42	42/1.069 = 39.29
	2	40	40/0.973 = 41.11
	3	42	42/0.954 = 44.03
	4	47	47/1.004 = 46.44
3	1	54	54/1.069 = 50.51
	2	53	53/0.973 = 54.47
	3	54	54/0.954 = 56.60
	4	62	62/1.004 = 61.75
4	1	70	70/1.069 = 65.48
	2	67	67/0.973 = 68.86
	3	70	70/0.954 = 73.38
	4	77	77/1.004 = 76.69
5	1	85	85/1.069 = 79.51
	2	82	82/0.973 = 84.28
	3	87	87/0.954 = 91.19
	4	99	99/1.004 = 98.61

Step 7: Determine a trend line of the deseasonalized data.

Now, label the periods $t = 1$ through $t = 20$ and use the regression trend analysis (see problem 2) to determine the following trend line describing the seasonally adjusted data over the 20 quarters:

$$T_t = 23.436 + 3.361t$$

Step 8: Determine the deseasonalized predictions.

Use this trend line to determine the trend predictions for the four quarters of year 6.

Step 9: Take into account the seasonality.

Then adjust these quarterly predictions by multiplying each by its seasonal adjustment factor.

Quarter	Period t	Trend Prediction ($T_t=23.436+3.361t$)	Seasonally Adjusted Forecast ($T_t S_t$)
1	21	94.02	(94.02)(1.069) = 100.50
2	22	97.38	(97.38)(0.973) = 94.72
3	23	100.74	(100.74)(0.954) = 96.15
4	24	104.10	(104.10)(1.004) = 104.53

PROBLEM 4

Business at Terry's Tie Shop can be viewed as falling into three distinct seasons: (1) Christmas (November-December); (2) Father's Day (late May - mid-June); and (3) all other times. Average weekly sales (in $'s) during each of these three seasons during the past four years has been as follows:

	Year			
Season	1	2	3	4
1	1856	1995	2241	2280
2	2012	2168	2306	2408
3	985	1072	1105	1120

Determine a forecast for the average weekly sales in year 5 for each of the three seasons.

SOLUTION 4

The table on the next page summarizes the computations in steps 1-6.

Step 1: Calculate the centered moving averages.

There are three distinct seasons in each year. Hence, take a three season moving average to eliminate seasonal and irregular factors. For example the first moving average is: (1856 + 2012 + 985)/3 = 1617.67.

Step 2: Center the CMAs on integer-valued periods.

The first moving average computed in step 1 (1617.67) will be centered on season 2 of year 1. Note that the moving averages from step 1 center themselves on integer-valued periods because n is an odd number.

Step 3: Determine the seasonal and irregular factors (S_t, I_t).

Isolate the trend and cyclical components. For each period t, this is given by Y_t/(Moving Average for period t).

Step 4: Determine the average seasonal factors.

Averaging all $S_t I_t$ values corresponding to that season:

    Season 1: (1.163 + 1.196 + 1.181) / 3   = 1.180
    Season 2: (1.244 + 1.242 + 1.224 + 1.244) / 4   = 1.238
    Season 3: (.592 + .587 + .582) / 3   = .587

Step 5: Scale the seasonal factors (S_t).

Average the seasonal factors = (1.180 + 1.238 + .587)/3 = 1.002. Divide each seasonal factor by the average of the seasonal factors.

Season 1: 1.180/1.002 = 1.178
Season 2: 1.238/1.002 = 1.236
Season 3: .587/1.002 = .586
          Total   = 3.000

Step 6: Determine the deseasonalized data.

Divide the data point values, Y_t, by S_t.

Year	Season	Dollar Sales (Y_t)	Moving Average	$S_t I_t$	Scaled S_t	Y_t/S_t
1	1	1856				1576
	2	2012	1617.67	1.244	1.236	1628
	3	985	1664.00	.592	.586	1681
2	1	1995	1716.00	1.163	1.178	1694
	2	2168	1745.00	1.242	1.236	1754
	3	1072	1827.00	.587	.586	1829
3	1	2241	1873.00	1.196	1.178	1902
	2	2306	1884.00	1.224	1.236	1866
	3	1105	1897.00	.582	.586	1886
4	1	2280	1931.00	1.181	1.178	1935
	2	2408	1936.00	1.244	1.236	1948
	3	1120			.586	1911

Step 7: Determine a trend line of the deseasonalized data.

Use the linear regression method illustrated in problem 2. For $t = 1, 2, ..., 12$, this gives:

$$T_t = 1580.11 + 33.96t.$$

Step 8: Determine the deseasonalized predictions for quarters (13, 14, 15).

Substitute $t = 13$, 14, and 15 into the above equation:

$T_{13} = 1580.11 + (33.96)(13) = 2022$
$T_{14} = 1580.11 + (33.96)(14) = 2056$
$T_{15} = 1580.11 + (33.96)(15) = 2090$

Step 9: Take into account the seasonality.

Multiply each deseasonalized prediction by its seasonal factor to give the following forecasts for year 5:

Season 1: (1.178)(2022) = 2382
Season 2: (1.236)(2056) = 2541
Season 3: (.586)(2090) = 1225

PROBLEM 5

Connie Harris, in charge of office supplies at First Capital Mortgage Corp., would like to predict the quantity of paper used in the office photocopying machines per month. She believes that the number of loans originated in a month influence the volume of photocopying performed. She has compiled the following recent monthly data:

Number of Loans Originated in Month	Sheets of Photocopy Paper Used (000's)
25	16
25	13
35	18
40	25
40	21
45	22
50	24
60	25

a) Develop the least-squares estimated regression equation that relates sheets of photocopy paper used to loans originated.

b) Use the regression equation developed in part (a) to forecast the amount of paper used in a month when 65 loan originations are expected.

SOLUTION 5

a) The regression equation is $y = b_0 + b_1 x$. The least squares method gives:

Month (i)	y_i	x_i	$x_i y_i$	x_i^2
1	16	25	400	625
2	13	25	325	625
3	18	35	630	1225
4	25	40	1000	1600
5	21	40	840	1600
6	22	45	990	2025
7	24	50	1200	2500
8	25	60	1500	3600
Totals	164	320	6885	13800

$$b_1 = \frac{\Sigma x_i y_i - (\Sigma x_i \Sigma y_i)/n}{\Sigma x_i^2 - (\Sigma x_i)^2/n}$$

$$= \frac{6885 - (320)(164)/8}{13800 - (320)^2/8}$$

$$= 325/1000 = 0.325$$

$b_0 = y - b_1 x = \Sigma y_i/n - b_1(\Sigma x_i/n) = 164/8 - .325(320/8) = 20.5 - 13 = 7.5$

Thus, the estimated regression equation is $y = 7.5 + .325x$

b) The forecast is $y = 7.5 + .325x = 7.5 + .325(65) = 28,625$ sheets.

ANSWERED PROBLEMS

PROBLEM 6

The monthly electricity bill at the Chez Paul Restaurant over the past 12 months has been as follows:

Month	Amount	Month	Amount
Jan	$271.90	Jul	$330.70
Feb	305.70	Aug	300.10
Mar	306.40	Sep	275.50
Apr	297.30	Oct	301.30
May	315.30	Nov	279.40
Jun	297.20	Dec	306.60

Paul is considering using exponential smoothing with $\alpha = .5$ or a four period weighted moving average with weights of .4, .3, .2, and .1 to forecast future electricity costs.

a) Which forecasting technique will give the smallest mean square error?

b) Give next January's forecast for each method.

PROBLEM 7

Sales (in thousands) of the new Thorton Model 506 convection oven over the eight week period since its introduction have been as follows:

Week	Sales
1	18.6
2	21.4
3	25.2
4	22.4
5	24.6
6	19.2
7	21.7
8	23.8

a) Which exponential smoothing model provides better forecasts, one using $\alpha = .6$ or $\alpha = .2$? Compare them using mean squared error.

b) Using the two forecast models in part (a), what are the forecasts for week 9?

PROBLEM 8

Forecast the sales of Jami Michelle skin cream for year 6 given the following quarterly sales (in thousands) over the past five years:

	\newline Year				
Quarter	1	2	3	4	5
1	34	38	43	47	49
2	27	33	37	39	45
3	49	51	60	68	72
4	27	28	29	32	40

PROBLEM 9

Weekly sales of the Weber La Guillotine food processor for the past ten weeks have been:

Week	Sales	Week	Sales
1	980	6	990
2	1040	7	1030
3	1120	8	1260
4	1050	9	1240
5	960	10	1100

a) Determine, on the basis of minimizing the mean square error, whether a three period or four period simple moving average model gives a better forecast for this problem.

b) For each model, forecast sales for week 11.

PROBLEM 10

Four months ago, the Bank Drug Company introduced Jeffrey William brand designer bandages. Advertised using the slogan, "What the best dressed cuts are wearing", weekly sales for this period (in 1000's) have been as follows:

Week	Sales	Week	Sales	Week	Sales
1	12.8	7	20.6	12	23.8
2	14.6	8	18.5	13	25.1
3	15.2	9	19.9	14	24.7
4	16.1	10	23.6	15	26.5
5	15.8	11	24.2	16	28.9
6	17.2				

a) Plot a graph of sales vs. weeks. Does linear trend appear reasonable?

b) Assuming linear trend, forecast sales for weeks 17, 18, 19, and 20.

PROBLEM 11

The number of haircuts performed each day at KwikKuts in the last four weeks are listed below:

Week	Monday	Tuesday	Wednesday	Thursday	Friday
1	122	122	103	133	98
2	127	130	106	137	97
3	126	131	111	151	104
4	135	135	110	146	107

a) Plot the sales data. Do you see both trend and seasonality components in the data?

b) Forecast the number of haircuts to be performed in each workday of week 6.

PROBLEM 12

At a local car dealership the following is a record of sales for the past 12 months:

Month	Sales	Month	Sales
Jan	36	Jul	25
Feb	34	Aug	22
Mar	28	Sep	26
Apr	30	Oct	22
May	27	Nov	21
Jun	24	Dec	19

a) Using the method of least squares, determine a trend line for forecasting future sales.

b) Using your model in part (a), determine how long it will be before zero sales are forecasted.

c) Consider your answer to part (b). What will be the forecasted sales for the month after that? Does this make sense? Comment on the validity of the model. What assumption about the model appears to be in error?

PROBLEM 13

A 24-hour coffee/donut shop makes donuts every eight hours. The manager must forecast donut demand so that the bakers have the fresh ingredients they need. Listed below is the actual number of glazed donuts (in dozens) sold in each of the preceding 13 eight-hour shifts.

Date	Shift	Demand (dozens)
June 3	Day	59
	Evening	47
June 4	Night	35
	Day	64
	Evening	43
June 5	Night	39
	Day	62
	Evening	46
June 6	Night	42
	Day	64
	Evening	50
June 7	Night	40
	Day	69

Forecast the demand for glazed donuts for the three shifts of June 8 and the three shifts of June 9.

PROBLEM 14

Scott Bell Builders would like to predict the total number of labor hours spent framing a house based on the square footage of the house. The following data has been compiled on ten houses recently built.

Square Footage (100's)	Framing Labor Hours	Square Footage (100's)	Framing Labor Hours
20	195	27	225
21	170	29	240
23	220	31	225
23	200	32	275
26	230	35	260

a) Develop the least-squares estimated regression equation that relates framing labor hours to house square footage.

b) Use the regression equation developed in part (a) to predict framing labor hours when the house size is 3350 square feet.

TRUE/FALSE

___ 15. If a time series has a trend component, then one should not use a moving average to forecast.

___ 16. In forecasting with trend and seasonal components using a multiplicative model, one computes moving averages in order to isolate the combined seasonal and irregular components.

___ 17. If the random variability in a time series is great, a high α value should be used to exponentially smooth out the fluctuations.

___ 18. In exponential smoothing, one typically chooses the smoothing constant as that value which minimizes the mean squared error.

___ 19. Forecasting errors are always less using exponential smoothing than a weighted moving average.

___ 20. A forecaster would choose trend projection using the least squares method over exponential smoothing if the data exhibited a trend component.

___ 21. To forecast using the multiplicative model, one must adjust the trend component by the seasonal factor.

___ 22. In a weighted moving average, the most recent occurrence is typically given the least weight.

___ 23. In using the Delphi technique, one attempts to obtain a group consensus.

___ 24. One advantage of exponential smoothing over moving averages is that fewer data points are used in the forecast.

___ 25. An α equal to 0.2 will cause an exponential smoothing forecast to react more quickly to a sudden drop in demand than will an α equal to 0.4.

___ 26. Exponential smoothing with $\alpha = .2$ and a moving average with $n = 5$ put the same weight on the actual value for the current period.

___ 27. The sum of the seasonal indexes should be adjusted, if necessary, to equal 1.

___ 28. With fewer periods in a moving average, it will take longer to adjust to a new level of demand.

___ 29. A causal forecasting method is most effective when demand data exhibit fluctuations caused by seasonal influences.

CHAPTER 7

Introduction to Linear Programming

KEY CONCEPTS

CONCEPT	ILLUSTRATED PROBLEMS	ANSWERED PROBLEMS
Formulation	7,8	15,16
Minimization	2,5	9,10,15
Standard Form	1	14
Slack/Surplus Variables	1	16
Equal-to Constraints	3,5	14
Redundant Constraints	5,7	12,13
Extreme Points	2,5	11,16
Alternative Optimal Solutions	7	10,11,15
Infeasibility	4	14
Unbounded	4	11
Spreadsheet Example	2	

REVIEW

1. A <u>mathematical programming</u> problem is one that seeks to maximize an objective function subject to constraints. If both the objective function and the constraints are linear, the problem is referred to as a <u>linear programming</u> problem.

2. <u>Linear functions</u> are functions in which each variable appears in a separate term raised to the first power and is multiplied by a constant (which could be 0).

3. <u>Linear constraints</u> are linear functions that are restricted to be "less than or equal to", "equal to", or "greater than or equal to" a constant.

4. The <u>maximization</u> or <u>minimization</u> of some quantity is the objective in all linear programming problems.

5. A <u>feasible solution</u> satisfies all the problem's constraints.

6. A linear program which is overconstrained so that no point satisfies all the constraints is said to be <u>infeasible</u>. Changes to the objective function coefficients do not affect the feasibility of the problem.

7. An <u>optimal solution</u> is a feasible solution that results in the largest possible objective function value, z, when maximizing or smallest possible z when minimizing.

8. A <u>graphical solution method</u> can be used to solve a linear program with two variables.

9. If a linear program possesses an optimal solution, then an <u>extreme point</u> will be optimal.

10. If a constraint can be removed without affecting the shape of the feasible region, the constraint is said to be <u>redundant</u>. If changes are anticipated to the linear programming model, constraints which were redundant in the original formulation may not be redundant in the revised formulation.

11. In the graphical method, if the objective function line is parallel to a boundary constraint in the direction of optimization, there are <u>alternative optimal solutions</u>, with all points on this line segment being optimal.

12. A feasible region may be <u>unbounded</u> and yet there may be optimal solutions. This is common in minimization problems and is possible in maximization problems.

13. The <u>feasible region</u> for a two-variable linear programming problem can be: a) nonexistent, b) a single point, c) a line, d) a polygon, or e) an unbounded area.

14. <u>Any linear program</u> either (a) is infeasible, (b) has a unique optimal solution or alternate optimal solutions, or (c) has an objective function that can be increased without bound.

15. A linear program in which all the variables are non-negative and all the constraints are equalities is said to be in <u>standard form</u>. Standard form is attained by adding <u>slack variables</u> to "less than or equal to" constraints, and by subtracting <u>surplus variables</u> from "greater than or equal to" constraints. They represent the difference between the left and right sides of the constraints.

16. A <u>nonbinding constraint</u> is one in which there is <u>positive slack or surplus</u> when evaluated at the optimal solution.

17. <u>Slack and surplus variables</u> have objective function coefficients equal to 0. If, however, extra resources could be sold at a profit, or if there were a penalty for surplus resources, the objective function coefficients would not be 0 and these variables would, in effect, become new decision variables.

GRAPHICAL SOLUTION PROCEDURE

1. Graph the constraints and shade in the feasible region, considering the feasible side of each constraint line.

2. Set the objective function equal to any arbitrary constant and graph it. If the line does not lie in the feasible region, move it (maintaining its slope) into the feasible region.

3. Move the objective function line parallel to itself in the direction that increases its value when maximizing (decreases its value when minimizing) until it touches the last point(s) of the feasible region.

4. If the optimal extreme point falls on an axis (say, x_2 axis), use the binding constraint equation to solve for the unknown x^* (in this case x_2^*, since x_1^* is zero). Otherwise, solve the two equations (binding constraints) in two unknowns (x_1^* and x_2^*) that determine the optimal extreme point.

5. Find z by substituting x_1* and x_2* in the objective function.

FLOW CHART OF
GRAPHICAL L.P. SOLUTION PROCEDURE

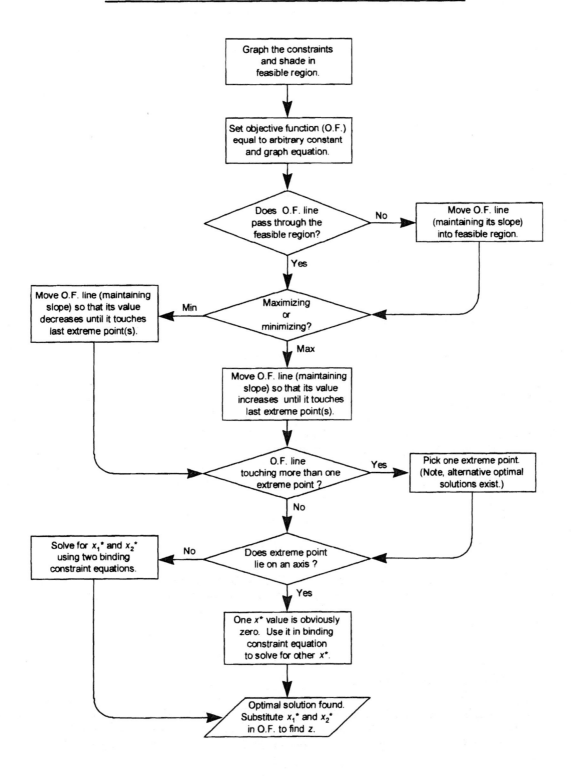

ILLUSTRATED PROBLEMS

PROBLEM 1

Given the following linear program:

$$\text{MAX} \quad 3x_1 + 4x_2$$
$$\text{s.t.} \quad 2x_1 + 3x_2 \leq 24$$
$$3x_1 + x_2 \leq 21$$
$$x_1 + x_2 \leq 9$$
$$x_1, x_2 \geq 0$$

a) Solve the problem graphically.

b) Write the problem in standard form.

c) Given your answer to (a), what are the optimal values of the slack variables.

> NOTE: Plotting an initial objective function line involves little more than reversing the objective coefficients for x_1 and x_2. Consider Problem 1 below. The objective line will cross the x_1 axis at 4 (x_2's coefficient) and the x_2 axis at 3 (x_1's coefficient). If the coefficients are too large (or small) for convenient graphing, scale them down (or up) in a consistent manner by dividing (or multiplying) both by, say, 10.

SOLUTION 1

a) (1) <u>Graph the constraints</u>. (See graph on next page.)
   Constraint 1: When $x_1 = 0$, then $x_2 = 8$; when $x_2 = 0$, then $x_1 = 12$.
   Connect (12,0) and (0,8). The "<" side is below the line.
   Constraint 2: When $x_1 = 0$, then $x_2 = 21$; when $x_2 = 0$, then $x_1 = 7$.
   Connect (7,0) and (0,21). The "<" side is below the line.
   Constraint 3: When $x_1 = 0$, then $x_2 = 9$; when $x_2 = 0$, then $x_1 = 9$.
   Connect (9,0) and (0,9). The "<" side is below the line.

   <u>Shade in the feasible region</u>.

   (2) <u>Graph the objective function</u> by setting the objective function equal to any arbitrary value (say 12) and graphing it. For $3x_1 + 4x_2 = 12$, when $x_2 = 0$, $x_1 = 4$; when $x_1 = 0$, $x_2 = 3$. Connect (4,0) and (0,3), the thick graphed line.

INTRO. TO LINEAR PROGRAMMING

(3) <u>Move the objective function line parallel to itself</u> in the direction that increases its value (upward) until it touches the last point of the feasible region. It is at the intersection of the first and third constraint lines.

(4) <u>Solve these two equations in two unknowns</u>:

$$2x_1 + 3x_2 = 24 \longrightarrow 2x_1 + 3x_2 = 24$$
$$x_1 + x_2 = 9 \longrightarrow 2x_1 + 2x_2 = 18$$
$$x_2 = 6$$

Substituting into $x_1 + x_2 = 9$, then $x_1 = 3$.

(5) <u>Solve for z</u>: $z = 3x_1 + 4x_2 = 3(3) + 4(6) = 33$. Thus the optimal solution is $x_1 = 3$, $x_2 = 6$, $z = 33$.

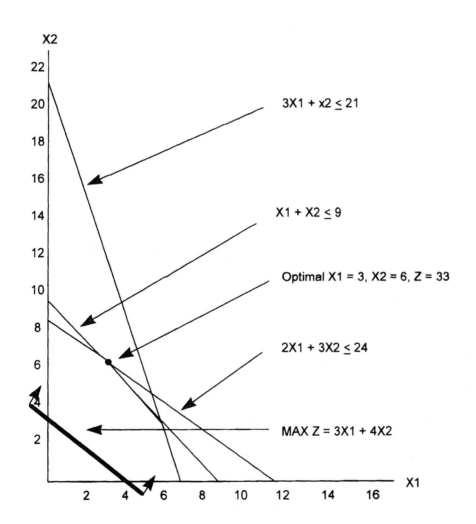

b) To write the problem in standard form, since each constraint is a "≤" constraint, add a slack variable to each constraint.

$$\text{MAX} \quad 3x_1 + 4x_2 + 0s_1 + 0s_2 + 0s_3$$

$$\text{s.t.} \quad 2x_1 + 3x_2 + s_1 = 24$$

$$3x_1 + x_2 + s_2 = 21$$

$$x_1 + x_2 + s_3 = 9$$

$$x_1, x_2, s_1, s_2, s_3 \geq 0$$

c) Since the optimal solution was $x_1 = 3$, $x_2 = 6$, then substituting these values into the above equations gives:

$$s_1 = 24 - 2(3) - 3(6) = 0$$

$$s_2 = 21 - 3(3) - 1(6) = 6$$

$$s_3 = 9 - 1(3) - 1(6) = 0$$

PROBLEM 2

Given the following linear program:

$$\text{MIN} \quad 5x_1 + 2x_2$$

$$\text{s.t.} \quad 2x_1 + 5x_2 \geq 10$$

$$4x_1 - x_2 \geq 12$$

$$x_1 + x_2 \geq 4$$

$$x_1, x_2 \geq 0$$

a) Solve graphically for the optimal solution.

b) How does one know that although $x_1 = 5$, $x_2 = 3$ is a feasible solution for the constraints, it will never be the optimal solution no matter what objective function is imposed?

c) Solve for the optimal solution using a spreadsheet.

SOLUTION 2

a) (1) <u>Graph the constraints</u>.

Constraint 1: When $x_1 = 0$, then $x_2 = 2$; when $x_2 = 0$, then $x_1 = 5$. Connect (5,0) and (0,2). The ">" side is above this line.

Constraint 2: When $x_2 = 0$, then $x_1 = 3$. But setting x_1 to 0 will yield $x_2 = -12$, which is not on the graph. Thus, to get a second point on this line, set x_1 to any number larger than 3 and solve for x_2: when $x_1 = 5$, then $x_2 = 8$. Connect (3,0) and (5,8). The ">" side is to the right.

Constraint 3: When $x_1 = 0$, then $x_2 = 4$; when $x_2 = 0$, then $x_1 = 4$. Connect (4,0) and (0,4). The ">" side is above this line.

<u>Shade in the feasible region</u>.

(2) <u>Graph the objective function</u> by setting the objective function equal to an arbitrary constant (say 20) and graphing it. For $5x_1 + 2x_2 = 20$, when $x_1 = 0$, then $x_2 = 10$; when $x_2 = 0$, then $x_1 = 4$. Connect (4,0) and (0,10).

(3) <u>Move the objective function line</u> in the direction which lowers its value until it touches the last point of the feasible region, determined by the last two constraints.

(4) <u>Solve these two equations in two unknowns</u>. $4x_1 - x_2 = 12$ and $x_1 + x_2 = 4$. Adding these two equations gives: $5x_1 = 16$ or $x_1 = 16/5$. Substituting this into $x_1 + x_2 = 4$ gives: $x_2 = 4/5$.

(5) <u>Solve for</u> $z = 5x_1 + 2x_2 = 5(16/5) + 2(4/5) = 88/5$. Thus the optimal solution is $x_1 = 16/5$; $x_2 = 4/5$; $z = 88/5$.

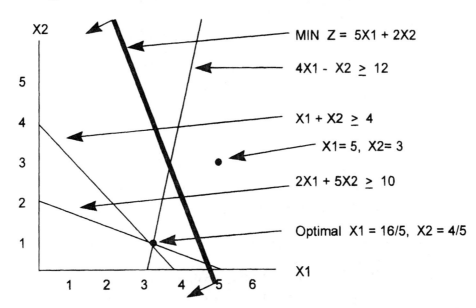

b) (5,3) lies in the feasible region, but it is not an extreme point and can never be optimal.

c) Spreadsheet showing data and formulas

	A	B	C	D
1		LHS Coefficients		
2	Constraints	X1	X2	RHS
3	#1	2	5	10
4	#2	4	-1	12
5	#3	1	1	4
6	Obj.Func.Coeff.	5	2	
7				
8		Decision Variables		
9		X1	X2	
10	Dec.Var.Values			
11				
12	Minimized Objective Function		=B6*B10+C6*C10	
13				
14	Constraints	Amount Used		Amount Avail.
15	#1	=B3*$B$10+C3*$C$10	>=	=D3
16	#2	=B4*$B$10+C4*$C$10	>=	=D4
17	#3	=B5*$B$10+C5*$C$10	>=	=D5

Steps in Using Excel Solver:

Step 1: Select the **Tools** pull-down menu.
Step 2: Select the **Solver** option.
Step 3: When the **Solver Parameters** dialog box appears:
　　Enter C12 in the **Set Target Cell** box.
　　Select the **Equal To: Min** option.
　　Enter B10:C10 in the **By Changing Cells** box.
　　Select **Add**.
Step 4: When the **Add Constraint** dialog box appears:
　　Enter B15:B17 in the **Cell Reference** box.
　　Select >=.
　　Enter D15:D17 in the **Constraint** box.
　　Click **OK**.
Step 5: When the **Solver Parameters** dialog box appears:
　　Choose **Options**.
Step 6: When the **Solver Options** dialog box appears:
　　Select **Assume Non-Negative**.
　　Click **OK**.
Step 7: When the **Solver Parameters** dialog box appears:
　　Choose **Solve**.
Step 8: When the **Solver Results** dialog box appears:
　　Select **Keep Solver Solution**.
　　Click **OK**.

	A	B	C	D
8		Decision Variables		
9		X1	X2	
10	Dec.Var.Values	3.20	0.800	
11				
12	Minimized Objective Function		17.600	
13				
14	Constraints	Amount Used		Amount Avail.
15	#1	10.4	>=	10
16	#2	12	>=	12
17	#3	4	>=	4

PROBLEM 3

Given the following linear program:

$$\text{MAX} \quad 4x_1 + 5x_2$$

$$\begin{aligned}
\text{s.t.} \quad & x_1 + 3x_2 \le 22 \\
& -x_1 + x_2 \le 4 \\
& x_2 \le 6 \\
& 2x_1 - 5x_2 \le 0 \\
& x_1, x_2 \ge 0
\end{aligned}$$

NOTE: If a constraint's righthand-side value is 0, the constraint line will pass through the origin ($x_1 = 0$, $x_2 = 0$). This is the case with the fourth constraint above.

a) Solve the problem by the graphical method.

b) What would be the optimal solution if the second constraint were $-x_1 + x_2 = 4$?

c) What would be the optimal solution if the first constraint were $x_1 + 3x_2 \ge 22$?

CHAPTER 7

SOLUTION 3

a) (1) <u>Graph the constraints</u>.

Constraint 1: When $x_1 = 0$, $x_2 = 22/3$; when $x_2 = 0$, then $x_1 = 22$. Connect (22,0) and (0,22/3). The "<" side is below this line.

Constraint 2: When $x_1 = 0$, then $x_2 = 4$. Setting x_2 to 0 would give $x_1 = -4$, which is outside the graph. Set x_2 to a number greater than 4 and solve for x_1. When $x_2 = 6$, then $x_1 = 2$. Connect (0,4) and (2,6). (0,0) is on the "<" side.

Constraint 3: This is a horizontal line through $x_2 = 6$.

Constraint 4: When $x_2 = 0$, then $x_1 = 0$; Set x_1 to any positive constant and solve for x_2. When $x_1 = 5$, then $x_2 = 2$. Connect the points (0,0) and (5,2). To determine the "<" side select any arbitrary point on one side of the line and substitute into the inequality. Arbitrarily choosing (0,5), this gives $2(0) - 5(5) = -25$. Thus the side containing (0,5) is the "<" side.

<u>Shade in the feasible region</u>.

(2) <u>Graph the objective function</u> by setting it to an arbitrary value, say 20. For $4x_1 + 5x_2 = 20$, when $x_1 = 0$, then $x_2 = 4$; when $x_2 = 0$, then $x_1 = 5$. Connect with a broken line the points (5,0) and (0,4).

(3) <u>Move the objective function line parallel to itself</u> in the direction which increases its value until it touches the last point of the feasible region. This is at the intersection of the first and fourth constraints.

(4) <u>Solve these two equations in two unknowns</u>:

$$x_1 + 3x_2 = 22 \quad \longrightarrow \quad 2x_1 + 6x_2 = 44$$
$$2x_1 - 5x_2 = 0 \quad \longrightarrow \quad 2x_1 - 5x_2 = 0$$

Subtracting the second equation from the first yields: $11x_2 = 44$ or $x_2 = 4$. Substituting $x_2 = 4$ into the first equation gives $x_1 = 10$.

(5) <u>Substitute for z</u> $= 4x_1 + 5x_2 = 4(10) + 5(4) = 60$. Thus the optimal solution is $x_1 = 10$; $x_2 = 4$; $z = 60$.

b) The feasible region is now the line segment of $-x_1 + x_2 = 4$ between (0,4) and (2,6). (2,6) now gives the optimal solution.

c) The feasible region is now the triangular section between (4,6), (15,6), and (10,4). (15,6) is now the optimal solution.

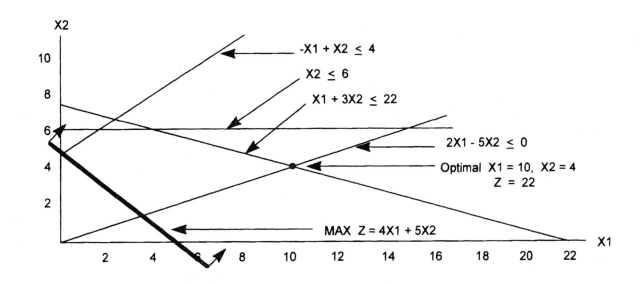

PROBLEM 4

Show graphically why the following two linear programs do not have optimal solutions and explain the difference between the two.

(a) MAX $2x_1 + 6x_2$

s.t. $4x_1 + 3x_2 \leq 12$

$2x_1 + x_2 \geq 8$

$x_1, x_2 \geq 0$

(b) MAX $3x_1 + 4x_2$

s.t. $x_1 + x_2 \geq 5$

$3x_1 + x_2 \geq 8$

$x_1, x_2 \geq 0$

SOLUTION 4

Refer to the graphs on the next page. Note that (a) has no points that satisfy both constraints, hence has no feasible region, and no optimal solution. (a) is infeasible.

Note that in (b) the feasible region is unbounded and the objective function line can be moved parallel to itself without bound so that z can be increased infinitely. (b) is unbounded.

(a)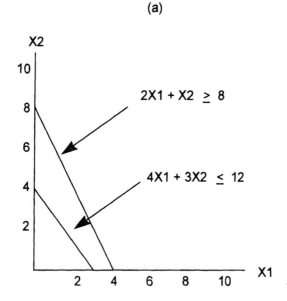

(b)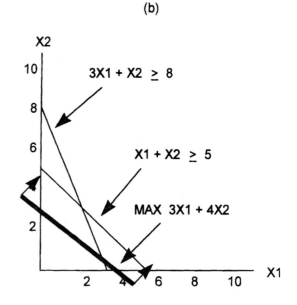

PROBLEM 5

Given the following linear program:

$$\text{MIN} \quad 150x_1 + 210x_2$$

$$\begin{aligned}
\text{s.t.} \quad 3.8x_1 + 1.2x_2 &\geq 22.8 \\
x_2 &\geq 6 \\
x_2 &\leq 15 \\
45x_1 + 30x_2 &= 630 \\
x_1, x_2 &\geq 0
\end{aligned}$$

a) Solve the problem graphically. How many extreme points exist for this problem?

b) What would be the optimal solution if the "=" in the fourth constraint was changed to "$\leq$"?

c) If the "=" in the fourth constraint was changed to "$\geq$", how would the problem be affected?

SOLUTION 5

a) (1) <u>Graph the constraints</u>.

Constraint 1: When $x_1 = 0$, $x_2 = 19$; when $x_2 = 0$, then $x_1 = 6$. Connect (6,0) and (0,9). The ">" side is to the right of this line.

Constraint 2: This is a horizontal line through $x_2 = 6$. The ">" side is above this line.

Constraint 3: This is a horizontal line through $x_2 = 15$. The "<" side is above this line.

Constraint 4: When $x_1 = 0$, $x_2 = 21$; when $x_2 = 0$, then $x_1 = 14$. Connect (14,0) and (0,21).

<u>Shade in the feasible region</u>.

> NOTE: The feasible region in this problem is limited to a segment of the line representing the "equal to" constraint. Only two extreme points exist.

(2) <u>Graph the objective function</u> by setting the objective function equal to an arbitrary constant as previously demonstrated or by using the following approach. Scale down the objective coefficients c_1 and c_2 (say, by dividing both by 10 to get 8 and 13, respectively). Now, use x_1's coefficient as a value to plot on the x_2 axis and use x_2's coefficient as a value to plot on the x_1 axis. Connect points (0,15) and (21,0).

(3) <u>Move the objective function line</u> in the direction that lowers its value until it touches the last point of the feasible region. The point is determined by the second and fourth constraints.

(4) <u>Solve for the unknown x</u> by substituting $x_2 = 6$ into $45x_1 + 30x_2 = 630$, yielding $x_1 = 10$.

(5) <u>Solve for z</u> = $150x_1 + 210x_2 = 150(10) + 210(6) = 2760$. Thus the optimal solution is $x_1 = 10$, $x_2 = 6$, and $z = 2760$. [See the graph on the next page.]

b) The feasible region is now shaped by all four constraints. The optimal extreme point is determined by the first and second constraints. Solving these two equations in two unknowns, the optimal solution is (4.105,6), point C on the graph.

c) The optimal solution is now (10,6), point B on the graph, and the first constraint is now redundant.

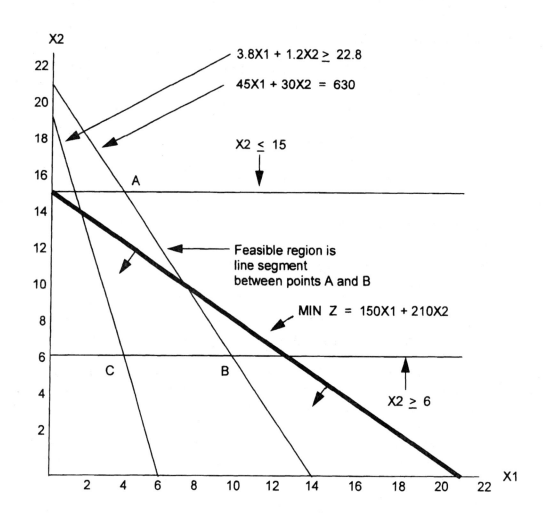

PROBLEM 6

Solve the following linear program graphically.

$$\begin{align}
\text{MAX} \quad & 5x_1 + 7x_2 \\
\text{s.t.} \quad & x_1 \leq 6 \\
& 2x_1 + 3x_2 \leq 19 \\
& x_1 + x_2 \leq 8 \\
& x_1, x_2 \geq 0
\end{align}$$

SOLUTION 6

From the graph below we see that the optimal solution occurs at $x_1 = 5$, $x_2 = 3$, and $z = 46$.

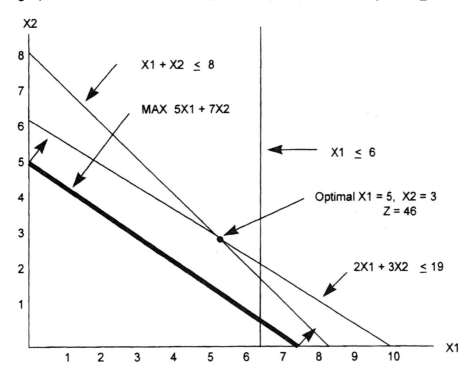

PROBLEM 7

A manager of a small fabrication plant must decide on a production schedule of two new products for the automobile industry. The profit on product 1 is $1(thousand) and on product 2 is $3(thousand).

The manufacture of these products depends largely on the availability of certain subassemblies the plant receives daily from a local distributor. It takes three of these subassemblies for each unit of product 1 and two for each unit of product 2. Twelve such subassemblies are delivered daily.

Further, it takes two hours to make a unit of product 1 and six hours to make a unit of product 2. The plant has assigned only three workers working 8-hour shifts for these new products. Due to limited demand, the manager does not want more than seven units of product 2 produced daily.

a) Formulate this problem as a linear program.

b) Solve graphically for the optimal solution. Describe the set of all optimal solutions. Identify any redundant constraints.

c) Give an optimal daily production schedule that manufactures exactly one unit of product 1.

d) Discuss the applicability of linear programming for this problem.

SOLUTION 7

a) (1) <u>Define variables</u>: x_1 and x_2 = the amount of product 1 and 2 produced daily.

   (2) <u>Define objective</u>:
   Maximize total daily profits: MAX $1x_1 + 3x_2$ (in thousands of dollars).

   (3) <u>Define constraints</u>:
   Subassemblies: Number used daily ≤ number available: $3x_1 + 2x_2 \leq 12$

   Labor: Number of hours used daily ≤ (3 men) x (8 hrs./day): $2x_1 + 6x_2 \leq 24$

   Product 2: Quantity produced daily ≤ specified limit: $x_2 \leq 7$

   Non-negativity of variables: $x_1, x_2 \geq 0$

   Summarizing,
   $$\text{MAX} \quad 1x_1 + 3x_2$$
   $$\text{s.t.} \quad 3x_1 + 2x_2 \leq 12$$
   $$2x_1 + 6x_2 \leq 24$$
   $$x_2 \leq 7$$
   $$x_1, x_2 \geq 0$$

b) Graphically,

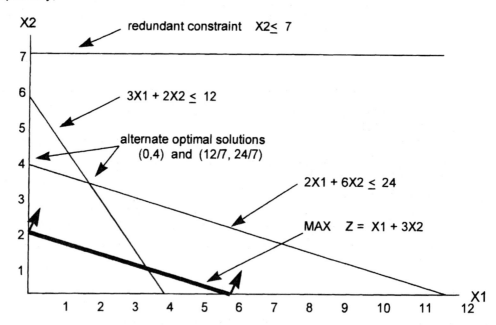

The optimal solution occurs at $x_1 = 0$, $x_2 = 4$ and at $x_1 = 12/7$, $x_2 = 24/7$, and at all points in between on the line $2x_1 + 6x_2 = 24$. At any point on this line, $z = 12$. The $x_2 \leq 7$ constraint does not help shape the feasible region and thus is redundant.

c) On the optimal solution line, $2x_1 + 6x_2 = 24$, when $x_1 = 1$, then $x_2 = 11/3$. Still, $z = 1(1) + 3(11/3) = 12$ (thousand).

d) One must consider whether these variables can be allowed to assume values that are not integers. For continuous production, frequently a fractional value can be considered as "work in progress"; products not finished on one day are simply completed the next day. Thus, LP appears to be appropriate for this problem.

PROBLEM 8

A small company will be introducing a new line of lightweight bicycle frames to be made from special aluminum and steel alloys. The frames will be produced in two models, deluxe and professional, with anticipated unit profits of $10 and $15, respectively.

The number of pounds of aluminum alloy and steel alloy needed per deluxe frame is 2 and 3, respectively. The number of pounds of aluminum alloy and steel alloy needed per professional frame is 4 and 2. A supplier delivers 100 pounds of the aluminum alloy and 80 pounds of the steel alloy weekly. What is the optimal weekly production schedule?

SOLUTION 8

Let x_1 and x_2 equal the number of deluxe and professional frames produced weekly.

$$\text{MAX} \quad 10x_1 + 15x_2$$
$$\text{s.t.} \quad 2x_1 + 4x_2 \leq 100$$
$$3x_1 + 2x_2 \leq 80$$
$$x_1, x_2 \geq 0$$

Solving graphically, the optimal production schedule is to produce $x_1 = 15$ deluxe frames weekly and $x_2 = 17.5$ professional frames weekly for an optimal weekly profit of $412.50.

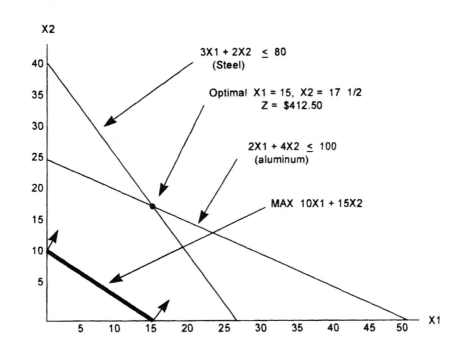

148   CHAPTER 7

ANSWERED PROBLEMS

PROBLEM 9

Solve graphically for the optimal solution to the following linear program:

$$\text{MIN} \quad 16x_1 + 12x_2$$

$$\begin{aligned}
\text{s.t.} \quad & 8x_1 + 4x_2 \le 36 \\
& x_1 + x_2 \le 7 \\
& 3x_1 + 12x_2 \ge 24 \\
& x_1 + 5x_2 \ge 20 \\
& x_1, x_2 \ge 0
\end{aligned}$$

PROBLEM 10

Given the following linear program:

$$\text{MAX} \quad 4x_1 + 2x_2$$

$$\begin{aligned}
\text{s.t.} \quad & x_1 \le 4 \\
& 3x_1 + 8x_2 \le 24 \\
& 2x_1 + x_2 \ge 6 \\
& x_1, x_2 \ge 0
\end{aligned}$$

a) Solve the problem graphically.

b) What would be the optimal solution(s) if the objective function were a minimization rather than a maximization objective?

INTRO. TO LINEAR PROGRAMMING

PROBLEM 11

Consider a linear programming problem with the following constraint set:

$$2x_1 + x_2 \geq 4$$

$$x_1 + 2x_2 \geq 5$$

$$x_1 - 2x_2 \leq 1$$

a) Graph the feasible region and note it is unbounded.

> NOTE: One might think an unbounded maximization problem would always have an unbounded objective function value. This problem proves the contrary.

b) Identify all extreme points.

c) Solve the problem with each of these objective functions: (1) MAX $z = 2x_1 - 5x_2$; (2) MAX $z = 2x_1 - 4x_2$; and (3) MAX $z = 2x_1 - 3x_2$ and discuss the results.

PROBLEM 12

Given the following linear programming problem:

$$\text{MAX} \quad 3x_1 + 5x_2$$

$$\text{s.t.} \quad 4x_1 + 3x_2 \geq 24$$

$$2x_1 + 3x_2 \leq 18$$

$$x_2 \geq 3$$

$$x_1, x_2 \geq 0$$

a) Solve the problem graphically.

> NOTE: The feasible region in this problem is limited to a single point. A common error is to mistake this situation for infeasibility.

b) What effect would changing the objective function to MAX $5x_1 + 4x_2$ have?

PROBLEM 13

Given the following linear programming problem:

$$\text{MAX} \quad 8x_1 + 10x_2$$
$$\begin{aligned} \text{s.t.} \quad & x_1 + x_2 \leq 35 \\ & 3x_1 + 2x_2 \leq 60 \\ & x_2 \leq 15 \\ & x_1, x_2 \geq 0 \end{aligned}$$

a) Solve for the optimal solution.

b) State why the first constraint is redundant.

c) Suppose the second constraint's right hand side is changed from 60 to 100. Solve for the new optimal solution and show that the first constraint is now binding and NOT redundant.

PROBLEM 14

Consider the following linear program:

$$\text{MAX} \quad 60x_1 + 43x_2$$
$$\begin{aligned} \text{s.t.} \quad & x_1 + 3x_2 \geq 9 \\ & 6x_1 - 2x_2 = 12 \\ & x_1 + 2x_2 \leq 10 \\ & x_1, x_2 \geq 0 \end{aligned}$$

a) Write the problem in standard form.

b) What is the feasible region for the problem?

c) Show that regardless of the values of the actual objective function coefficients, the optimal solution will occur at one of two points. Solve for these points and then determine which one maximizes the current objective function.

INTRO. TO LINEAR PROGRAMMING

PROBLEM 15

A businessman is considering opening a small specialized trucking firm. To make the firm profitable, it is estimated that it must have a daily trucking capacity of at least 84,000 cu. ft. Two types of trucks are appropriate for the specialized operation. Their characteristics and costs are summarized in the table below. Note that truck 2 requires 3 drivers for long haul trips. There are 41 potential drivers available and there are facilities for at most 40 trucks.

The businessman's objective is to minimize the total cost outlay for trucks.

Truck	Cost	Capacity (Cu. ft.)	Drivers Needed
x_1	$18,000	2,400	1
x_2	$45,000	6,000	3

Solve the problem graphically and note there are alternate optimal solutions. Which optimal solution:

a) uses only one type of truck?

b) utilizes the minimum total number of trucks?

c) uses the same number of truck x_1 as truck x_2?

PROBLEM 16

A baseball glove manufacturer has 1200 linear feet of cowhide and 800 linear feet of synthetic material. It makes two styles of baseball gloves: child's and adult's. Requirements and profit PER DOZEN are summarized below:

	Cowhide	Synthetic	Profit
Child's	4	4	$60
Adult's	12	6	$95

a) Solve for the optimal number of dozen of each model to manufacture. What are the values of the slack variables?

b) Suppose the company could make $1 on each unused linear foot of cowhide and $.25 on each unused linear foot of synthetic material. Reformulate the linear programming model. The new optimal solution is to make 200 dozen child models and no adult models and sell 400 linear feet of cowhide. Locate this new point on your graph and show it is not the optimal extreme point of part (a).

TRUE/FALSE

___ 17. A problem formulation that includes a term that is the product of two variables would not be a linear program.

___ 18. A nonbinding constraint, like a binding constraint, helps form the shape (boundaries) of the feasible region.

___ 19. If a linear program has an optimal solution, then an extreme point must be optimal.

___ 20. All optimal solutions are extreme points.

___ 21. A redundant constraint lies entirely within the feasible region.

___ 22. It is possible to have exactly two optimal solutions to a linear programming problem.

___ 23. A linear programming problem can be both unbounded and infeasible.

___ 24. If a problem has a constraint which is parallel to the objective function, then there must be alternative optimal solutions.

___ 25. An infeasible problem is one in which the objective function can be increased to infinity.

___ 26. A slack variable is a variable that represents the difference between the amount of a resource that was available and the actual amount used by the solution.

___ 27. In a feasible problem, an equal-to constraint cannot be redundant.

___ 28. A variable in a linear programming problem must be allowed to assume fractional values.

___ 29. Any change to an objective function coefficient of a variable that is positive in the optimal solution will change the optimal solution.

___ 30. An unbounded feasible region might not result in an unbounded solution for a minimization or maximization problem.

___ 31. Increasing the right-hand side of a nonbinding constraint will not cause a change in the optimal solution.

CHAPTER 8

Linear Programming: Sensitivity Analysis and Interpretation of Solution

CHAPTER 8

KEY CONCEPTS

CONCEPT	ILLUSTRATED PROBLEMS	ANSWERED PROBLEMS
Sensitivity Analysis: Graphical Solution	1,2	12-14
Sensitivity Analysis: Computer Solution	3-6	7-11,15
Changes to Objective Function Coefficients		
Reduced Cost	6	
Range of Optimality	3-6	7-11
Changes to Right Hand Side Values		
Dual Price	3-6	8-11
Range of Feasibility	3-5	8-11
Sunk/Relevant Costs	5	8
100% Rule	3-6	8-11

REVIEW

1. <u>Sensitivity analysis</u> is used to determine how the optimal solution is affected by changes, within specified ranges, in the objective function coefficients and the right-hand side (RHS) values.

2. <u>Sensitivity analysis</u> is important to the manager who must operate in a dynamic environment with imprecise estimates of the coefficients. Sensitivity analysis allows him to ask certain <u>what-if questions</u> about the problem.

3. A <u>reduced cost</u> for a decision variable whose value is 0 in the optimal solution is the amount the variable's objective coefficient would have to improve (increase for maximization problems, decrease for minimization problems) before this variable could assume a positive value. Thus, the reduced cost for a decision variable with a positive value is 0.

4. A <u>range of optimality</u> of an <u>objective function coefficient</u> is found by determining an interval for the objective function coefficient in which the original solution remains optimal while keeping all other data of the problem constant. The value of z might change in this range.

5. Graphically, the limits of a <u>range of optimality</u> are found by changing the slope of the objective function line within the limits of the slopes of the binding constraint lines. This would also apply to simultaneous changes in the objective coefficients. The slope of an objective function line, MAX $c_1x_1 + c_2x_2$, is $-c_1/c_2$, and the slope of a constraint i, $a_{i1}x_1 + a_{i2}x_2 = b_i$, is $-a_{i1}/a_{i2}$.

6. The <u>100% rule</u> states that <u>simultaneous changes in objective function coefficients</u> will not change the optimal solution as long as the sum of the percentages of the change divided by the corresponding maximum allowable change in the range of optimality for each coefficient does not exceed 100%.

7. A <u>dual price</u> for a <u>right-hand side</u> (or resource limit) is the amount the objective function will improve per unit increase in the right-hand side value of a constraint.

8. A <u>dual price</u> reflects the value of an additional unit of the resource if the <u>resource cost</u> is <u>sunk</u>. It represents the extra value over the normal cost of the resource when the resource cost is <u>relevant</u>.

9. A <u>resource cost</u> is <u>relevant</u> if the amount paid for it is dependent upon the amount of the resource used by the decision variables. Relevant costs are reflected in the objective function coefficients.

10. A <u>resource cost</u> is <u>sunk</u> if it must be paid regardless of the amount of the resource actually used by the decision variables. Sunk resource costs are not reflected in the objective function coefficients.

11. Graphically, a <u>dual price</u> is determined by adding +1 to the right hand side value in question and then resolving for the optimal solution in terms of the same two binding constraints. The dual price is equal to the difference in the values of the objective functions between the new and original problems.

12. A <u>nonbinding constraint</u> is one in which there is <u>positive slack or surplus</u> when evaluated at the optimal solution. The dual price for a nonbinding constraint is zero.

13. The <u>range of feasibility</u> for a change in a right-hand side value is the range of values for this parameter in which the original <u>dual price</u> remains constant.

14. Graphically, the <u>range of feasibility</u> is determined by finding the values of a right hand side coefficient such that the same two lines that determined the original optimal solution continue to determine the optimal solution for the problem.

15. The <u>100% rule</u> also states that <u>simultaneous changes in right hand sides</u> will not change the dual prices as long as the sum of the percentages of the changes divided by the corresponding maximum allowable change in the range of feasibility for each right-hand side does not exceed 100%.

16. **Computer software packages** (such as *The Management Scientist* or *Microsoft Excel*) that solve linear programming problems all give five elements of relevant information about the optimal solution:
    1. The optimal value of the objective function;
    2. Information about the decision variables: (a) their values, (b) their reduced costs;
    3. Information about the constraints: (a) amount of slack or surplus, (b) dual prices;
    4. Ranges of optimality for objective function coefficients: (a) lower limit, (b) upper limit;
    5. Ranges of feasibility for righthand-side values: (a) lower limit, (b) upper limit.

ILLUSTRATED PROBLEMS

PROBLEM 1

Given the following linear program:

$$\text{MAX} \quad 5x_1 + 7x_2$$
$$\begin{aligned} \text{s.t.} \quad x_1 &\leq 6 \\ 2x_1 + 3x_2 &\leq 19 \\ x_1 + x_2 &\leq 8 \\ x_1, x_2 &\geq 0 \end{aligned}$$

The graphical solution to the problem is shown below. From the graph we see that the optimal solution occurs at $x_1 = 5$, $x_2 = 3$, and $z = 46$.

a) Calculate the range of optimality for each objective function coefficient.

b) Calculate the dual price for each resource.

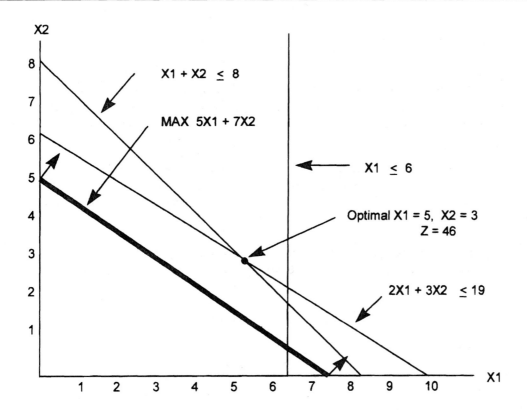

SOLUTION 1

a) The slope of the objective function line is $-c_1/c_2$. The slope of the first binding constraint, $x_1 + x_2 = 8$, is -1 and the slope of the second binding constraint, $2x_1 + 3x_2 = 19$ is $-2/3$.

Range of optimality for c_1:
Find the range of values for c_1 (with c_2 staying 7) such that the objective function line slope lies between that of the two binding constraints:

$$-1 \leq -c_1/7 \leq -2/3.$$

Multiplying through by -7 (and reversing the inequalities):

$$14/3 \leq c_1 \leq 7.$$

Range of optimality for c_2:
Find the range of values for c_2 (with c_1 staying 5) such that the objective function line slope lies between that of the two binding constraints:

$$-1 \leq -5/c_2 \leq -2/3.$$

Multiplying by -1: $\qquad\qquad\qquad 1 \geq 5/c_2 \geq 2/3.$

Inverting, $\qquad\qquad\qquad\qquad 1 \leq c_2/5 \leq 3/2.$

Multiplying by 5: $\qquad\qquad\qquad 5 \leq c_2 \leq 15/2.$

158 CHAPTER 8

b) <u>Dual prices:</u>

   Constraint 1: Since $x_1 \leq 6$ is not a binding constraint, its dual price is 0.

   Constraint 2: Change the right hand side of the second constraint to 20 and resolve for the optimal point determined by the last two constraints: $2x_1 + 3x_2 = 20$ and $x_1 + x_2 = 8$. The solution is $x_1 = 4$, $x_2 = 4$, $z = 48$. Hence, the dual price = $z_{new} - z_{old} = 48 - 46 = 2$.

   Constraint 3: Change the right hand side value of the third constraint to 9 and resolve for the optimal point determined by the last two constraints: $2x_1 + 3x_2 = 19$ and $x_1 + x_2 = 9$. The solution is: $x_1 = 8$, $x_2 = 1$, $z = 47$. Hence, the dual price is $z_{new} - z_{old} = 47 - 46 = 1$.

   Summarizing, the dual price for the first resource is 0, for the second resource is 2, and for the third is 1. Note that these dual prices are only valid in the range of feasibility for each resource.

PROBLEM 2

A small company will be introducing a new line of lightweight bicycle frames to be made from special aluminum and steel alloys. The frames will be produced in two models, deluxe and professional, with anticipated unit profits of $10 and $15, respectively.

The number of pounds of aluminum alloy and steel alloy needed per deluxe frame is 2 and 3, respectively. The number of pounds of aluminum alloy and steel alloy needed per professional frame is 4 and 2, respectively. A supplier delivers 100 pounds of the aluminum alloy and 80 pounds of the steel alloy weekly.

The problem of determining the optimal weekly production schedule can be formulated as follows.

Let: x_1 = the number of deluxe frames produced weekly
      x_2 = the number of professional frames produced weekly

$$\begin{aligned} \text{MAX} \quad & 10x_1 + 15x_2 \\ \text{s.t.} \quad & 2x_1 + 4x_2 \leq 100 \\ & 3x_1 + 2x_2 \leq 80 \\ & x_1, x_2 \geq 0 \end{aligned}$$

Solving graphically, the optimal production schedule is to produce $x_1 = 15$ deluxe frames weekly and $x_2 = 17.5$ professional frames weekly for an optimal weekly profit of $412.50.

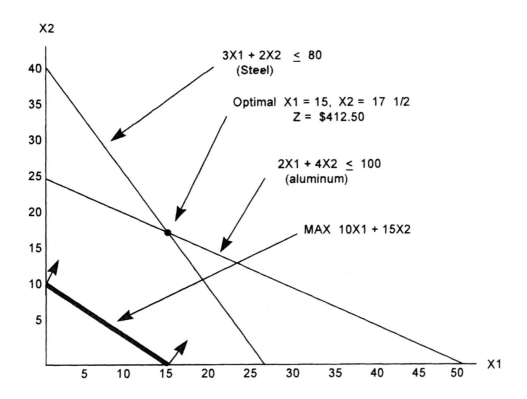

a) Within what limits must the unit profits lie for each of the frames for this solution to remain optimal?

b) Suppose the unit profits assumed all aluminum purchased would be used and hence the profit figures did not include a unit cost for the aluminum. Now extra aluminum can be purchased at $2.50 per pound. Should the company purchase additional pounds of aluminum at that price?

SOLUTION 2

a) Note that the binding constraints are the aluminum and the steel constraints, with slopes -1/2 and -3/2 respectively.
Range of optimality for deluxe profits (c_1):

$$-3/2 \leq -c_1/15 \leq -1/2 \quad \text{or} \quad 15/2 \leq c_1 \leq 45/2.$$

Range of optimality for professional profits (c_2):

$$-3/2 \leq -10/c_2 \leq -1/2 \quad \text{or} \quad 20/3 \leq c_2 \leq 20.$$

b) The aluminum costs are then considered <u>sunk</u> costs and the dual price for aluminum would yield the maximum worth for additional aluminum. Resolve the two equations and two unknown with the right hand side of the aluminum constraint changed to 101.

This results in $x_1 = 59/4$, $x_2 = 143/8$, $z = \$415.625$. Hence the dual price for aluminum is $\$415.625 - \$412.50 = \$3.125$. Since this is greater than the selling price of $\$2.50$ per pound of aluminum, additional aluminum should be purchased at this price.

PROBLEM 3

Consider the following linear program:

$$\text{MAX } 3x_1 + 4x_2 \text{ (\$ Profit)}$$
$$\text{s.t. } x_1 + 3x_2 \leq 12$$
$$2x_1 + x_2 \leq 8$$
$$x_1 \leq 3$$
$$x_1, x_2 \geq 0$$

The Management Scientist provided the following solution output:

OBJECTIVE FUNCTION VALUE = 20.000

VARIABLE	VALUE	REDUCED COST
X1	2.400	0.000
X2	3.200	0.000

CONSTRAINT	SLACK/SURPLUS	DUAL PRICES
1	0.000	1.000
2	0.000	1.000
3	0.600	0.000

OBJECTIVE COEFFICIENT RANGES

VARIABLE	LOWER LIMIT	CURRENT VALUE	UPPER LIMIT
X1	1.333	3.000	8.000
X2	1.500	4.000	9.000

RIGHT HAND SIDE RANGES

CONSTRAINT	LOWER LIMIT	CURRENT VALUE	UPPER LIMIT
1	9.000	12.000	24.000
2	4.000	8.000	9.000
3	2.400	3.000	NO UPPER LIMIT

a) What is the optimal solution including the optimal value of the objective function?

b) Suppose the profit on x_1 is increased to $7. Is the above solution still optimal? What is the value of the objective function when this unit profit is increased to $7?

c) If the unit profit on x_2 was $10 instead of $4, would the optimal solution change?

> NOTE: A "change in the optimal solution" refers to a change in the optimal values of the decision variables, not a change in the optimal value of the objective function alone.

d) If simultaneously the profit on x_1 was raised to $5.5 and the profit on x_2 was reduced to $3, would the current solution still remain optimal?

SOLUTION 3

a) According to the output $x_1 = 2.4$ and $x_2 = 3.2$, and $z = \$20.00$.

b) The output states that the solution remains optimal as long as the objective function coefficient of x_1 is between 1.333 and 8.0. Since 7 is within this range, the optimal solution will not change. However, the optimal profit will be affected: $z = 7x_1 + 4x_2 = 7(2.4) + 4(3.2) = \$29.60$.

c) The output states that the solution remains optimal as long as the objective function coefficient of x_2 is between 1.5 and 9.0. Since 10 is outside this range, the optimal solution would change.

d) Use the 100% rule for simultaneous changes. If $c_1 = 5.5$, the amount c_1 changed is 5.5 - 3 = 2.5. The maximum allowable increase is 8 - 3 = 5, so this is a 2.5/5 = a 50% change. If $c_2 = 3$, the amount that c_2 changed is 4 - 3 = 1. The maximum allowable decrease is 4 - 1.5 = 2.5, so this is a 1/2.5 = a 40% change. The sum of the change percentages is 50% + 40% = 90%. Since this does not exceed 100% the optimal solution would not change.

> NOTE: For the 100% rule, a reduction does not offset an increase. For example, increasing c_1 by 70% of its allowed maximum increase and <u>decreasing</u> c_2 by 50% of its allowed maximum decrease does <u>not</u> result in a combined change of 20%. The combined change is 120%, which means the problem should be solved again because the simultaneous changes to c_1 and c_2 might have (probably) changed the optimal solution (x_1 and x_2 values).

PROBLEM 4

Consider the following linear program:

$$\text{MIN} \quad 6x_1 + 9x_2 \quad (\$ \text{ cost})$$
$$\text{s.t.} \quad x_1 + 2x_2 \leq 8$$
$$10x_1 + 7.5x_2 \geq 30$$
$$x_2 \geq 2$$
$$x_1, x_2 \geq 0$$

The Management Scientist provided the following solution output:

OBJECTIVE FUNCTION VALUE = 27.000

VARIABLE	VALUE	REDUCED COST
X1	1.500	0.000
X2	2.000	0.000

CONSTRAINT	SLACK/SURPLUS	DUAL PRICES
1	2.500	0.000
2	0.000	-0.600
3	0.000	-4.500

OBJECTIVE COEFFICIENT RANGES

VARIABLE	LOWER LIMIT	CURRENT VALUE	UPPER LIMIT
X1	0.000	6.000	12.000
X2	4.500	9.000	NO UPPER LIMIT

RIGHT HAND SIDE RANGES

CONSTRAINT	LOWER LIMIT	CURRENT VALUE	UPPER LIMIT
1	5.500	8.000	NO UPPER LIMIT
2	15.000	30.000	55.000
3	0.000	2.000	4.000

a) What is the optimal solution including the optimal value of the objective function?

b) Suppose the unit cost of x_1 is decreased to $4. Is the above solution still optimal? What is the value of the objective function when this unit cost is decreased to $4?

c) How much can the unit cost of x_2 be decreased without concern for the optimal solution changing?

LP: SOLUTION AND SENSITIVITY   163

d) If simultaneously the cost of x_1 was raised to $7.5 and the cost of x_2 was reduced to $6, would the current solution still remain optimal?

e) If the right-hand side of constraint 3 is increased by 1, what will be the effect on the optimal solution?

SOLUTION 4

a) According to the output $x_1 = 1.5$ and $x_2 = 2.0$, and the objective function value = 27.00.

b) The output states that the solution remains optimal as long as the objective function coefficient of x_1 is between 0 and 12. Since 4 is within this range, the optimal solution will not change. However, the optimal total cost will be affected: $z = 6x_1 + 9x_2 = 4(1.5) + 9(2.0) = \$24.00$.

c) The output states that the solution remains optimal as long as the objective function coefficient of x_2 does not fall below 4.5.

d) Use the 100% rule for simultaneous changes. If $c_1 = 7.5$, the amount c_1 changed is 7.5 - 6 = 1.5. The maximum allowable increase is 12 - 6 = 6, so this is a 1.5/6 = 25% change. If $c_2 = 6$, the amount that c_2 changed is 9 - 6 = 3. The maximum allowable decrease is 9 - 4.5 = 4.5, so this is a 3/4.5 = 66.7% change. The sum of the change percentages is 25% + 66.7% = 91.7%. Since this does not exceed 100% the optimal solution would not change.

e) A dual price represents the improvement in the objective function value per unit increase in the right-hand side. A negative dual price indicates a deterioration (negative improvement) in the objective, which in this problem means an increase in total cost because we're minimizing. Since the right-hand side remains within the range of feasibility, there is no change in the optimal solution. However, the objective function value increases by $4.50.

PROBLEM 5

A small company will be introducing a new line of lightweight bicycle frames to be made from special aluminum and steel alloys. The frames will be produced in two models, deluxe and professional. The anticipated unit profits are currently $10 for a deluxe frame and $15 for a professional frame.

The number of pounds of each alloy needed per frame is summarized in the table below. A supplier delivers 100 pounds of the aluminum alloy and 80 pounds of the steel alloy weekly.

	Aluminum Alloy	Steel Alloy
Deluxe	2	3
Professional	4	2

Use a spreadsheet to determine how many frames of each type should be produced?

a) What is the optimal solution including the optimal value of the objective function?

b) Suppose the profit on deluxe frames is increased to $20. Is the above solution still optimal? What is the value of the objective function when this unit profit is increased to $20?

c) If the unit profit on deluxe frames were $6 instead of $10 would the optimal solution change?

d) If simultaneously the profit on deluxe frames were raised to $16 and the profit on professional frames were raised to $17, would the current solution still remain optimal?

e) Given that aluminum is a sunk cost, what is the maximum amount the company should pay for 50 extra pounds of aluminum? (Constraint 1 pertains to aluminum availability.)

f) How would your answer to (e) change if aluminum were a relevant cost?

SOLUTION 5

The steps to follow in setting up and using a spreadsheet are outlined on the next page.

Spreadsheet showing data and formulas

	A	B	C	D
1		Material Requirements		Amount
2	Material	Deluxe	Profess.	Available
3	Aluminum	2	4	100
4	Steel	3	2	80
5	Profit/Bike	10	15	
6		Decision Variables		
7		Deluxe	Professional	
8	Bikes Made			
9				
10	Maximized Total Profit		=B5*B8+C5*C8	
11				
12	Constraints	Amount Used		Amount Avail.
13	Aluminum	=B3*B8+C3*C8	<=	=D3
14	Steel	=B4*B8+C4*C8	<=	=D4

LP: SOLUTION AND SENSITIVITY

Steps in Using Excel Solver:

Step 1: Select the **Tools** pull-down Menu.
Step 2: Select the **Solver** option.
Step 3: When the **Solver Parameters** dialog box appears:
   Enter C10 in the **Set Target Cell** box.
   Select the **Equal To: Max** option.
   Enter B8:C8 in the **By Changing Cells** box.
   Select **Add**.
Step 4: When the **Add Constraint** dialog box appears:
   Enter B13:B14 in the **Cell Reference** box.
   Select <=.
   Enter D13:D14 in the **Constraint** box.
   Click **OK**.
Step 5: When the **Solver Parameters** dialog box appears:
   Choose **Options**.
Step 6: When the **Solver Options** dialog box appears:
   Select **Assume Non-Negative**.
   Click **OK**.
Step 7: When the **Solver Parameters** dialog box appears:
   Choose **Solve**.
Step 8: When the **Solver Results** dialog box appears:
   Select **Keep Solver Solution**.
   Select **Sensitivity** in the Reports box.
   Click **OK**.

Sensitivity Report

Adjustable Cells

Cell	Name	Final Value	Reduced Cost	Objective Coefficient	Allowable Increase	Allowable Decrease
$B$8	Deluxe	15	0	10	12.5	2.5
$C$8	Profess.	17.500	0.000	15	5	8.333333333

Constraints

Cell	Name	Final Value	Shadow Price	Constraint R.H. Side	Allowable Increase	Allowable Decrease
$B$13	Aluminum	100	3.125	100	60	46.66666667
$B$14	Steel	80	1.25	80	70	30

a) According to the output x_1 (deluxe frames) = 15, and x_2 (professional frames) = 17.5, and this yields an objective function value of $412.50.

b) The output states that the solution remains optimal as long as the objective function coefficient of x_1 is between 7.5 and 22.5. Since 20 is within this range, the optimal solution will not change. However the optimal profit will be affected: $z = 20x_1 + 15x_2 = 20(15) + 15(17.5) = \$562.50$.

c) The output states that the solution remains optimal as long as the objective function coefficient of x_1 is between 7.5 and 22.5. Since 6 is outside this range, the optimal solution would change.

d) Use the 100% rule for simultaneous changes. If $c_1 = 16$, the amount $c1$ changed is $16 - 10 = 6$. The maximum allowable increase is $22.5 - 10 = 12.5$, so this is a $6/12.5 = 48\%$ change. If $c_2 = 17$, the amount that c_2 changed is $17 - 15 = 2$. The maximum allowable increase is $20 - 15 = 5$ so this is a $2/5 = 40\%$ change. The sum of the change percentages is 88%. This is less than 100%, so the optimal solution would not change.

e) Since the cost for aluminum is a sunk cost, the dual price provides the value of extra aluminum. The dual price for aluminum is $3.125 per pound. Thus, the value of 50 additional pounds is = $156.25.
   This analysis is valid only if the change is within the range of feasibility for aluminum. From the output we can see that the maximum allowable increase for aluminum is 60. Since 50 is in this range, then the $156.25 is valid.

f) If aluminum were a relevant cost, the dual price would be the amount above the normal price of aluminum the company would be willing to pay. Thus if initially aluminum cost $4 per pound, then additional units in the range of feasibility would be worth $4 + $3.125 = $7.125 per pound.

PROBLEM 6

Comfort Plus Inc. (CPI) manufactures a standard dining chair used in restaurants. The demand forecasts for quarter 1 (January-March) and quarter 2 (April-June) are 3700 chairs and 4200 chairs, respectively. CPI has a policy of satisfying all demand in the quarter in which it occurs.

The chair contains an upholstered seat that can be produced by CPI or purchased from DAP, a subcontractor. DAP currently charges $12.50 per seat, but has announced a new price of $13.75 effective April 1. CPI can produce the seat at a cost of $10.25. CPI can produce up to 3800 seats per quarter.

Seats that are produced or purchased in quarter 1 and used to satisfy demand in quarter 2 cost CPI $1.50 each to hold in inventory, but maximum inventory cannot exceed 300 seats.

The problem was formulated as follows:

x_1 = number of seats produced by CPI in quarter 1,
x_2 = number of seats purchased from DAP in quarter 1,
x_3 = number of seats carried in inventory from quarters 1 to 2,
x_4 = number of seats produced by CPI in quarter 2, and
x_5 = number of seats purchased from DAP in quarter 2.

MIN $10.25x_1 + 12.5x_2 + 1.5x_3 + 10.25x_4 + 13.75x_5$ (costs)

s.t. $x_1 + x_2 - x_3 \geq 3700$ (quarter 1 demand)

$x_3 + x_4 + x_5 \geq 4200$ (quarter 2 demand)

$x_1 \leq 3800$ (CPI's production capacity in quarter 1)

$x_4 \leq 3800$ (CPI's production capacity in quarter 2)

$x_3 \leq 300$ (inventory capacity)

$x_1, x_2, x_3, x_4, x_5 \geq 0$

The Management Scientist provided the following solution output:

OBJECTIVE FUNCTION VALUE = 82175.000

VARIABLE	VALUE	REDUCED COST
X1	3800.000	0.000
X2	0.000	0.250
X3	100.000	0.000
X4	3800.000	0.000
X5	300.000	0.000

CONSTRAINT	SLACK/SURPLUS	DUAL PRICES
1	0.000	-12.250
2	0.000	-13.750
3	0.000	2.000
4	0.000	3.500
5	200.000	0.000

OBJECTIVE COEFFICIENT RANGES

VARIABLE	LOWER LIMIT	CURRENT VALUE	UPPER LIMIT
X1	NO LOWER LIMIT	10.250	12.250
X2	12.250	12.500	NO UPPER LIMIT
X3	1.250	1.500	3.500
X4	NO LOWER LIMIT	10.250	13.750
X5	11.750	13.750	14.000

RIGHT HAND SIDE RANGES

CONSTRAINT	LOWER LIMIT	CURRENT VALUE	UPPER LIMIT
1	3500.000	3700.000	3800.000
2	3900.000	4200.000	NO UPPER LIMIT
3	3700.000	3800.000	4000.000
4	0.000	3800.000	4100.000
5	100.000	300.000	NO UPPER LIMIT

a) What is the optimal solution including the optimal value of the objective function?

b) If the per-unit inventory cost increased from $1.50 to $2.50, would the optimal solution change? Would the optimal value of the objective function change?

c) If in quarter 2 CPI's per-seat production cost increased by $1.25 and DAP changed its mind about the announced price increase (thus leaving it at $12.50 per seat), would the optimal solution change?

d) If DAP reduced its per-seat selling price in quarter 1 from $12.50 to $12.25, should CPI purchase any seats in quarter 1?

e) How much is it worth to CPI to increase its inventory capacity from 300 seats to 400?

f) If CPI increased its production capacity by 100 seats in both quarters 1 and 2, what would be the savings for CPI (ignoring the capacity-expansion expense)?

SOLUTION 6

a) CPI will produce 3800 seats in quarter 1 and another 3800 in quarter 2. 100 seats will be carried in inventory from quarter 1 to quarter 2. 300 seats will be purchased from DAP in quarter 2. Total cost for this plan is $82,175.

b) The optimal solution will not change as a result of a change in the per-unit inventory cost as long as the cost remains in the range of $1.25 to $3.50. The objective function value <u>will</u> change; it will increase by $100 to $82,275.

c) Use the 100% rule for simultaneous changes. The amount c_4 changed is 1.25. The maximum allowable increase is 13.75 - 10.25 = 3.50, so this is a 1.25/3.50 = 35.7% change. If c_5 = 12.50, the amount that c_5 changed is 13.75 - 12.50 = 1.25. The maximum allowable decrease is 13.75 - 11.75 = 2.00, so this is a 1.25/2.00 = 62.5% change. The sum of the percentages is 98.2%; the optimal solution would not change.

d) x_2 is the number of seats purchased from DAP in quarter 1 and its current value is 0. Its reduced cost value is 0.25, indicating that if c_2 improved (decreased in this case) by 0.25 or more, x_2 would have a positive value in the new optimal solution. c_2 = $12.25 represents an improvement of exactly 0.25, so the answer is yes.

e) Increasing the inventory capacity (without other changes) will not benefit CPI. Actually, a <u>decrease</u> as great as 200 seats will not change the optimal solution. This is indicated by the slack value of 200 for constraint 5.

f) Based on the dual prices for constraints 3 and 4, the objective function value (total cost) will decrease (remember, we're minimizing) by $5.50 (2.00 + 3.50) for each unit increase in CPI's production capacity in quarters 1 and 2. We can conclude that a 100-unit capacity increase will reduce the total cost by $550 <u>if</u> the 100% rule has not been violated. It has not. (For constraint 3 the percent of allowed increase is 50% (100/200) and for constraint 4 the percent of allowed increase is 33%, for a total of 83%.)

ANSWERED PROBLEMS

PROBLEM 7

Regal Investments has just received instructions from a client to invest in two stocks, one an airline stock, the other an insurance stock. The total maximum appreciation in stock value over the next year is to be maximized subject to the following restrictions: (1) the total investment shall not exceed $100,000., (2) at most $40,000 is to be invested in the insurance stock, and (3) quarterly dividends must total at least $2,600.

The airline stock is currently selling for $40 per share and its quarterly dividend is $1 per share. The insurance stock is currently selling for $50 per share and the quarterly dividend is $1.50 per share. Regal's analysts predict that over the next year the airline stock will increase $2 per share and the insurance stock will increase $3 per share.

The Management Scientist provided the following solution output:

OBJECTIVE FUNCTION VALUE = 5400.000

VARIABLE	VALUE	REDUCED COST
X1	1500.000	0.000
X2	800.000	0.000

CONSTRAINT	SLACK/SURPLUS	DUAL PRICES
1	0.000	0.050
2	0.000	0.010
3	100.000	0.000

OBJECTIVE COEFFICIENT RANGES

VARIABLE	LOWER LIMIT	CURRENT VALUE	UPPER LIMIT
X1	0.000	2.000	2.400
X2	2.500	3.000	NO UPPER LIMIT

RIGHT HAND SIDE RANGES

CONSTRAINT	LOWER LIMIT	CURRENT VALUE	UPPER LIMIT
1	96000.000	100000.000	NO UPPER LIMIT
2	20000.000	40000.000	100000.000
3	NO LOWER LIMIT	2600.000	2700.000

---

a) How should the client's money be invested to satisfy his restrictions?

b) Suppose Regal's estimate of the airline stock's appreciation is in error. Within what limits must the actual appreciation lie for the answer in (a) to remain optimal?

PROBLEM 8

A company produces two products made from aluminum and copper. The table below gives the unit requirements, the unit production man-hours required, the unit profit and the availability of the resources (in tons).

	Aluminum	Copper	Man-hours	Unit Profit
Product 1	1	0	2	50
Product 2	1	1	3	60
Available	10	6	24	

The Management Scientist provided the following solution output:

OBJECTIVE FUNCTION VALUE = 540.000

VARIABLE	VALUE	REDUCED COST
X1	6.000	0.000
X2	4.000	0.000

CONSTRAINT	SLACK/SURPLUS	DUAL PRICES
1	0.000	30.000
2	2.000	0.000
3	0.000	10.000

OBJECTIVE COEFFICIENT RANGES

VARIABLE	LOWER LIMIT	CURRENT VALUE	UPPER LIMIT
X1	40.000	50.000	60.000
X2	50.000	60.000	75.000

RIGHT HAND SIDE RANGES

CONSTRAINT	LOWER LIMIT	CURRENT VALUE	UPPER LIMIT
1	9.000	10.000	12.000
2	4.000	6.000	NO UPPER LIMIT
3	20.000	24.000	26.000

a) What is the optimal production schedule?

b) Within what range for the profit on product 2 will the solution in (a) remain optimal? What is the optimal profit when $c_2 = 70$?

c) Suppose that simultaneously the unit profits on x_1 and x_2 changed from 50 to 55 and 60 to 65 respectively. Would the optimal solution change?

d) Explain the meaning of the "DUAL PRICES" column. Given the optimal solution, why should the dual price for copper be 0?

e) What is the increase in the value of the objective function for an extra unit of aluminum?

f) Man-hours were not figured into the unit profit as it must pay three workers for eight hours of work regardless of the number of man-hours used. What is the dual price for man-hours? Interpret.

g) On the other hand, aluminum and copper are resources that are ordered as needed. The unit profit coefficients were determined by: (selling price per unit) - (cost of the resources per unit). The 10 units of aluminum cost the company $100. What is the most the company should be willing to pay for extra aluminum?

PROBLEM 9

A small company produces only two sizes of frames for stereo receivers: standard size and slim-line. The accounting department has provided the following analysis of the unit profit:

	Standard	Slim-Line
Selling Price	$6.00	$4.25
Raw Materials	$0.75	$0.50
	(1.5 units @ .50/unit)	(1 unit @ .50/unit)
Packaging	$0.25	$0.25
Labor*	0.40 hours	0.25 hours
Profit (excluding labor costs)	$5.00	$3.50

• Labor is considered a fixed cost as it is performed by salaried workers at the plant.

There are 350 units of the raw material and 300 packing boxes available daily. (Both products utilize the same packing boxes.) At most 10 workers (at 8 hrs./day) will be assigned to this project.

The Management Scientist provided the following output:

OBJECTIVE FUNCTION VALUE = 1100.000

VARIABLE	VALUE	REDUCED COST
X1	33.333	0.000
X2	266.667	0.000

CONSTRAINT	SLACK/SURPLUS	DUAL PRICES
1	33.333	0.000
2	0.000	1.000
3	0.000	10.000

172 CHAPTER 8

OBJECTIVE COEFFICIENT RANGES

VARIABLE	LOWER LIMIT	CURRENT VALUE	UPPER LIMIT
X1	3.500	5.000	5.600
X2	3.125	3.500	5.000

RIGHT HAND SIDE RANGES

CONSTRAINT	LOWER LIMIT	CURRENT VALUE	UPPER LIMIT
1	316.667	350.000	NO UPPER LIMIT
2	200.000	300.000	320.000
3	75.000	80.000	90.000

a) What is the optimal daily production plan?

b) What is the maximum profit for standard models that will keep the same optimal solution as (a)?

c) What is the new optimal solution and optimal profit if an additional worker (8 additional hours) is assigned to the project? (Hint: The output states the current values for the right-hand sides of the constraints. Compare them with the resource data provided in the problem description to determine the constraint that corresponds with each resource.)

PROBLEM 10

Consider Problem 9 further.

a) Suppose c_1 was changed from 5 to 5.5. Would the optimal solution change?

b) Suppose c_2 was changed from 3.5 to 4. Would the optimal solution change?

c) Suppose simultaneously c_1 changed to 5.5 and c_2 changed to 4. Would the optimal solution change?

d) What is the dual price for man-hours? Interpret.

e) What is the dual price for the packaging? Interpret.

f) Suppose simultaneously the amount of material available increased from 350 to 500, the number of boxes available increased from 300 to 310 and the number of man-hours increased from 80 to 84. What conclusion can be drawn regarding the dual prices?

Note: If the range of optimality for an objective function coefficient (or the range of feasibility for a right-hand side) involved in a simultaneous change is unlimited in the direction of the change, the change to this coefficient (or constant) is of no concern.

PROBLEM 11

A client of an investment firm has $10,000 available for investment. He has instructed that his money be invested in three stocks so that no more than $5,000 is invested in any one stock but at least $1,000 is invested in each stock. He has further instructed the firm to use its current data and invest in a manner that maximizes his expected overall gain during a one-year period.

The stocks, the current price per share, and the firm's projected stock price a year from now are summarized in the following table.

Stock	Current Price	Projected Price 1 Year Hence
James Industries	$25	$35
QM Inc.	$50	$60
Delicious Candy Co.	$100	$125

This problem was formulated as follows:

x_1 = number of shares of James Industries to purchase,
x_2 = number of shares of QM Inc. to purchase, and
x_3 = number of shares of Delicious Candy Co. to purchase.

$$\text{MAX} \quad 10x_1 + 10x_2 + 25x_3$$

$$\begin{align}
\text{s.t.} \quad 25x_1 + 50x_2 + 100x_3 &\leq 10000 \quad \text{(\$ available)} \\
25x_1 &\leq 5000 \quad \text{(max. \$ stock 1)} \\
50x_2 &\leq 5000 \quad \text{(max. \$ stock 2)} \\
100x_3 &\leq 5000 \quad \text{(max. \$ stock 3)} \\
25x_1 &\geq 1000 \quad \text{(min. \$ stock 1)} \\
50x_2 &\geq 1000 \quad \text{(min. \$ stock 2)} \\
100x_3 &\geq 1000 \quad \text{(min. \$ stock 3)} \\
x_1, x_2, x_3 &\geq 0
\end{align}$$

> Use the computer output on the next page to help answer the following questions.

a) What is the optimal solution including the optimal value of the objective function?

b) How should the dual price for constraint 6 be interpreted?

c) If the client had an additional $1000 available for investing how much would the expected overall one-year gain increase?

d) If the client increased the allowed maximum investment amount to $6000 for just one stock, should it be Delicious Candy (the stock with the greatest gain)? Why?

e) Based on your stock choice in (d) above, how much would the objective function increase?

f) For your stock choice in (d) above, how much could the allowed maximum investment amount be raised before the optimal investment mix might change?

g) If the expected one-year gains on James Industries and QM Inc. each increased by $1.00, would the optimal investment mix change?

The Management Scientist provided the following solution output:

OBJECTIVE FUNCTION VALUE = 3200.000

VARIABLE	VALUE	REDUCED COST
X1	200.000	0.000
X2	20.000	0.000
X3	40.000	0.000

CONSTRAINT	SLACK/SURPLUS	DUAL PRICES
1	0.000	0.250
2	0.000	0.150
3	4000.000	0.000
4	1000.000	0.000
5	4000.000	0.000
6	0.000	-0.050
7	3000.000	0.000

OBJECTIVE COEFFICIENT RANGES

VARIABLE	LOWER LIMIT	CURRENT VALUE	UPPER LIMIT
X1	6.250	10.000	NO UPPER LIMIT
X2	NO LOWER LIMIT	10.000	12.500
X3	20.000	25.000	40.000

RIGHT HAND SIDE RANGES

CONSTRAINT	LOWER LIMIT	CURRENT VALUE	UPPER LIMIT
1	7000.000	10000.000	11000.000
2	4000.000	5000.000	8000.000
3	1000.000	5000.000	NO UPPER LIMIT
4	4000.000	5000.000	NO UPPER LIMIT
5	NO LOWER LIMIT	1000.000	5000.000
6	0.000	1000.000	4000.000
7	NO LOWER LIMIT	1000.000	4000.000

PROBLEM 12

Given the following linear program:

$$\text{MAX} \quad 6x_1 + 5x_2$$

$$\begin{aligned}
\text{s.t.} \quad x_1 + x_2 &\leq 6 \\
2x_1 + x_2 &\leq 8 \\
x_1 &\leq 3 \\
x_1, x_2 &\geq 0
\end{aligned}$$

a) Solve graphically for the optimal solution.

b) Calculate the range of optimality for both c_1 and c_2.

c) What is the new optimal point when c_2 exceeds the upper limit determined in (a)?

d) Determine and interpret the dual price for iron (resource of second constraint).

e) For what values of zinc, the third resource, will its dual price be 0?

PROBLEM 13

The Asia Import Company (AIC) has 600 cu. ft. of excess cargo space on its ships and has decided to import two new items: jade figurines and linen placemats. Each container of jade figurines is 4 cu. ft. and will net a profit of $80 per container. Each container of linen placemats is 2 cu. ft. and will realize a profit of $60 per container. AIC expects no more than 140 containers of jade figurines available on any trip.

Additionally, AIC wishes to use no more than 480 man-hours for loading, storing, and processing the items through customs. The normal estimate is that a container requires 2 man-hours. However, because of special agricultural restrictions, an extra 2 man-hours can be expected for the linen products.

a) Using the graphical method, determine the number of containers of each item that should be shipped.

b) What is the range of profit on jade containers for which the solution in (a) remains optimal?

c) Determine the value of: (1) an extra man-hour; (2) an extra cubic foot of cargo space; and, (3) the availability of an extra jade container.

PROBLEM 14

Tom manages Leisure Time Motors, a dealership selling minivans and large travel trailers. He is trying to decide how to allocate 90,000 square feet of outside display space to his two products. The products differ in terms of required display space, monthly upkeep, and generated monthly profit as summarized below on a per-unit basis:

	Space Requirement	Monthly Upkeep	Monthly Profit
Minivan	300 sq. ft.	2.0 man-hours	$3200
Trailer	500	3.2	4500

Leisure Motors has three yard men, each working a 150-hour month, keeping the minivans and trailers clean. Tom feels he needs a minimum of 50 minivans on display. The manufacturer of his trailers requires that he display at least 75 trailers.

a) Graphically solve for the numbers of minivans and trailers on display that will maximize Leisure Time's profit.

b) Calculate the range of optimality for both c_1 and c_2.

c) Determine the dual price for yard men, the resource of the second constraint. Interpret.

PROBLEM 15

Harvey owns a Harley motorcycle and a Hauler pickup. The Harley gets an average of 45 miles per gallon (mpg) using 93 octane gasoline that sells for $1.35 per gallon. The Hauler averages 26 mpg using 89 octane that sells for $1.17 per gallon.

The Harley requires 15 hours of maintenance work per 5000 miles ridden. The Hauler requires 10 hours of maintenance per 5000 miles. Harvey does his own maintenance work, but he cannot devote more than 100 hours annually to the task.

Harvey predicts he will have to travel 45,000 miles in the upcoming year. He would like to ride his Harley a minimum of 5,000 miles annually in order to stay in practice.

a) How should Harvey divide his mileage among his Harley and Hauler so that his annual fuel expense is minimized.

b) What is the value of an additional hour of Harvey's time per year for maintenance?

c) By how much will Harvey's annual fuel expense increase for each mile that he travels in excess of 45,000?

TRUE/FALSE

___ 16. Any change to an objective function coefficient of a variable which is positive in the optimal solution will change the optimal value of the objective function.

___ 17. The dual price for labor hours is $25. The profit coefficients c_1 and c_2 take into account a $10 per hour labor cost. Then the maximum value of an overtime hour is $35.

___ 18. Ranges of optimality or feasibility are calculated for a single change only, and they assume no other coefficients in the problem have been changed.

___ 19. Regarding the 100 percent rule, it is possible for the optimal solution to not change even though changes in the objective function coefficients exceed 100 percent.

___ 20. If the range of feasibility for b_1 is between 16 and 37, then if $b_1 = 22$, the optimal solution will not change from the original optimal solution.

___ 21. The 100 percent rule can be applied to changes in both objective function coefficients and right-hand sides at the same time.

___ 22. Relevant costs should be reflected in the objective function, but sunk costs should not.

___ 23. Degeneracy occurs when the dual price equal zero for one of the binding constraints.

___ 24. For any constraint, either its slack/surplus value must be zero or its dual price must be zero.

___ 25. For any decision variable, either its value must be zero or its reduced cost must be zero.

___ 26. If the dual price for the right-hand side of a $\leq$ constraint is zero, there is no upper limit on its range of feasibility.

___ 27. A $\leq$ constraint cannot have a negative dual price, regardless of whether it is a maximization or minimization problem.

___ 28. Slack corresponds to $\leq$ constraints.

___ 29. The Management Scientist can only be used to solve linear programs in two variables.

___ 30. Dual prices are only valid within the range of feasibility.

CHAPTER 9

Linear Programming Applications

KEY CONCEPTS

CONCEPT	ILLUSTRATED PROBLEMS	ANSWERED PROBLEMS
Programming Applications:		
Blending	1,4	15
Equipment Acquisition	2	11,16
Multiperiod Planning	3	12
Product Mix	4	13,14,15
Staff Scheduling	5	21
Portfolio Selection	6	17
Data Envelopment Analysis	7	20
Media Selection	8	18
Transportation	9	19
Revenue Management	10	

L.P. APPLICATIONS **181**

REVIEW

1. <u>Linear programming applications</u> include problems in production, marketing, finance, and numerous other business-related areas.

2. To develop a <u>good formulation</u>, one should strive to understand the problem thoroughly. Then one should: (1) define the decision variables (the inputs over which you have direct control); (2) define the objective (the goal that you wish to maximize or minimize); and, (3) define the constraints (the restrictions that deter you from even better values for the objective function.) Frequently it is advisable to write the objective function and constraints in English first before translating them into mathematical notation.

3. <u>Computer packages</u>, such as *The Management Scientist* or *Microsoft Excel* may be used to solve linear programming problems giving the optimal solution and appropriate sensitivity analyses.

4. <u>Data Envelopment Analysis (DEA)</u> is an application of linear programming used to determine the relative operating efficiency of units with the same goals and objectives (e.g. banks, schools, company divisions, etc.)

5. DEA creates a fictitious composite unit made up of an optimal weighted average (W_1, W_2, etc.) of existing units. Then an individual unit, k, can be compared by determining E, the fraction of this unit's input resources required by the "optimal" composite unit.

6. The DEA model is given by:

    MIN E

    S.T. Weighted outputs $\geq$ Unit k's output   . . . for each measured output

    Weighted inputs $\leq E$ (Unit k's input)   . . . for each measured input

    The sum of the weights = 1

    E, weights ≥ 0

7. In DEA, if the optimal value of E is less than 1, unit k is less efficient than the composite unit and can be judged relatively inefficient. If $E = 1$, there is no evidence that unit k is inefficient, but one cannot conclude that unit k is absolutely efficient.

8. <u>Revenue management</u> involves the managing of short-term demand for a fixed perishable inventory in order to maximixe the revenue potential for an organization.

ILLUSTRATED PROBLEMS

> NOTE: Students often confuse a constraint for the objective in a problem. If there is a limit imposed on an entity, it probably represents a constraint. (We are solving for the objective function's limit.)

> NOTE: A common dilemma is whether to use "=" or an inequality sign ($\leq$ or $\geq$) in a constraint. The use of "=" when not necessary might result in an infeasible problem. Quite often, if you are choosing between "$\leq$" and "=", "$\leq$" will do no harm if you are maximizing ("$\geq$" if you are minimizing); equality is likely to be achieved if it is possible. Of course, using "$\leq$" when you should be using "$\geq$" is very harmful!

PROBLEM 1

Frederick's Feed Company receives four raw grains from which it blends its dry pet food. The pet food advertises that each 8-ounce can meets the minimum daily requirements for vitamin C, protein and iron. The cost of each raw grain as well as the vitamin C, protein, and iron units per pound of each grain are summarized below.

Grain	Vitamin C Units/lb	Protein Units/lb	Iron Units/lb	Cost/lb
1	9	12	0	.75
2	16	10	14	.90
3	8	10	15	.80
4	10	8	7	.70

Frederick's is interested in producing the 8-ounce mixture at minimum cost while meeting the minimum daily requirements of 6 units of vitamin C, 5 units of protein, and 5 units of iron.

a) Formulate this problem as a linear program.

b) Solve for the optimal solution using a program such as *The Management Scientist*.

c) If the mixture costs 5.4 cents to can and Frederick's puts a 50% markup on the package to its retailers, how much will it charge its retailers for an 8-ounce can?

SOLUTION 1

a) <u>Define the decision variables</u>

x_j = the pounds of grain j (j = 1,2,3,4) used in the 8-ounce mixture

<u>Define the objective function</u>

Minimize the total cost for an 8-ounce mixture:
Min $.75x_1 + .90x_2 + .80x_3 + .70x_4$

<u>Define the constraints</u>

The total weight of the mixture ($x_1 + x_2 + x_3 + x_4$) is 8-ounces (.5 pounds):
(1) $x_1 + x_2 + x_3 + x_4 = .5$

Total amount of Vitamin C in the mixture is at least 6 units:
(2) $9x_1 + 16x_2 + 8x_3 + 10x_4 \geq 6$

Total amount of protein in the mixture is at least 5 units:
(3) $12x_1 + 10x_2 + 10x_3 + 8x_4 \geq 5$

Total amount of iron in the mixture is at least 5 units:
(4) $14x_2 + 15x_3 + 7x_4 \geq 5$

Nonnegativity of variables: $x_j \geq 0$ for all j

b) When solved by *The Management Scientist* the following output was generated:

OBJECTIVE FUNCTION VALUE = 0.406

VARIABLE	VALUE	REDUCED COSTS
X1	0.099	0.000
X2	0.213	0.000
X3	0.088	0.000
X4	0.099	0.000

Thus, the optimal blend is about .10 lb. of grain 1, .21 lb. of grain 2, .09 lb. of grain 3, and .10 lb. of grain 4.

c) The mixture costs Frederick's 40.6 cents. With 5.4 cents for packaging, this brings their costs to 46 cents. A 50% markup would mean it would charge retailers 1.5(46) = 69 cents per can.

PROBLEM 2

Floataway Tours has $400,000 that may be used to purchase new rental boats for hire during the summer. The boats can be purchased from two different manufacturers. Pertinent data concerning the boats are summarized below:

Boat	Manufacturer	Cost	Maximum Seating	Expected Daily Profit
Speedhawk	Sleekboat	$6000	3	$ 70
Silverbird	Sleekboat	$7000	5	$ 80
Catman	Racer	$5000	2	$ 50
Classy	Racer	$9000	6	$110

Floataway Tours would like to purchase at least 50 boats and would like to purchase the same number from Sleekboat as from Racer to maintain goodwill. Also, Floataway Tours wishes to have a total seating capacity of at least 200.

a) Formulate this as an LP.

b) Solve the problem using a spreadsheet.

SOLUTION 2

a) The problem can be formulated as follows:

Define the decision variables

x_1 = number of Speedhawks ordered
x_2 = number of Silverbirds ordered
x_3 = number of Catmans ordered
x_4 = number of Classys ordered

Define the objective function

Maximize total expected daily profit:

Maximize (Expected daily profit per unit)(Number of units)
MAX $70x_1 + 80x_2 + 50x_3 + 110x_4$

Define the constraints

Spend no more than $400,000:

(1) $6000x_1 + 7000x_2 + 5000x_3 + 9000x_4 \leq 400,000$

Purchase at least 50 boats:

(2) $x_1 + x_2 + x_3 + x_4 \geq 50$

Number of boats from Sleekboat equals number of boats from Racer:

(3) $x_1 + x_2 = x_3 + x_4$ or $x_1 + x_2 - x_3 - x_4 = 0$

Capacity at least 200:

(4) $3x_1 + 5x_2 + 2x_3 + 6x_4 \geq 200$

Nonnegativity of variables:

$x_j \geq 0$, for $j = 1,2,3,4$

b) Refer to Chapters 2 and 3 of this study guide for coverage of the steps to follow in setting up and using a spreadsheet to solve an LP.

Spreadsheet showing the optimal solution

	A	B	C	D	E	F
1			LHS Coefficients			
2	Constraints	X1	X2	X3	X4	RHS
3	#1	6	7	5	9	420
4	#2	1	1	1	1	50
5	#3	1	1	-1	-1	0
6	#4	3	5	2	6	200
7	Obj.Func.	70	80	50	110	
8						
9			Decision Variable Values			
10			X1	X2	X3	X4
11	No. of Boats		28	0	0	28
12						
13	Maximum Total Profit			5040		
14						
15	Constraints			LHS		RHS
16	Spending Max.			420.0	<=	420
17	Min. # Boats			56.0	>=	50
18	Equal Sourcing			0.0	=	0
19	Min. Seating			252.0	>=	200

Sensitivity Report

Adjustable Cells						
		Final	Reduced	Objective	Allowable	Allowable
Cell	Name	Value	Cost	Coefficient	Increase	Decrease
$D$12	X1	28	0	70	45	1.875
$E$12	X2	0	-2	80	2	1E+30
$F$12	X3	0	-12	50	12	1E+30
$G$12	X4	28	0	110	1E+30	16.36363636
Constraints						
		Final	Shadow	Constraint	Allowable	Allowable
Cell	Name	Value	Price	R.H. Side	Increase	Decrease
$E$17	#1	420.0	12.0	420	1E+30	45
$E$18	#2	56.0	0.0	50	6	1E+30
$E$19	#3	0.0	-2.0	0	70	30
$E$20	#4	252.0	0.0	200	52	1E+30

PROBLEM 3

Burt Wheeler is the production manager of Wheeler Wheels, Inc. Burt has just received orders for 1,000 standard wheels and 1,250 deluxe wheels next month and for 800 standard and 1,500 deluxe wheels the following month. All orders are to be filled.

The cost of producing standard wheels is $10 and deluxe wheels is $16. Overtime rates are 50% higher. There are 1,000 hours of regular time and 500 hours of overtime available each month. The cost of storing one wheel from one month to the next is $2.

Develop a two-month production schedule for Burt of standard and deluxe wheel production if it takes .5 hour to make a standard wheel and .6 hour to make a deluxe wheel and solve using a program such as *The Management Scientist*.

L.P. APPLICATIONS   187

SOLUTION 3

Define the decision variables

We must determine how many of each type of wheel to make each month during regular time and overtime; also we must determine the number of each wheel stored from one month to the next. Thus we want to determine the production levels, x_j, as follows:

	Month 1		Month 2	
	Reg. Time	Overtime	Reg. Time	Overtime
Standard	x_1	x_2	x_5	x_6
Deluxe	x_3	x_4	x_7	x_8

Also let,
y_1 = number of standard wheels stored from month 1 to month 2
y_2 = number of deluxe wheels stored from month 1 to month 2.

Define the objective function

Minimize total production and storage costs:
Min (cost per wheel)(number of wheels produced) + $2y_1 + 2y_2$
Min $10x_1 + 15x_2 + 16x_3 + 24x_4 + 10x_5 + 15x_6 + 16x_7 + 24x_8 + 2y_1 + 2y_2$

Define the constraints

Standard Wheel Production Month 1 = (Requirements) + (Amount Stored)

(1) $x_1 + x_2 = 1{,}000 + y_1 \longrightarrow x_1 + x_2 - y_1 = 1{,}000$

Deluxe Wheel Production Month 1 = (Requirements) + (Amount Stored)

(2) $x_3 + x_4 = 1{,}250 + y_2 \longrightarrow x_3 + x_4 - y_2 = 1{,}250$

Standard Wheel Production Month 2 = (Requirements) - (Amount Stored)

(3) $x_5 + x_6 = 800 - y_1 \longrightarrow x_5 + x_6 + y_1 = 800$

Deluxe Wheel Production Month 2 = (Requirements) - (Amount Stored)

(4) $x_7 + x_8 = 1{,}500 - y_2 \longrightarrow x_7 + x_8 + y_2 = 1{,}500$

Regular Hours Used Month 1 $\leq$ Regular Hours Available Month 1:

(5) $.5x_1 + .6x_3 \leq 1000$

Overtime Hours Used Month 1 $\leq$ Overtime Hours Available Month 1:

(6) $.5x_2 + .6x_4 \leq 500$

Regular Hours Used Month 2 $\leq$ Regular Hours Available Month 2:

(7) $.5x_5 + .6x_7 \leq 1000$

Overtime Hours Used Month 2 ≤ Overtime Hours Available Month 2:

(8) $.5x_6 + .6x_8 \leq 500$

Nonnegativity of variables:

$x_j \geq 0, j = 1,,8$ and $y_j \geq 0\ j = 1,2$

The Management Scientist provided the following solution:

OBJECTIVE FUNCTION VALUE = 67500.000

VARIABLE	VALUE	REDUCED COSTS
X1	500.000	0.000
X2	500.000	0.000
X3	1250.000	0.000
X4	0.000	2.000
X5	200.000	0.000
X6	600.000	0.000
X7	1500.000	0.000
X8	0.000	2.000
Y1	0.000	2.000
Y2	0.000	2.000

Thus, the recommended production schedule is:

	Month 1		Month 2	
	Reg. Time	Overtime	Reg. Time	Overtime
Standard	500	500	200	600
Deluxe	1250	0	1500	0

No wheels are stored and the minimum total cost is $67,500.

PROBLEM 4

Target Shirt Company makes three varieties of shirts: Collegiate, Traditional and European. These shirts are made from different combinations of cotton and polyester.

The cost per yard of unblended cotton is $5 and for unblended polyester is $4. Target can receive up to 4,000 yards of raw cotton and 3,000 yards of raw polyester fabric weekly. The table below pertinent data concerning the manufacture of the shirts.

Shirt	Total Yards	Fabric Requirements	Weekly Contracts	Weekly Demand	Selling Price
Collegiate	1.00	At least 50% cotton	500	600	$14.00
Traditional	1.20	No more than 20% polyester	650	850	$15.00
European	.90	As much as 80% polyester	280	675	$18.00

Formulate and solve a linear program that would give a manufacturing policy for Target Shirt Company.

SOLUTION 4

Define the decision variables

Not only must we decide how many shirts to make and how much fabric to purchase, we also need to decide how much of each fabric is blended into each shirt.

Let, s_j = the total number of shirt style j produced
f_i = the number of yards of material i purchased
x_{ij} = yards of fabric i blended into shirt style j

where i = 1 (cotton) or 2 (polyester) and
j = 1 (collegiate), 2 (traditional), or 3 (European)

Define the objective function

Maximize the overall profit.
To determine the profit function, subtract the cost of purchasing the fabric from the shirt sales revenue. Thus the objective function is:

MAX $14s_1 + 15s_2 + 18s_3 - 5f_1 - 4f_2$

Define the constraints

Definition of Total Number of Shirts of Each Style

Total Number of each style =
(Total Yardage used in making the style) / (Yardage/Shirt)

(1) Collegiate: $s_1 = (x_{11} + x_{21}) / 1$ $\longrightarrow$ $s_1 - x_{11} - x_{21} = 0$
(2) Traditional: $s_2 = (x_{12} + x_{22}) / 1.2$ $\longrightarrow$ $1.2s_2 - x_{12} - x_{22} = 0$
(3) European: $s_3 = (x_{13} + x_{23}) / .9$ $\longrightarrow$ $.95s_3 - x_{13} - x_{23} = 0$

Definition of Total Yardage of Materials

(4) Cotton: $f_1 = x_{11} + x_{12} + x_{13}$ $\longrightarrow$ $f_1 - x_{11} - x_{12} - x_{13} = 0$
(5) Polyester: $f_2 = x_{21} + x_{22} + x_{23}$ $\longrightarrow$ $f_2 - x_{21} - x_{22} - x_{23} = 0$

Weekly Availability of the Resources

(6) Cotton: $f_1 \leq 4000$
(7) Polyester: $f_2 \leq 3000$

Meet Weekly Contracts

(8) Collegiate: $s_1 \geq 500$
(9) Traditional: $s_2 \geq 650$
(10) European: $s_3 \geq 280$

Do Not Exceed Weekly Demand

(11) Collegiate: $s_1 \leq 600$
(12) Traditional: $s_2 \leq 850$
(13) European: $s_3 \leq 675$

Fabric Requirements

Collegiate At Least 50% Cotton:
(Total yds. of Cotton in Colleg. Shirts) $\geq [(.5(1.00)$ yds./shirt)(number of colleg. shirts)]
(14) $x_{11} \geq .5s_1$ $\longrightarrow$ $x_{11} - .5s_1 \geq 0$

Traditional At Most 20% Polyester:
(Total yds. Polyester in Tradit. Shirts) $\leq [.2(1.20)$ yds./shirt)(number of tradit. shirts)]
(15) $x_{22} \leq .24s_2$ $\longrightarrow$ $x_{22} - .24s_2 \leq 0$

European At Most 80% Polyester:
(Total yds. Polyester in Europ. Shirts) $\leq [.8(.90)$ yds./shirt)(number of Europ. shirts)]
(16) $x_{23} \leq .72s_3$ $\longrightarrow$ $x_{23} - .72s_3 \leq 0$

Nonnegativity of variables

$s_j, f_i, x_{ij} \geq 0$ for $i = 1,2$ and $j = 1,2,3$.

The Management Scientist provided the following solution:

OBJECTIVE FUNCTION VALUE =	23152.500	
VARIABLE	VALUE	REDUCED COSTS
X11	300.000	0.000
X12	816.000	0.000
X13	121.500	0.000
X21	300.000	0.000
X22	204.000	0.000
X23	486.000	0.000
S1	600.000	0.000
S2	850.000	0.000
S3	675.000	0.000
F1	1237.500	0.000
F2	990.000	0.000

PROBLEM 5

The Accounting Department at Lenny's Restaurant requires information on the total number of employees it will be hiring for its new restaurant located across the street from a major university. Lenny's has broken down its requirements into 4-hour periods.

Time Period	Employees Required
7am-11am	12
11am- 3pm	20
3pm- 7pm	18
7pm-11pm	22
11pm- 7am	Closed

Staffing is done by hiring personnel for eight-hour shifts commencing at 7am, 11am, and 3pm. Also, there are enough students willing to work the before and after school eight-hour shift covering 7am-11am and 7pm-11pm.

a) Formulate and solve for the minimum number of personnel the Accounting Department at Lenny's should expect for this new restaurant.

b) Based on the solution in part (a), determine the timing and amount of overstaffing that will occur.

SOLUTION 5

a) Define the decision variables

x_1 = number of workers hired for 7am-3pm shift
x_2 = number of workers hired for 11am-7pm shift
x_3 = number of workers hired for 3pm-11pm shift
x_4 = number of workers hired for 7am-11am and 7pm-11pm shift

Define the objective function

Minimize the total number of personnel hired:
Min $x_1 + x_2 + x_3 + x_4$

Define the constraints

Total personnel working during each 4-hour period must be greater than or equal to the number of employees required:

(1) $x_1 + x_4 \geq 12$
(2) $x_1 + x_2 \geq 20$
(3) $x_2 + x_3 \geq 18$
(4) $x_3 + x_4 \geq 22$

Nonnegativity of variables:

$x_j \geq 0$ for $j = 1,2,3,4$

The Management Scientist provided the following solution:

OBJECTIVE FUNCTION VALUE = 42.000

VARIABLE	VALUE	REDUCED COSTS
X1	20.000	0.000
X2	0.000	0.000
X3	18.000	0.000
X4	4.000	0.000

CONSTRAINT	SLACK/SURPLUS	DUAL PRICES
1	12.000	0.000
2	0.000	-1.000
3	0.000	0.000
4	0.000	-1.000

There are alternate optimal solutions, one of which is: $x_1 = 12$, $x_2 = 8$, $x_3 = 22$, $x_4 = 0$.

b) The timing and amount of overstaffing corresponds to the constraint surplus values shown in the computer output. The only four-hour period that is overstaffed is 7am-11am. It is overstaffed by 12 workers. Thus, the total man-hours of overstaffing is 4(12) = 48.

PROBLEM 6

Winslow Savings has $20 million available for investment. It wishes to invest over the next four months in such a way that it will maximize the total interest earned over the four month period as well as have at least $10 million available at the start of the fifth month for a high rise building venture in which it will be participating.

For the time being, Winslow wishes to invest only in 2-month government bonds (earning 2% over the 2-month period) and 3-month construction loans (earning 6% over the 3-month period). Each of these is available each month for investment. Funds not invested in these two investments are liquid and earn 3/4 of 1% per month when invested locally.

Formulate and solve a linear program that will help Winslow Savings determine how to invest over the next four months if at no time does it wish to have more than $8 million in either government bonds or construction loans.

SOLUTION 6

Define the decision variables

g_j = amount of <u>new</u> investment in government bonds in month j
c_j = amount of <u>new</u> investment in construction loans in month j
l_j = amount invested locally in month j, where j = 1,2,3,4

Define the objective function

Maximize the total interest earned over the four-month period:

MAX (interest rate on investment)(amount invested)
MAX $.02g_1 + .02g_2 + .02g_3 + .02g_4 + .06c_1 + .06c_2 + .06c_3 + .06c_4$
    $+ .0075l_1 + .0075l_2 + .0075l_3 + .0075l_4$

Define the constraints

Month 1's total investment amount limited to $20 million:

(1) $g_1 + c_1 + l_1 = 20,000,000$

Month 2's total investment amount limited to principle and interest invested locally in Month 1:

(2) $g_2 + c_2 + l_2 = 1.0075l_1$ or $g_2 + c_2 - 1.0075l_1 + l_2 = 0$

Month 3's total investment amount limited to principle and interest invested in government bonds in Month 1 and locally invested in Month 2:

(3) $g_3 + c_3 + l_3 = 1.02g_1 + 1.0075l_2$ or $-1.02g_1 + g_3 + c_3 - 1.0075l_2 + l_3 = 0$

Month 4's total investment amount limited to principle and interest invested in construction loans in Month 1, goverment bonds in Month 2, and locally invested in Month 3:

(4) $g_4 + c_4 + l_4 = 1.06c_1 + 1.02g_2 + 1.0075l_3$
    or $-1.02g_2 + g_4 - 1.06c_1 + c_4 - 1.0075l_3 + l_4 = 0$

$10 million must be available at start of Month 5:

(5) $1.06c_2 + 1.02g_3 + 1.0075l_4 \geq 10{,}000{,}000$

No more than $8 million in government bonds at any time:

(6) $g_1 \leq 8{,}000{,}000$
(7) $g_1 + g_2 \leq 8{,}000{,}000$
(8) $g_2 + g_3 \leq 8{,}000{,}000$
(9) $g_3 + g_4 \leq 8{,}000{,}000$

No more than $8 million in construction loans at any time:

(10) $c_1 \leq 8{,}000{,}000$
(11) $c_1 + c_2 \leq 8{,}000{,}000$
(12) $c_1 + c_2 + c_3 \leq 8{,}000{,}000$
(13) $c_2 + c_3 + c_4 \leq 8{,}000{,}000$

Nonnegativity:

$g_j, c_j, l_j \geq 0$ for $j = 1,2,3,4$

The Management Scientist provided the following solution:

OBJECTIVE FUNCTION VALUE = 1429213.7987

VARIABLE	VALUE	REDUCED COSTS
G1	8000000.000	0.000
G2	0.000	0.000
G3	5108613.923	0.000
G4	2891386.077	0.000
C1	8000000.000	0.000
C2	0.000	0.045
C3	0.000	0.008
C4	8000000.000	0.000
I1	4000000.000	0.000
I2	4030000.000	0.000
I3	7111611.077	0.000
I4	4753562.083	0.000

PROBLEM 7

    The Langley County School District is trying to determine the relative efficiency of its three high schools. In particular, it wants to evaluate Roosevelt High School.
    The district is evaluating performances on SAT scores, the number of seniors finishing high school, and the number of students who enter college as a function of the number of teachers teaching senior classes, the prorated budget for senior instruction, and the number of students in the senior class.

Input	Roosevelt	Lincoln	Washington
Senior Faculty	37	25	23
Budget ($100,000's)	6.4	5.0	4.7
Senior Enrollments	850	700	600

Output	Roosevelt	Lincoln	Washington
Average SAT Score	800	830	900
High School Graduates	450	500	400
College Admissions	140	250	370

Solve the linear program using *The Management Scientist*. Comment on the relative efficiency of Roosevelt High School for each of the output measures.

SOLUTION 7

a) The goal of data envelopment analysis is to compare Roosevelt High School to a fictitious <u>composite</u> high school developed by determining an optimal set of weights of all high schools in the Langley School District.

This set of weights minimizes the fraction, E, of Roosevelt High School's input resources required by the composite school.

If $E = 1$, the composite high school requires the same input resources as Roosevelt's and there would be no evidence that Roosevelt High School is inefficient. If $E < 1$, Roosevelt High School is operating inefficiently with its given inputs and further study of Roosevelt should be made to determine the reasons for this inefficiency.

Define the decision variables

E = Fraction of Roosevelt High School's input resources required by the composite high school
w_1 = Weight applied to Roosevelt High School's input/output resources by the composite high school
w_2 = Weight applied to Lincoln High School's input/output resources by the composite high school
w_3 = Weight applied to Washington High School's input/output resources by the composite high school

Define the objective

Minimize the fraction of Roosevelt High School's input resources required by the composite high school:

Min E

Define the constraints

Sum of the Weights is 1:

(1) $w_1 + w_2 + w_3 = 1$

Output Constraints:
Since $w_1 = 1$ is possible, each output of the composite school must be at least as great as that of Roosevelt:

(2) $800w_1 + 830w_2 + 900w_3 \geq 800$ (SAT Scores)
(3) $450w_1 + 500w_2 + 400w_3 \geq 450$ (Graduates)
(4) $140w_1 + 250w_2 + 370w_3 \geq 140$ (College Admissions)

Input Constraints:
The input resources available to the composite school is a fractional multiple, E, of the resources available to Roosevelt. Since the composite high school cannot use more input than that available to it, the input constraints are:

(5) $37w_1 + 25w_2 + 23w_3 \leq 37E$ (Faculty)
(6) $6.4w_1 + 5.0w_2 + 4.7w_3 \leq 6.4E$ (Budget)
(7) $850w_1 + 700w_2 + 600w_3 \leq 850E$ (Seniors)

Nonnegativity of variables:

$E, w_1, w_2, w_3 \geq 0$

b) The problem was solved by *The Management Scientist* and the output is below.

OBJECTIVE FUNCTION VALUE = 0.765

VARIABLE	VALUE	REDUCED COSTS
E	0.765	0.000
W1	0.000	0.235
W2	0.500	0.000
W3	0.500	0.000

CONSTRAINT	SLACK/SURPLUS	DUAL PRICES
1	0.000	-0.235
2	65.000	0.000
3	0.000	-0.001
4	170.000	0.000
5	4.294	0.000
6	0.044	0.000
7	0.000	0.001

The output shows that the composite high school is made up of equal weights of Lincoln High School and Washington High School. Roosevelt High School is 76.5% efficient compared to this composite high school when measured by college admissions (because of the 0 slack on this constraint (#4)). It is less than 76.5% efficient when using measures of SAT scores and high school graduates (there is positive slack in constraints 2 and 3.)

PROBLEM 8

The SMM Company, which is manufacturing a new instant salad machine, has $280,000 to spend on advertising. The product is only to be test marketed initially in the Dallas area. The money is to be spent on an advertising blitz during one weekend (Friday, Saturday, and Sunday) in January, and SMM is limited to television advertising.

The company has three options available: day time advertising, evening news advertising and the Super Bowl. Even though the Super Bowl is a national telecast, the Dallas Cowboys will be playing in it, and hence, the viewing audience will be especially large in the Dallas area. A mixture of one-minute TV spots is desired. The table below gives pertinent data:

	Cost Per Ad	Estimated New Audience Reached With Each Ad
Day Time	$5,000	3000
Evening News	$7,000	4000
Super Bowl	$100,000	75,000

SMM has decided to take out at least one ad in each option. Further, there are only two Super Bowl ad spots available. There are 10 day time spots and 6 evening news spots available daily. SMM wants to have at least 5 ads per day, but spend no more than $50,000 on Friday and no more than $75,000 on Saturday. Formulate and solve a linear program to help SMM decide how the company should advertise over the weekend.

SOLUTION 8

Define the decision variables

x_1 = the number of day ads on Friday
x_2 = the number of day ads on Saturday
x_3 = the number of day ads on Sunday
x_4 = the number of evening ads on Friday
x_5 = the number of evening ads on Saturday
x_6 = the number of evening ads on Sunday
x_7 = the number of Super Bowl ads

Define the objective function

Maximize the estimated total new audience reached:

MAX (new audience reached per ad of each type)(number of ads of each type)
MAX $3000x_1 + 3000x_2 + 3000x_3 + 4000x_4 + 4000x_5 + 4000x_6 + 75000x_7$

Define the constraints

Take out at least one ad of each type:

(1) $x_1 + x_2 + x_3 \geq 1$
(2) $x_4 + x_5 + x_6 \geq 1$
(3) $x_7 \geq 1$

10 daytime spots available:

(4) $x_1 \leq 10$
(5) $x_2 \leq 10$
(6) $x_3 \leq 10$

6 evening news spots available:

(7) $x_4 \leq 6$
(8) $x_5 \leq 6$
(9) $x_6 \leq 6$

Only two Super Bowl ad spots available:

(10) $x_7 \leq 2$

At least 5 ads per day:

(11) $x_1 + x_4 \geq 5$
(12) $x_2 + x_5 \geq 5$
(13) $x_3 + x_6 + x_7 \geq 5$

Spend no more than $50,000 on Friday:

(14) $5000x_1 + 7000x_4 \leq 50000$

Spend no more than $75,000 on Saturday:

(15) $5000x_2 + 7000x_5 \leq 75000$

Spend no more than $350,000 in total:

(16) $5000x_1 + 5000x_2 + 5000x_3 + 7000x_4 + 7000x_5 + 7000x_6 + 100000x_7 \leq 280000$

Nonnegativity:

$x_j \geq 0 \quad j = 1,...,7$

The Management Scientist provided the following solution:

OBJECTIVE FUNCTION VALUE =   197800.000

VARIABLE	VALUE	REDUCED COSTS
X1	7.600	0.000
X2	5.000	0.000
X3	2.000	0.000
X4	0.000	0.000
X5	0.000	0.000
X6	1.000	0.000
X7	2.000	0.000

PROBLEM 9

The Navy has 9,000 pounds of material in Albany, Georgia which it wishes to ship to three installations: San Diego, Norfolk, and Pensacola. They require 4,000, 2,500, and 2,500 pounds respectively. The following gives the shipping costs per pound for truck, railroad, and airplane transit.

Mode	San Diego	Norfolk	Pensacola
Truck	$12	$6	$5
Railroad	20	11	9
Airplane	30	26	28

Destination

Government regulations require equal distribution of shipping among the three carriers. Formulate and solve a linear program to determine the shipping arrangements (mode, destination, and quantity) that will minimize the total shipping cost.

SOLUTION 9

Define the decision variables

We want to determine the pounds of material, x_{ij}, to be shipped by mode i to destination j. The following table summarizes the decision variables:

Mode	San Diego	Norfolk	Pensacola
Truck	x_{11}	x_{12}	x_{13}
Railroad	x_{21}	x_{22}	x_{23}
Airplane	x_{31}	x_{32}	x_{33}

Destination

Define the objective function

Minimize the total shipping cost:

MIN (shipping cost per pound for each mode/destination pairing)
 X (number of pounds shipped by mode/destination pairing):
MIN $12x_{11} + 6x_{12} + 5x_{13} + 20x_{21} + 11x_{22} + 9x_{23} + 30x_{31} + 26x_{32} + 28x_{33}$

Define the constraints

Equal use of transportation modes:

(1) $x_{11} + x_{12} + x_{13} = 3000$
(2) $x_{21} + x_{22} + x_{23} = 3000$
(3) $x_{31} + x_{32} + x_{33} = 3000$

Destination material requirements:

(4) $x_{11} + x_{21} + x_{31} = 4000$
(5) $x_{12} + x_{22} + x_{32} = 2500$
(6) $x_{13} + x_{23} + x_{33} = 2500$

Nonnegativity of variables:

$x_{ij} \geq 0$, $i = 1,2,3$ and $j = 1,2,3$

The Management Scientist provided the following solution:

OBJECTIVE FUNCTION VALUE = 142000.000

VARIABLE	VALUE	REDUCED COSTS
X11	1000.000	0.000
X12	2000.000	0.000
X13	0.000	1.000
X21	0.000	3.000
X22	500.000	0.000
X23	2500.000	0.000
X31	3000.000	0.000
X32	0.000	2.000
X33	0.000	6.000

To summarize: San Diego will receive 1000 lbs. by truck and 3000 lbs. by airplane; Norfolk will receive 2000 lbs. by truck and 500 lbs. by railroad; Pensacola will receive 2500 lbs. by railroad. The total shipping cost will be $142,000.

> NOTE: This problem is referred to as a transportation problem. Transportation problems have a mathematical structure that has enabled management scientists to develop efficient, specialized procedures for solving them. These procedures are covered in Chapter 10.

PROBLEM 10

LeapFrog Airways provides passenger service for Indianapolis, Baltimore, Memphis, Austin, and Tampa. LeapFrog has two WB828 airplanes, one based in Indianapolis and the other in Baltimore. Each morning the Indianapolis based plane flies to Austin with a stopover in Memphis, and the Baltimore based plane flies to Tampa with a stopover in Memphis. Both planes have a coach section with a 120-seat capacity.

LeapFrog uses two fare classes: a discount fare D class and a full fare F class. Leapfrog's products, each referred to as an origin destination itinerary fare (ODIF), are listed below with their fares and forecasted demand.

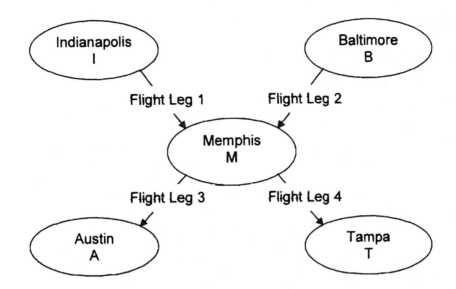

ODIF	Origin	Destination	Fare Class	ODIF Code	Fare	Demand
1	Indianapolis	Memphis	D	IMD	175	44
2	Indianapolis	Austin	D	IAD	275	25
3	Indianapolis	Tampa	D	ITD	285	40
4	Indianapolis	Memphis	F	IMF	395	15
5	Indianapolis	Austin	F	IAF	425	10
6	Indianapolis	Tampa	F	ITF	475	8
7	Baltimore	Memphis	D	BMD	185	26
8	Baltimore	Austin	D	BAD	315	50
9	Baltimore	Tampa	D	BTD	290	42
10	Baltimore	Memphis	F	BMF	385	12
11	Baltimore	Austin	F	BAF	525	16
12	Baltimore	Tampa	F	BTF	490	9
13	Memphis	Austin	D	MAD	190	58
14	Memphis	Tampa	D	MTD	180	48
15	Memphis	Austin	F	MAF	310	14
16	Memphis	Tampa	F	MTF	295	11

Develop a linear programming model for LeapFrog's problem situation and determine how many seats LeapFrog should allocate to each ODIF.

SOLUTION 9

Define the decision variables

There are 16 variables, one for each ODIF.

IMD = number of seats allocated to Indianapolis-Memphis-Discount class
IAD = number of seats allocated to Indianapolis-Austin-Discount class
ITD = number of seats allocated to Indianapolis-Tampa-Discount class
IMF = number of seats allocated to Indianapolis-Memphis-Full Fare class
IAF = number of seats allocated to Indianapolis-Austin-Full Fare class
ITF = number of seats allocated to Indianapolis-Tampa-Full Fare class
BMD = number of seats allocated to Baltimore-Memphis-Discount class
BAD = number of seats allocated to Baltimore-Austin-Discount class
BTD = number of seats allocated to Baltimore-Tampa-Discount class
BMF = number of seats allocated to Baltimore-Memphis-Full Fare class
BAF = number of seats allocated to Baltimore-Austin-Full Fare class
BTF = number of seats allocated to Baltimore-Tampa-Full Fare class
MAD = number of seats allocated to Memphis-Austin-Discount class
MTD = number of seats allocated to Memphis-Tampa-Discount class
MAF = number of seats allocated to Memphis-Austin-Full Fare class
MTF = number of seats allocated to Memphis-Tampa-Full Fare class

Define the objective function

Maximize total revenue:

Max (fare per seat for each ODIF) x (number of seats allocated to the ODIF).
Max 175IMD + 275IAD + 285ITD + 395IMF + 425IAF + 475ITF
    + 185BMD + 315BAD + 290BTD + 385BMF + 525BAF
       + 490BTF + 190MAD + 180MTD + 310MAF + 295MTF

Define the constraints

There are 4 capacity constraints, one for each flight leg:

(1) IMD + IAD + ITD + IMF + IAF + ITF $\leq$ 120 (Indianapolis-Memphis leg)
(2) BMD + BAD + BTD + BMF + BAF + BTF $\leq$ 120 (Baltimore-Memphis leg)
(3) IAD + IAF + BAD + BAF + MAD + MAF $\leq$ 120 (Memphis-Austin leg)
(4) ITD + ITF + BTD + BTF + MTD + MTF $\leq$ 120 (Memphis-Tampa leg)

There are 16 demand constraints, one for each ODIF:

(5) IMD $\leq$ 44   (11) BMD $\leq$ 26   (17) MAD $\leq$ 58
(6) IAD $\leq$ 25   (12) BAD $\leq$ 50   (18) MTD $\leq$ 48
(7) ITD $\leq$ 40   (13) BTD $\leq$ 42   (19) MAF $\leq$ 14
(8) IMF $\leq$ 15   (14) BMF $\leq$ 12   (20) MTF $\leq$ 11
(9) IAF $\leq$ 10   (15) BAF $\leq$ 16
(10) ITF $\leq$ 8   (16) BTF $\leq$ 9

Nonnegativity constraints: IMD, IAD, ITD, ..., MTF $\geq$ 0

The problem was solved by *The Management Scientist* and the output is below.

OBJECTIVE FUNCTION VALUE =   94735.000

VARIABLE	VALUE	REDUCED COSTS
IMD	44.000	0.000
IAD	3.000	0.000
ITD	40.000	0.000
IMF	15.000	0.000
IAF	10.000	0.000
ITF	8.000	0.000
BMD	26.000	0.000
BAD	50.000	0.000
BTD	7.000	0.000
BMF	12.000	0.000
BAF	16.000	0.000
BTF	9.000	0.000
MAD	27.000	0.000
MTD	45.000	0.000
MAF	14.000	0.000
MTF	11.000	0.000

CONSTRAINT	SLACK/SURPLUS	DUAL PRICES
1	0.000	85.000
2	0.000	110.000
3	0.000	190.000
4	0.000	180.000
5	0.000	90.000
6	22.000	0.000
7	0.000	20.000
8	0.000	310.000
9	0.000	150.000
10	0.000	210.000
11	0.000	75.000
12	0.000	15.000
13	35.000	0.000
14	0.000	275.000
15	0.000	225.000
16	0.000	200.000
17	31.000	0.000
18	3.000	0.000
19	0.000	120.000
20	0.000	115.000

ANSWERED PROBLEMS

PROBLEM 11

Fullerton Trucking has $550,000 allocated to purchase at least 40 trucks. It will buy both Japanese and American-built trucks, but keeping with its "BUY-AMERICAN" image, it wishes to purchase at least twice as many American as Japanese-built trucks. Fullerton has narrowed its choices to three American and three Japanese models. It wishes to purchase at least 20 two-seat models. The following table summarizes the data for each model:

Truck	Country	Cost	Capacity	Seats
Hauler	U.S.	$20,000	2.00 tons	2
Mauler	U.S.	$16,000	1.00	2
Bruiser	U.S.	$13,000	.75	1
Econotruck	Japan	$ 9,000	.50	1
T-150	Japan	$12,000	.75	2
Maxitruck	Japan	$17,000	1.50	2

a) Formulate and solve this problem as a linear program with the objective of maximizing the overall trucking capacity.

b) Why is linear programming not the technically correct procedure to use to solve this problem?

PROBLEM 12

National Wing Company (NWC) is gearing up for the new B-48 contract. Currently NWC has 100 equally qualified workers. Over the next three months NWC has made the following commitments for wing production:

Month	Contract
May	20
June	24
July	30

Each worker can either be placed in production or can train new recruits. A new recruit can be trained to be an apprentice in one month. The next month, he, himself, becomes a qualified worker (after two months from the start of training). Each trainer can train two recruits. The production rate and salary per employee is estimated below.

Employee	Production Rate (Wings/Month)	Salary Per Month
Production	.6	$3,000
Trainer	.3	$3,300
Apprentice	.4	$2,600
Recruit	.05	$2,200

At the end of July, NWC wishes to have no recruits or apprentices but have at least 140 full-time workers. Formulate and solve a linear program for NWC to accomplish this at minimum total cost.

PROBLEM 13

Triumph Trumpet Company makes two styles each of both trumpets and cornets: deluxe and professional models. Its unit profit on deluxe trumpets is $80 and on deluxe cornets is $60. The professional models realize twice the profit of the deluxe models.

Trumpets and cornets are made basically from two mixtures of two different brass alloys. The amount of each alloy (in pounds) required to produce each type of horn is summarized in the following table along with the monthly availability to Triumph of the alloys.

	Trumpets		Cornets		Monthly
	Deluxe	Pro.	Deluxe	Pro.	Availability
Alloy 1	2	1.5	1.5	1	2000
Alloy 2	1	1.5	1	1.5	1800

Triumph must fulfill contracts calling for at least 500 deluxe trumpets and 300 deluxe cornets monthly. Monthly demand for professional trumpets is not expected to exceed 150 and for professional cornets is not expected to exceed 100. Production set-ups are such that the company will produce exactly twice as many trumpets as cornets.

Formulate and solve this problem as a linear program.

PROBLEM 14

Millard Construction is contemplating building a planned community with the help of federal funds. These funds are to be distributed only if Millard meets federal standards for low cost housing. There are three types of units -- houses, town-houses, and high rise condominiums. There are three styles each of the houses and town-houses -- low cost, standard, and deluxe. The high rise condominiums will have standard and deluxe models.

The amount of total ground space (including allowances for parking and green belts) is given in the following table:

	Ground Area (Sq. Ft.)		
	Low Cost	Standard	Deluxe
Houses	1800	2200	3000
Town-Houses	740	1600	2230
Condominiums	X	1000	1500

The profit to Millard per unit is summarized in the following table:

	Profit ($1000's)		
	Low Cost	Standard	Deluxe
Houses	5	12	25
Town-Houses	4	10	18
Condominiums	X	9	16

Millard has 300,000 square feet for construction. To make the project "work", Millard wants houses and town-houses each to occupy between 25% and 40% of the total area while condominiums only need to occupy 10% to 25% of the total area.

The federal government requires that at least 25% of the total units built in the complex be low cost units.

a) Formulate this problem as a linear program.

b) Solve for the optimal solution using a computer package such as *The Management Scientist*.

c) Why is linear programming not the technically correct formulation procedure for this problem?

PROBLEM 15

Delicious Candy Company manufactures three types of candy bars-- Chompers, Smerks, and Delicious Chocolate. All three candies come in a one-ounce size while Delicious Chocolate also comes in a one-pound mini-bar bag.

The basic ingredients used are chocolate, peanuts, and caramel. Delicious Chocolate is all chocolate, while Chompers consists of chocolate and caramel, and Smerks consists of chocolate, caramel and peanuts. Chompers' recipe allows for the amount of caramel to be anywhere between 18% and 28% of the candy bar's weight with chocolate making up the rest. Smerks' recipe calls for an equal amount of caramel and peanuts, with chocolate making up between 20% and 40% of the bar's weight.

For each one-ounce bar, labor and packaging costs $.012, while labor and packaging for the one-pound bag costs $.039. The company has production facilities for making up to 20,000 one-ounce bars and up to 1000 one-pound bags daily.

Delicious has contracts to produce at least 3000 one-ounce bars of each type of candy daily. Also, the difference between the number of Chompers and the number of Smerks produced must be less than 10% of the total number of Chompers and Smerks made. The present prices for chocolate, caramel, and peanuts are $1.60, $.95, and $1.40 per pound respectively. The company has contracts which will supply it with at least 1,000 pounds of chocolate, exactly 350 pounds of caramel, and at most 500 pounds of peanuts daily.

The company currently sells Chompers one-ounce bars for $.14, Smerks one-ounce bars for $.16, Delicious Chocolate one- ounce bars for $.15, and Delicious Chocolate one-pound bags for $2.30. Formulate and solve a linear program that would determine the optimal daily production schedule and ingredients required. (HINT: Variables must be established for each product type and the amount of each ingredient in each product.)

PROBLEM 16

WeBuild Construction must decide how many small bulldozers to purchase or lease for the coming year. Bulldozers may be purchased for $40,000 each and their salvage value at the end of a year is $20,000. WeBuild can also lease bulldozers for $8,000 per year payable in advance.

WeBuild has $1,000,000 available in its budget for purchase and/or lease of bulldozers. Any monies not invested in purchasing or leasing bulldozers will be invested at 8%.

There are 60 projects each requiring four bulldozers throughout the year. Because of timing and reliability, each new bulldozer will be available for eight projects, whereas each leased bulldozer will be available for five projects.

Formulate and solve for the number of bulldozers WeBuild should purchase and lease to minimize its total annual cost.

PROBLEM 17

John Sweeney is an investment advisor who is attempting to construct an "optimal portfolio" for a client who has $400,000 cash to invest. There are ten different investments, falling into four broad categories that John and his client have identified as potential candidates for this portfolio.

The following table lists the investments and their important characteristics. Note that Unidyde Equities (stocks) and Unidyde Debt (bonds) are two separate investments, whereas First General REIT is a single investment that is considered both an equities and a real estate investment.

Category	Investment	Exp. Annual After Tax Return	Liquidity Factor	Risk Factor
Equities	Unidyde Corporation	15.0%	100	60
	Col. Must. Restaurants	17.0%	100	70
	First General REIT	17.5%	100	75
Debt	Metropolitan Electric	11.8%	95	20
	Unidyde Corporation	12.2%	92	30
	Lemonville Transit	12.0%	79	22
Real Estate	Fairview Apartment Partn.	22.0%	0	50
	First General REIT		(See above)	
Money	T-Bill Account	9.6%	80	0
	Money Market Fund	10.5%	100	10
	All Saver's Certificate	12.6%	0	0

Formulate and solve a linear program to accomplish John's objective as an investment advisor which is to construct a portfolio that maximizes his client's total expected after-tax return over the next year, subject to a number of constraints placed upon him by the client for the portfolio:

1. Its (weighted) average liquidity factor must be at least 65.
2. The (weighted) average risk factor must be no greater than 55.
3. At most, $60,000 is to be invested in Unidyde stocks or bonds.
4. No more than 40% of the investment can be in any one category except the money category.
5. No more than 20% of the investment can be in any one investment except the money market fund.
6. At least $1,000 must be invested in the money market fund.
7. The maximum investment in All Saver's Certificates is $15,000.
8. The minimum investment desired for debt is $90,000.
9. At least $10,000 must be placed in a T-Bill account.

PROBLEM 18

BP Cola must decide how much money to allocate for new soda and traditional soda advertising over the coming year. The advertising budget is $10,000,000.

Because BP wants to push its new sodas, at least one-half of the advertising budget is to be devoted to new soda advertising. However, at least $2,000,000 is to be spent on its traditional sodas. BP estimates that each dollar spent on traditional sodas will translate into 100 cans sold, whereas, because of the harder sell needed for new products, each dollar spent on new sodas will translate into 50 cans sold. To attract new customers BP has lowered its profit margin on new sodas to 2 cents per can as compared to 4 cents per can for traditional sodas.

How should BP allocate its advertising budget if it wants to maximize its profits while selling at least 750 million cans?

PROBLEM 19

Maybury Public School System has three high schools to serve a territory divided into five districts. The capacity of each high school, the student population in each district, and the distance (in miles) between each school and the center of each district are listed in the table below:

District	McHale	McCallum	McBride	Student Population
Northeast	1.5	2.5	0.5	700
Southeast	4	1.5	3	1100
Southwest	2.5	3	3.5	900
Northwest	0.5	4	1.5	600
Central	1	2	1	800
Capacity	1500	1800	1100	

Formulate and solve a linear program to determine the school-student assignment that minimizes the total student-miles traveled per day.

PROBLEM 20

The June Company is a department store chain serving three states in the South: Georgia, Alabama, and Mississippi. Recently management has been concerned about the relative efficiency of its Alabama store.

The June Company tabulates the monthly gross sales of home appliances, clothing, home entertainment, and all other divisions and measures these outputs against inputs of store size, number of sales personnel for the store, and the population service area for each store (defined as the number of adults over 16 living within a 35 mile radius of the store.)

The tables below give the average monthly values for each of the inputs and outputs based on last year's data. Based on last year's data, use data envelopment analysis to formulate and solve a linear program to help June Company determine the relative efficiency of its Alabama store.

	Georgia	Alabama	Mississippi
Inputs			
Store Size (1000's sq. ft.)	25	24	18
Sales Personnel	250	210	180
Service Area Population (1000's)	600	750	375
Outputs: Average Monthly Sales ($mil.)			
Appliances	1.2	0.8	0.6
Clothing	2.2	1.4	1.5
Home Entertainment	1.6	1.5	1.6
All Others	2.7	2.0	2.1

PROBLEM 21

Niteton Power and Light Company (NPLC) wants to develop an efficient work schedule for its full- and part-time customer service clerks. The number of clerks needed to provide adequate service during each hour the office is open on a weekday is given below:

Hour	8-9am	9-10	10-11	11-12pm	12-1	1-2	2-3	3-4	4-5	5-6
Clerks	5	4	6	8	10	9	7	4	7	5

A full-time clerk works 3 hours, has a 1-hour break, and then works another 3 hours. Part-time clerks work 4 consecutive hours. Full-timers get paid for their break. All clerks start work on the hour.

NPLC's office manager insists that at least one full-time clerk be on duty during all open hours and that a minimum of four full-time clerks are on the payroll. A full-time clerk costs NPLC $9.00 per hour, and a part-timer costs $6.50 per hour.

Formulate and solve a linear program that will provide a schedule that will meet NPLC's customer service needs at a minimum labor cost. (Hint: there are 4 different full-time shifts and 7 different part-time shifts.)

TRUE/FALSE

___ 22. Sunk costs should be viewed as relevant costs when developing the objective function for a linear programming model.

___ 23. Double-subscript notation for decision variables should be avoided unless the number of decision variables exceeds nine.

___ 24. Generating the data for large-scale LP models can be more time consuming than either the formulation of the model or the development of the computer solution.

___ 25. Using minutes as the unit of measurement on the left-hand side of a constraint and using hours on the right-hand side is acceptable since both are a measure of time.

___ 26. No real-world problem, when correctly formulated, has an unbounded solution.

___ 27. A company makes two products from steel; one requires 2 tons of steel and the other requires 3 tons. There are 100 tons of steel available daily. A constraint on daily production could be written as: $2x_1 + 3x_2 \leq 100$.

___ 28. Compared to the problems in the textbook, real world problems generally require more variables and constraints.

___ 29. For the multiperiod production scheduling problem in the textbook, period $n - 1$'s ending inventory variable was also used as period n's beginning inventory variable.

___ 30. To reliably answer "what-if" questions about the effects of changes in the parameters used in an LP formulation, one should always resolve the problem.

___ 31. If a real-world problem is correctly formulated, it is impossible to have alternative optimal solutions.

___ 32. A company makes two products, A and B. A sells for $100 and B sells for $90. The variable production costs are $30 per unit for A and $25 for B. The company's objective could be written as: MAX $190x_1 - 55x_2$.

___ 33. The primary limitation of linear programming's applicability is the requirement that all decision variables be nonnegative.

___ 34. If an LP problem is not correctly formulated, the computer will indicate it is infeasible when trying to solve it.

___ 35. The wide-spread use of linear programming is due largely to the fact that the nature of most business functions is linear.

___ 36. A decision maker would be wise to not deviate from the optimal solution found by an LP model because it is the best solution.

CHAPTER 10

Transportation, Assignment, and Transshipment Problems

CHAPTER 10

KEY CONCEPTS

CONCEPT	ILLUSTRATED PROBLEMS	ANSWERED PROBLEMS
Transportation Problems		
Network Representation	1	8
Dummy Sources	2,3	9,10
Maximization Problems	3	11
Unacceptable Routes	3	11
Spreadsheet Solution	2	
Assignment Problem		
Network Representation	4	13
Maximization Problems	5	12,14
Unacceptable Assignments	5	14
Transshipment Problem		
Network Representation	6,7	15
LP Formulation	6	15
Production and Inventory Application	7	

REVIEW

1. A <u>network model</u> is one which can be represented by a set of nodes, a set of arcs, and functions (e.g. costs, supplies, demands, etc.) associated with the arcs and/or nodes.

2. Transportation, assignment, and transshipment problems of this chapter, and PERT/CPM problems in Chapter 12 are all examples of <u>network problems</u>.

3. Each of the three models of this chapter (transportation, assignment, and transshipment models) can be formulated as <u>linear programs</u> and solved by general purpose linear programming codes. However, there are many computer packages (including The Management Scientist) which contain separate computer codes for these models which take advantage of their network structure. This radically enhances the speed of solving particularly large problems.

4. For each of the three models in this chapter, if the right-hand side of the linear programming formulations are all integers, the optimal solution will be in terms of <u>integer values</u> for the decision variables.

TRANSPORTATION PROBLEM

1. The <u>transportation problem</u> seeks to minimize the total shipping costs of transporting goods from m origins (each with a supply s_i) to n destinations (each with a demand d_j), when the unit shipping cost from an origin, i, to a destination, j, is c_{ij}.

2. The <u>network representation</u> for a transportation problem with two sources and three destinations is given below:

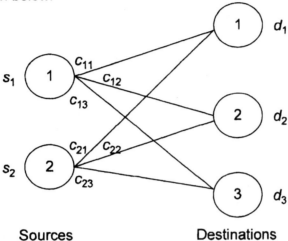

3. Transportation problems are special cases of linear programs. The <u>linear programming formulation</u> in terms of the amounts shipped from the origins to the destinations, x_{ij}, can be written as:

$$\text{MIN } \Sigma_i \Sigma_j c_{ij} x_{ij}$$

s. t. $\Sigma_j x_{ij} \leq s_i$ for each origin i

$\Sigma_i x_{ij} = d_j$ for each destination j

$x_{ij} \geq 0$ for all i and j.

4. The following <u>special-case modifications</u> to the linear programming formulation can be made:
   (a) Minimum shipping guarantees from i to j: $x_{ij} \geq L_{ij}$
   (b) Maximum route capacity from i to j: $x_{ij} \leq L_{ij}$
   (c) Unacceptable routes: delete the variable

5. To solve the transportation problem as it is formulated above, it is required that the sum of the supplies at the origins equals or exceeds the sum of the demands at the destinations. If total supply is less than total demand, a <u>dummy origin</u> is added. The objective function coefficient associated with any route stemming from a dummy origin has a value of zero.

ASSIGNMENT PROBLEM

1. An <u>assignment problem</u> seeks to minimize the total cost assignment of m workers to m jobs, given that the cost of worker i performing job j is c_{ij}. It assumes all workers are assigned and each job is performed.

2. The <u>network representation</u> of an assignment problem with three workers and three jobs is:

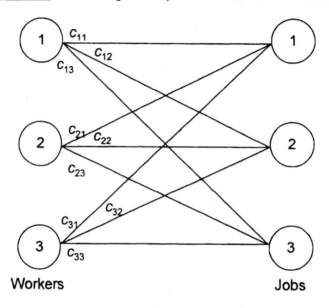

3. The <u>linear programming formulation</u> of the assignment problem using $x_{ij} = 0$ or 1 denoting whether worker i is assigned to job j is:

$$\text{MIN} \quad \sum_i \sum_j c_{ij} x_{ij}$$

$$\text{s. t.} \quad \sum_j x_{ij} = 1 \quad \text{for each worker } i$$

$$\sum_I x_{ij} = 1 \quad \text{for each job } j$$

$$x_{ij} = 0 \text{ or } 1 \quad \text{for all } i \text{ and } j.$$

4. An assignment problem is a <u>special case of a transportation problem</u> in which all supplies and all demands are equal to 1; hence assignment problems may be solved as linear programs.

5. A modification to the <u>right-hand side</u> of the linear program can be made if a worker is permitted to work more than 1 job.

TRANSSHIPMENT PROBLEM

1. **Transshipment problems** are transportation problems in which a shipment may move through intermediate nodes (transshipment nodes) before reaching a particular destination node.

2. The **network representation** for a transshipment problem with two sources, three intermediate nodes, and two destinations is given below:

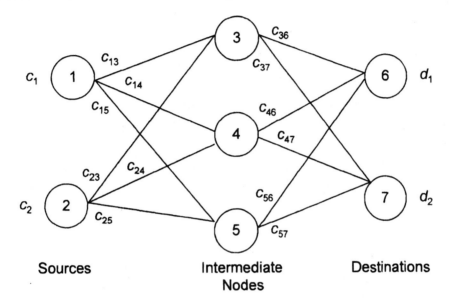

Sources   Intermediate Nodes   Destinations

3. The **linear programming formulation** for the transshipment problem with x_{ij} representing the shipment from node i to node j is:

$$\text{MIN} \sum_i \sum_j c_{ij} x_{ij}$$

s. t. $\sum_j x_{ij} \leq s_i$ for each origin i

$\sum_I x_{ik} - \sum_j x_{kj} = 0$ for each intermediate node k

$\sum_I x_{ij} = d_j$ for each destination j

$x_{ij} \geq 0$ for all i and j.

4. A **capacitated transshipment problem** is a transshipment problem involving one or more routes (arcs) that have a shipping capacity (upper limit).

218 CHAPTER 10

ILLUSTRATED PROBLEMS

PROBLEM 1

Given the following minimum cost transportation network:

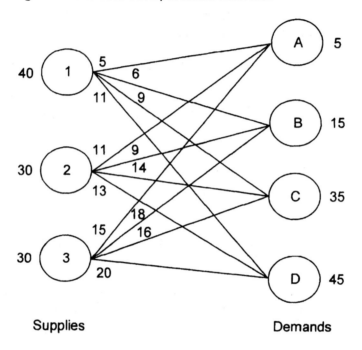

Supplies   Demands

a) Formulate this problem as a linear program.

b) Solve the problem using *The Management Scientist* or similar package.

SOLUTION 1

a) <u>Linear Programming Formulation</u>

Define the decision variables:
$Q_{1A}, Q_{1B}, Q_{1C}, Q_{1D}$ = Quantity shipped from Source 1 to Destinations A, B, C, D
$Q_{2A}, Q_{2B}, Q_{2C}, Q_{2D}$ = Quantity shipped from Source 2 to Destinations A, B, C, D
$Q_{3A}, Q_{3B}, Q_{3C}, Q_{3D}$ = Quantity shipped from Source 3 to Destinations A, B, C, D

Define objective: Minimize overall shipping costs:
 Min $5Q_{1A} + 6Q_{1B} + 9Q_{1C} + 11Q_{1D} + 11Q_{2A} + 9Q_{2B} + 14Q_{2C} + 13Q_{2D}$
  $+ 15Q_{3A} + 18Q_{3B} + 16Q_{3C} + 20Q_{3D}$

Define the constraints:
 Amount out of Origin 1: $Q_{1A} + Q_{1B} + Q_{1C} + Q_{1D} \leq 40$
 Amount out of Origin 2: $Q_{2A} + Q_{2B} + Q_{2C} + Q_{2D} \leq 30$
 Amount out of Origin 3: $Q_{3A} + Q_{3B} + Q_{3C} + Q_{3D} \leq 30$

Amount into Destination A: $Q_{1A} + Q_{2A} + Q_{3A} \leq 5$
Amount into Destination B: $Q_{1B} + Q_{2B} + Q_{3B} \leq 15$
Amount into Destination C: $Q_{1C} + Q_{2C} + Q_{3C} \leq 35$
Amount into Destination D: $Q_{1D} + Q_{2D} + Q_{3D} \leq 45$
Non-negativity of variables: $Q_{1A}, Q_{1B}, Q_{1C}, Q_{1D} \geq 0$
$Q_{2A}, Q_{2B}, Q_{2C}, Q_{2D} \geq 0$
$Q_{3A}, Q_{3B}, Q_{3C}, Q_{3D} \geq 0$

b) *The Management Scientist* provided the following solution:

OBJECTIVE FUNCTION VALUE = 1195.000

VARIABLE	VALUE	REDUCED COSTS
Q_{1A}	5.000	0.000
Q_{1B}	15.000	0.000
Q_{1C}	5.000	0.000
Q_{1D}	15.000	0.000
Q_{2A}	0.000	4.000
Q_{2B}	0.000	1.000
Q_{2C}	0.000	3.000
Q_{2D}	30.000	0.000
Q_{3A}	0.000	3.000
Q_{3B}	0.000	5.000
Q_{3C}	30.000	0.000
Q_{3D}	0.000	2.000

CONSTRAINT	SLACK/SURPLUS	DUAL PRICES
1	0.000	7.000
2	0.000	5.000
3	0.000	0.000
4	0.000	-12.000
5	0.000	-13.000
6	0.000	-16.000
7	0.000	-18.000

PROBLEM 2

Building Brick Company (BBC) has orders for 80 tons of bricks at three suburban locations as follows: Northwood -- 25 tons, Westwood -- 45 tons, and Eastwood -- 10 tons. BBC has two plants, each of which can produce 50 tons per week. How should end of week shipments be made to fill the above orders given the following delivery cost per ton:

	Northwood	Westwood	Eastwood
Plant 1	24	30	40
Plant 2	30	40	42

a) Formulate this problem as a linear program.

b) Solve the problem using a spreadsheet.

SOLUTION 2

a) Let T_{1N} = Number of tons of brick shipped from Plant 1 to Northwood
T_{1W} = Number of tons of brick shipped from Plant 1 to Westwood
T_{1E} = Number of tons of brick shipped from Plant 1 to Eastwood
T_{2N} = Number of tons of brick shipped from Plant 2 to Northwood
T_{2W} = Number of tons of brick shipped from Plant 2 to Westwood
T_{2E} = Number of tons of brick shipped from Plant 2 to Eastwood

Min $24T_{1N} + 30T_{1W} + 40T_{1E} + 30T_{2N} + 40T_{2W} + 42T_{2E}$

s.t. $T_{1N} + T_{1W} + T_{1E} \leq 50$
$T_{2N} + T_{2W} + T_{2E} \leq 50$
$T_{1N} + T_{2N} = 25$
$T_{1W} + T_{2W} = 45$
$T_{1E} + T_{2E} = 10$
$T_{1N}, T_{1W}, T_{1E}, T_{2N}, T_{2W}, T_{2E} \geq 0$

b) *The Management Scientist* provided the following solution:

OBJECTIVE FUNCTION VALUE = 2490.000

VARIABLE	VALUE	REDUCED COSTS
T_{1N}	5.000	0.000
T_{1W}	45.000	0.000
T_{1E}	0.000	4.000
T_{2N}	20.000	0.000
T_{2W}	0.000	4.000
T_{2E}	10.000	0.000

CONSTRAINT	SLACK/SURPLUS	DUAL PRICES
1	0.000	6.000
2	20.000	0.000
3	0.000	-30.000
4	0.000	-36.000
5	0.000	-42.000

Thus the optimal solution is:

From	To	Amount	Cost
Plant 1	Northwood	5	120
Plant 1	Westwood	45	1,350
Plant 2	Northwood	20	600
Plant 2	Eastwood	10	420

Total Cost = $2,490

b) Refer to Chapters 7 and 8 of this study guide for coverage of the steps to follow in setting up and using a spreadsheet to solve an LP.

Spreadsheet showing optimal solution

	A	B	C	D	E	F	G
1		LHS Coefficients					
2	Constraint	X11	X12	X13	X21	X22	X23
3	#1	1	1	1			
4	#2				1	1	1
5	#3	1			1		
6	#4		1			1	
7	#5			1			1
8	Obj.Coefficients	24	30	40	30	40	42
9							
10		X11	X12	X13	X21	X22	X23
11	Dec.Var.Values	5	45	0	20	0	10
12	Minimized Total Shipping Cost			2490			
13							
14	Constraints			LHS		RHS	
15	P1.Cap.			50	<=	50	
16	P2.Cap.			30	<=	50	
17	N.Dem.			25	=	25	
18	W.Dem.			45	=	45	
19	E.Dem.			10	=	10	

Sensitivity Report

Adjustable Cells						
Cell	Name	Final Value	Reduced Cost	Objective Coefficient	Allowable Increase	Allowable Decrease
$B$11	X11	5	0	24	4	4
$C$11	X12	45	0	30	4	1E+30
$D$11	X13	0	4	40	1E+30	4
$E$11	X21	20	0	30	4	4
$F$11	X22	0	4	40	1E+30	4
$GH$1	X23	10.000	0.000	42	4	1E+30

Constraints						
Cell	Name	Final Value	Shadow Price	Constraint R.H. Side	Allowable Increase	Allowable Decrease
$D$16	P2.Cap	30.0	0.0	50	1E+30	20
$D$17	N.Dem	25.0	30.0	25	20	20
$D$18	W.Dem	45.0	36.0	45	5	20
$D$19	E.Dem	10.0	42.0	10	20	10
$D$15	P1.Cap	50.0	-6.0	50	20	5

PROBLEM 3

Telly's Toy Company produces three kinds of dolls called Bertha, Holly, and Terri. Maximum production quantities for the dolls are 1,000, 2,000, and 2,000 per week, respectively. These dolls are purchased by three large department stores: Shears, Nichols and Words. Each department store wishes 1,500 total dolls per week form Telly's. However, Words does not want any Bertha dolls.

Because of past commitments and the sizes of other orders from Telly's, unit profits per doll vary from store to store for each style of doll. These are summarized as follows:

	Shears	Nichols	Words
Bertha	$5	$10	X
Holly	$16	$8	$9
Terri	$12	$9	$11

a) Formulate this maximization transportation problem as a linear program.

b) Solve the problem using *The Management Scientist* or similar package.

SOLUTION 3

a) The origins are the dolls, the destinations are the stores.
Let: D_{BS}, D_{BN} = Number of Bertha dolls shipped to Shears, Nichols
D_{HS}, D_{HN}, D_{HW} = Number of Holly dolls shipped to Shears, Nichols, Words
D_{TS}, D_{TN}, D_{TW} = Number of Terri dolls shipped to Shears, Nichols. Words

MAX $5D_{BS} + 10D_{BN} + 16D_{HS} + 8D_{HN} + 9D_{HW} + 12D_{TS} + 9D_{TN} + 11D_{TW}$

s.t.
$D_{BS} + D_{BN} \leq 1000$
$D_{HS} + D_{HN} + D_{HW} \leq 2000$
$D_{TS} + D_{TN} + D_{TW} \leq 2000$
$D_{BS} + D_{HS} + D_{TS} = 1500$
$D_{BN} + D_{HN} + D_{TN} = 1500$
$D_{HW} + D_{TW} = 1500$
$D_{BS}, D_{BN}, D_{HS}, D_{HN}, D_{HW}, D_{TS}, D_{TN}, D_{TW} \geq 0$

b) *The Management Scientist* provided the following solution:

OBJECTIVE FUNCTION VALUE = 55000.000

VARIABLE	VALUE	REDUCED COSTS
D_{BS}	0.000	12.000
D_{BN}	1000.000	0.000
D_{HS}	1500.000	0.000
D_{HN}	0.000	1.000
D_{HW}	0.000	2.000
D_{TS}	0.000	4.000
D_{TN}	500.000	0.000
D_{TW}	1500.000	0.000

CONSTRAINT	SLACK/SURPLUS	DUAL PRICES
1	0.000	1.000
2	500.000	0.000
3	0.000	0.000
4	0.000	16.000
5	0.000	9.000
6	0.000	11.000

The optimal solution is:

Dolls	Store	Shipment	Profit
Bertha	Nichols	1000	$10,000
Holly	Shears	1500	$24,000
Shari	Nichols	500	$ 4,500
Shari	Words	1500	$16,500

Total Profit = $55,000

PROBLEM 4

A contractor pays his subcontractors a fixed fee plus mileage for work performed. On a given day the contractor is faced with three electrical jobs associated with various projects. He has four electrical subcontractors which are located at various places throughout the area. Given below are the distances between the subcontractors and the projects.

		Project		
		A	B	C
	Westside	50	36	16
Subcontractors	Federated	28	30	18
	Goliath	35	32	20
	Universal	25	25	14

a) Draw a network representation of this problem.

b) Formulate and solve this problem as a linear program.

SOLUTION 4

a) This is an assignment problem.

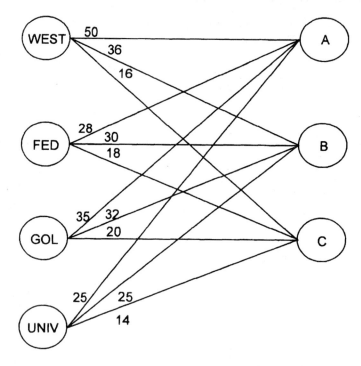

b) Let: $A_{ij} = 1$ if Subcontractor i is assigned to Project j;
= 0 otherwise
where: i is Westside, Federated, Universal, Goliath
j is A, B, C

MIN $\quad 50A_{WA} + 36A_{WB} + 16A_{WC} + 28A_{FA} + 30A_{FB} + 18A_{FC}$
$\quad\quad\quad + 35A_{UA} + 32A_{UB} + 20A_{UC} + 25A_{GA} + 25A_{GB} + 14A_{GC}$

s.t. $\quad A_{WA} + A_{WB} + A_{WC} \leq 1$
$\quad\quad A_{FA} + A_{FB} + A_{FC} \leq 1$
$\quad\quad A_{UA} + A_{UB} + A_{UC} \leq 1$
$\quad\quad A_{GA} + A_{GB} + A_{GC} \leq 1$
$\quad\quad A_{WA} + A_{FA} + A_{UA} + A_{GA} = 1$
$\quad\quad A_{WB} + A_{FB} + A_{UB} + A_{GB} = 1$
$\quad\quad A_{WC} + A_{FC} + A_{UC} + A_{GC} = 1$
$\quad\quad A_{WA}, A_{WB}, A_{WC}, A_{FA}, A_{FB}, A_{FC}, A_{UA}, A_{UB}, A_{UC}, A_{GA}, A_{GB}, A_{GC} \geq 0$

The optimal assignment is:

Subcontractor	Project	Distance
Westside	C	16
Federated	A	28
Universal	B	25
Goliath	(unassigned)	---

Total distance is 69 miles

PROBLEM 5

A foreman is trying to assign crews to produce the maximum number of parts per hour of a certain product. He has three crews and four possible work centers. The estimated number of parts per hour for each crew at each work center is summarized below. Solve for the optimal assignment of crews to work centers.

	Work Center			
	WC1	WC2	WC3	WC4
Crew A	15	20	18	30
Crew B	20	22	26	30
Crew C	25	26	27	30

SOLUTION 5

This is a maximization assignment problem with fewer crews than work centers. A dummy crew is created to make the problem feasible.

The Management Scientist provided the following solution:

OBJECTIVE FUNCTION VALUE = 55000.000

VARIABLE	VALUE	REDUCED COSTS
A_{A1}	0.000	12.000
A_{A2}	0.000	0.000
A_{A3}	0.000	0.000
A_{A4}	1.000	1.000
A_{B1}	0.000	2.000
A_{B2}	0.000	4.000
A_{B3}	1.000	0.000
A_{B4}	0.000	0.000
A_{C1}	0.000	12.000
A_{C2}	1.000	0.000
A_{C3}	0.000	0.000
A_{C4}	0.000	1.000
A_{D1}	1.000	2.000
A_{D2}	0.000	4.000
A_{D3}	0.000	0.000
A_{D4}	0.000	0.000

CONSTRAINT	SLACK/SURPLUS	DUAL PRICES
1	0.000	18.000
2	0.000	23.000
3	0.000	24.000
4	0.000	-1.000
5	0.000	1.000
6	0.000	2.000
7	0.000	3.000
8	0.000	12.000

An optimal solution is:

Crew	Work Center	Parts/Hour
Crew A	WC4	30
Crew B	WC3	26
Crew C	WC2	26
----------	WC1	---
	Total Parts Per Hour =	82

PROBLEM 6

Thomas Industries and Washburn Corporation supply three firms (Zrox, Hewes, Rockwright) with customized shelving for its offices. They both order shelving from the same two manufacturers, Arnold Manufacturers and Supershelf, Inc. Because of long standing contracts based on past orders, unit costs from the manufacturers to the suppliers are given below:

	Thomas	Washburn
Arnold	5	8
Supershelf	7	4

The chart below gives the cost to install the shelving at the various locations:

	Zrox	Hewes	Rockwright
Thomas	1	5	8
Washburn	3	4	4

Currently weekly demands by the users are 50 for Zrox, 60 for Hewes, and 40 for Rockwright. Both Arnold and Supershelf can supply at most 75 units to its customers.

a) Draw the network representation for this problem.

b) Formulate and solve this problem as a transshipment linear program.

SOLUTION 6

a) Arnold Manufacturers and Supershelf are manufacturers (origins) and Zrox, Hewes, and Rockwright are customers (destinations). Thomas and Washburn represent transshipment points (middlemen) supplying Arnold and Supershelf products to their customers.

TRANSPORTATION, ASSIGNMENT, TRANSSHIPMENT

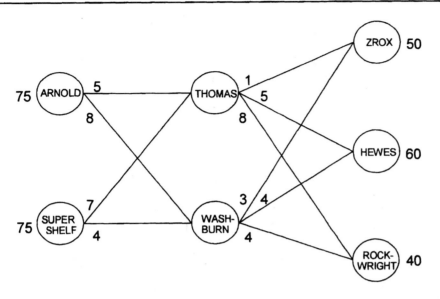

b) <u>Linear Programming Formulation and Solution</u>

Define the decision variables:

x_{ij} = amount shipped from manufacturer i to supplier j
x_{jk} = amount shipped from supplier j to customer k
   i = 1 (Arnold), 2 (Supershelf)
   j = 3 (Thomas), 4 (Washburn)
   k = 5 (Zrox), 6 (Hewes), 7 (Rockwright)

Define objective: Minimize overall shipping costs:

MIN $5x_{13} + 8x_{14} + 7x_{23} + 4x_{24} + 1x_{35} + 5x_{36} + 8x_{37} + 3x_{45} + 4x_{46} + 4x_{47}$

Define the constraints:

Amount out of Arnold: $x_{13} + x_{14} \leq 75$
Amount out of Supershelf: $x_{23} + x_{24} \leq 75$

Amount through Thomas: $x_{13} + x_{23} - x_{35} - x_{36} - x_{37} = 0$
Amount through Washburn: $x_{14} + x_{24} - x_{45} - x_{46} - x_{47} = 0$

Amount into Zrox: $x_{35} + x_{45} = 50$
Amount into Hewes: $x_{36} + x_{46} = 60$
Amount into Rockwright: $x_{37} + x_{47} = 40$

Non-negativity of variables: $x_{ij} \geq 0$, for all i and j.

The Management Scientist provided the following solution:

OBJECTIVE FUNCTION VALUE =   1150.000

VARIABLE	VALUE	REDUCED COSTS
X13	75.000	0.000
X14	0.000	2.000
X23	0.000	4.000
X24	75.000	0.000
X35	50.000	0.000
X36	25.000	0.000
X37	0.000	3.000
X45	0.000	3.000
X46	35.000	0.000
X47	40.000	0.000

Thus the optimal solution is:

From	To	Amount	Cost
Arnold	Thomas	75	375
Thomas	Zrox	50	50
Thomas	Hewes	25	125
Supershelf	Washburn	75	300
Washburn	Hewes	35	140
Washburn	Rockwright	40	160

Total Cost = $1,150

PROBLEM 7

Fodak must schedule its production of camera film for the first four months of the year. Film demand (in 1,000s of rolls) in January, February, March and April is expected to be 300, 500, 650 and 400, respectively. Fodak's production capacity is 500 thousand rolls of film per month. The film business is highly competitive, so Fodak cannot afford to lose sales or keep its customers waiting. Meeting month i's demand with month $i+1$'s production is unacceptable.

Film produced in month i can be used to meet demand in month i or can be held in inventory to meet demand in month $i+1$ or month $i+2$ (but not later due to the film's limited shelflife). There is no film in inventory at the start of January.

The film's production and delivery cost per thousand rolls will be $500 in January and February. This cost will increase to $600 in March and April due to a new labor contract. Any film put in inventory requires additional transport costing $100 per thousand rolls. It costs $50 per thousand rolls to hold film in inventory from one month to the next.

a) Modeling this problem as a transshipment problem, draw the network representation.

b) Formulate and solve this problem as a linear program.

c) How might the network in (a) be changed if there was no limit on the length of time film could be held in inventory?

SOLUTION 7

a) The source nodes in this transshipment problem are the four months of production. The destination nodes are the four months of demand. The intermediate nodes are the first three months' ending inventory (there is no reason to be holding any inventory at the end of April).

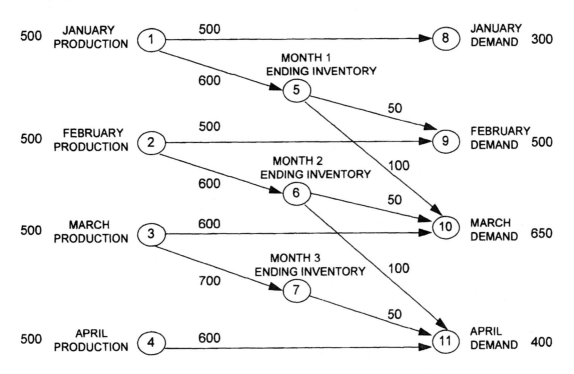

b) <u>Linear Programming Formulation and Solution</u>

Define the decision variables:

x_{ij} = amount of film "moving" between node i and node j

Define objective: Minimize total production, transportation, and inventory holding cost:

MIN $600x_{15} + 500x_{18} + 600x_{26} + 500x_{29} + 700x_{37} + 600x_{310} + 600x_{411} + 50x_{59}$
$+ 100x_{510} + 50x_{610} + 100x_{611} + 50x_{711}$

Define the constraints:

Amount (1000's of rolls) of film produced in January:	$x_{15} + x_{18} \leq 500$
Amount (1000's of rolls) of film produced in February:	$x_{26} + x_{29} \leq 500$
Amount (1000's of rolls) of film produced in March:	$x_{37} + x_{310} \leq 500$
Amount (1000's of rolls) of film produced in April:	$x_{411} \leq 500$

Amount (1000's of rolls) of film in/out of January inventory: $x_{15} - x_{59} - x_{510} = 0$
Amount (1000's of rolls) of film in/out of February inventory: $x_{26} - x_{610} - x_{611} = 0$
Amount (1000's of rolls) of film in/out of March inventory: $x_{37} - x_{711} = 0$

230 CHAPTER 10

Amount (1000's of rolls) of film satisfying January demand: x_{18} = 300
Amount (1000's of rolls) of film satisfying February demand $x_{29} + x_{59}$ = 500
Amount (1000's of rolls) of film satisfying March demand: $x_{310} + x_{510} + x_{610}$ = 650
Amount (1000's of rolls) of film satisfying April demand: $x_{411} + x_{611} + x_{711}$ = 400

Non-negativity of variables: $x_{ij} \geq 0$, for all i and j.

The Management Scientist provided the following solution:

OBJECTIVE FUNCTION VALUE = 1045000.000

VARIABLE	VALUE	REDUCED COSTS
X15	150.000	0.000
X18	300.000	0.000
X26	0.000	100.000
X29	500.000	0.000
X37	0.000	250.000
X310	500.000	0.000
X411	400.000	0.000
X59	0.000	0.000
X510	150.000	0.000
X610	0.000	0.000
X611	0.000	150.000
X711	0.000	0.000

c)

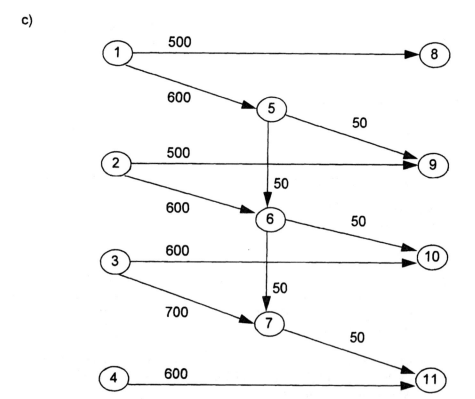

ANSWERED PROBLEMS

PROBLEM 8

The Navy has 9,000 pounds of material in Albany, Georgia which it wishes to ship to three installations: San Diego, Norfolk, and Pensacola. They require 4,000, 2,500, and 2,500 pounds, respectively. Laws require equal allotment of shipping among the 3 carriers. The following gives the shipping costs per pound for truck, railroad, and airplane transit.

	San Diego	Norfolk	Pensacola
Truck	$12	$6	$5
Railroad	$20	$11	$9
Airplane	$30	$26	$28

a) Draw the network representation of this problem.

b) Formulate this problem as a linear program.

c) Solve for the optimal shipping strategy.

PROBLEM 9

There are four marketing research firms (MR1, MR2, MR3, MR4) that Hairways has faith in to advertise its products. Hairways has just come out with a new hairspray and they wish to have 30 newspaper ads, 15 television ads, and 25 radio ads available within three months. Given the size of the firms it is expected that MR1 can produce 15 total ads, MR2 can produce 25 total ads, MR3 can produce 10 total ads, and MR4 can produce 20 total ads. The bids submitted (in thousands of dollars per ad) are:

	MR1	MR2	MR3	MR4
Newspaper	16	10	12	12
Television	26	20	30	21
Radio	22	15	23	14

a) Formulate this problem as a linear program and show that it fits the structure of a transportation problem.

b) Find a solution with six media-firm combinations.

c) Give another solution with five media-firm combinations.

d) Give another solution with seven media-firm combinations.

PROBLEM 10

The city of Francene has 25 contracts up for bids in each of four different departments: Sanitation, Police Services, Parks Department, and Administration.

Three different consulting firms are bidding on the contracts: Ace Consulting, Band Corporation, and QM Associates. Ace has personnel for 40 contracts, Band for 40, and QM for 30. Because contracts are similar within each department, each firm is able to set a fixed bid per contract for each department. The bid price (in $1000s per contract) is summarized below:

	Sanitation	Police	Parks	Administration
Ace	10	15	14	16
Band	15	18	8	10
QM	12	12	12	12

How should the contracts be awarded and how much will the city spend?

PROBLEM 11

Independent Auditors (IA) has committed 100 of its auditors in three locations (35 from its Los Angeles branch, 30 from its Chicago branch, and 35 from its New York branch) to audit firms in three cities: 25 for Denver, 35 for Atlanta, and 40 for New York. Because of possible charges of conflicts of interest, no New York based IA auditor will audit a New York firm. Taking into account all costs and revenues, and the following profit table giving the average profit per auditor (in $1000's), solve for IA's optimal distribution of auditors.

	Denver	Atlanta	New York
Los Angeles	25	18	10
Chicago	30	35	20
New York	14	24	X

PROBLEM 12

Emily Rodd, manager of Camp Pinnacle, must assign her five head counselors to cabins for the summer. The counselors have cabin assignment preferences (based on cabin size, location, condition and other factors). The counselors' cabin preference ratings on a 1-to-9 scale (9 being most favorable) are listed below:

	Cabin 1	2	3	4	5
Abbey	3	9	7	5	5
Babbs	6	8	3	5	7
Carla	2	8	5	2	7
Diane	7	4	2	2	9
Ellsa	6	8	5	2	5

Help Emily make the counselor-cabin assignments that will maximize the sum of the preference ratings achieved.

TRANSPORTATION, ASSIGNMENT, TRANSSHIPMENT

PROBLEM 13

In addition to Pine City's established microcomputer firm, Local Computer, there are three new microcomputer firms that have opened up in the area. In an effort to establish good relations, Manny's Manufacturing plans on buying one computer system from each of the new firms and two from Local Computer. Manny's has five plants, each with different needs, and hence Manny needs to install five different systems. The bids (in $1,000's) from the firms are:

	_____Plant_____				
	P1	P2	P3	P4	P5
Computer Town	10	12	14	18	20
Computer World	11	13	15	14	18
Universal Comp	6	14	24	20	19
Local Computer	14	15	16	22	25

a) Give a network representation for this problem.

b) Solve for the optimal assignment strategy.

c) Suppose Computer World did not have the system needed by P5. How would this have affected the optimal solution?

d) Suppose Computer World did not have the systems needed by P4 or P5. How would this have affected the optimal solution?

PROBLEM 14

A plant manager for a sporting goods manufacturer is in charge of assigning the manufacture of four new aluminum products to four different departments. Because of varying expertise and workloads, the different departments can produce the new products at various rates. If only one product is to be produced by each department and the daily output rates are given in the table below, which department should manufacture which product to maximize total daily product output? (Note: Department 1 does not have the facilities to produce golf clubs.)

Department	Bats	Tennis Rackets	Golf Clubs	Racquetball Rackets
1	100	60	X	80
2	100	80	140	100
3	110	75	150	120
4	85	50	100	75

PROBLEM 15

RVW (Restored Volkswagens) buys 15 used VW's at each of two car auctions each week held at different locations. It then transports the cars to repair shops it contracts with. When they are restored to RVW's specifications, RVW sells 10 each to three different used car lots.

There are various costs associated with the average purchase and transportation prices from each auction to each repair shop. Also there are transportation costs from the repair shops to the used car lots. RVW is concerned with minimizing its total cost given the costs in the table below.

a) Given the costs below, draw a network representation for this problem.

	Repair Shops			Used Car Lots		
	S1	S2		L1	L2	L3
Auction 1	550	500	S1	250	300	500
Auction 2	600	450	S2	350	650	450

b) Formulate and solve this problem as a transshipment linear programming model.

TRUE/FALSE

___ 16. Flow in a transportation network is limited to one direction.

___ 17. If a transportation problem has four origins and five destinations, the LP formulation of the problem will have nine constraints.

___ 18. In a transportation problem, if the total supply is less than the total demand, a dummy source is included in the linear programming formulation.

___ 19. The transportation, assignment, and transshipment problems are all special cases of linear programming models known as network models and can be solved by the simplex method.

___ 20. In the general assignment problem, one agent can be assigned to several tasks.

___ 21. In a transportation problem with total supply equal to total demand, if there are four origins and seven destinations, and there is a unique optimal solution, the optimal solution will utilize 11 shipping routes.

___ 22. The assignment problem can be considered to be a transportation problem, but the opposite is not true.

___ 23. In assignment problems, dummy agents or tasks are created when the number of agents and tasks is not equal.

___ 24. A transshipment problem is a generalization of the transportation problem in which certain nodes are neither supply nodes nor destination nodes.

___ 25. The assignment problem is a special case of the transportation problem in which all supply and demand values equal one.

___ 26. To compensate for a transshipment route with limited capacity, add a constraint to the LP.

___ 27. A transportation problem with 3 sources and 4 destinations will have 7 variables in the objective function.

___ 28. All of the transportation costs associated with a dummy origin are zero, regardless of whether you are maximizing or minimizing.

___ 29. The optimal solution to a transportation problem will never include a shipment from a dummy origin.

___ 30. Transshipment problem allows shipments both in and out of some nodes while transportation problems do not.

CHAPTER 11

Integer Linear Programming

KEY CONCEPTS

CONCEPT	ILLUSTRATED PROBLEMS	ANSWERED PROBLEMS
All-Integer Linear Program	1,2	6,7,8,11,12,13
Mixed-Integer Linear Program	2	11,13
ILP: Maximization	1,3	6,7
ILP: Minimization	2	11
Rounding LP Solution	1	6,10,12
Graphical Solution	1,2	6
0-1 Integer Linear Program	3,4,5	9,10
Special 0-1 Constraints	3	8,9
Spreadsheet Example	1,3	

REVIEW

1. A linear program in which all the variables are restricted to be integers is called an <u>integer linear program (ILP)</u>. If only a subset of the variables are restricted to be integers, the problem is called a <u>mixed integer linear program (MILP)</u>.

2. <u>Binary variables</u> are variables whose values are restricted to be 0 or 1. If all variables are restricted to be 0 or 1, the problem is called a <u>0-1 or binary integer program</u>.

3. Many practical applications involve only binary integer variables and hence many computer codes are written only for this case (such as LINDO/PC). <u>Binary expansion</u> is a mathematical technique which may be used to convert any integer variable into the sum of binary variables.

4. <u>Rounding</u> the values of the variables obtained by solving the linear programming problem may yield an infeasible solution or a solution that might be feasible but not optimal for the ILP or MILP.

5. For problems in which x_i and and x_j represent binary variables designating whether projects i and j have been completed, the following <u>special constraints</u> may be formulated:

    (a) At most <u>k out of n</u> projects will be completed: $\quad \Sigma_j x_j \leq k$

    (b) Project j is <u>conditional</u> on project i: $\quad x_j - x_i \leq 0$

    (c) Project i is a <u>corequisite</u> for project j: $\quad x_j - x_i = 0$

    (d) Projects i and j are <u>mutually exclusive</u>: $\quad x_i + x_j \leq 1$.

6. <u>Sensitivity analysis</u> information for integer linear programming does not have the same connotation as that for linear programming and should be either discarded or used with great caution. In fact, small changes in the coefficients of an integer program can cause large changes in its optimal solution or even cause the problem to be infeasible.

ILLUSTRATED PROBLEMS

PROBLEM 1

Given the following all-integer linear program:

$$\text{MAX} \quad 3x_1 + 2x_2$$
$$\text{s. t.} \quad 3x_1 + x_2 \leq 9$$
$$x_1 + 3x_2 \leq 7$$
$$-x_1 + x_2 \leq 1$$
$$x_1, x_2 \geq 0 \text{ and integer}$$

a) Solve the problem as a linear program ignoring the integer constraints. Show that the optimal solution to the linear program gives fractional values for both x_1 and x_2.

b) What is the solution obtained by rounding fractions greater than of equal to 1/2 to the next larger number? Show that this solution is not a feasible solution.

c) What is the solution obtained by rounding down all fractions? Is it feasible?

d) Enumerate all points in the linear programming feasible region in which both x_1 and x_2 are integers, and show that the feasible solution obtained in (c) is not optimal and that in fact the optimal integer is not obtained by any form of rounding.

e) Solve the problem as an integer linear program using a computer spreadsheet.

SOLUTION 1

a) From the graph on the next page, the optimal solution to the linear program is $x_1 = 2.5$, $x_2 = 1.5$, $z = 10.5$.

b) By rounding the optimal solution of $x_1 = 2.5$, $x_2 = 1.5$ to $x_1 = 3$, $x_2 = 2$, this point lies outside the feasible region.

c) By rounding the optimal solution down to $x_1 = 2$, $x_2 = 1$, we see that this solution indeed is an integer solution within the feasible region, and substituting in the objective function, it gives $z = 8$.

d) There are eight feasible integer solutions in the linear programming feasible region with z values as follows:

INTEGER LINEAR PROGRAMMING

	x_1	x_2	z	
1.	0	0	0	
2.	1	0	3	
3.	2	0	6	
4.	3	0	9	← optimal
5.	0	1	2	
6.	1	1	5	
7.	2	1	8	← part (c) solution
8.	1	2	7	

$x_1 = 3$, $x_2 = 0$ is the optimal solution. Rounding the LP solution ($x_1 = 2.5$, $x_2 = 1.5$) would not have been optimal.

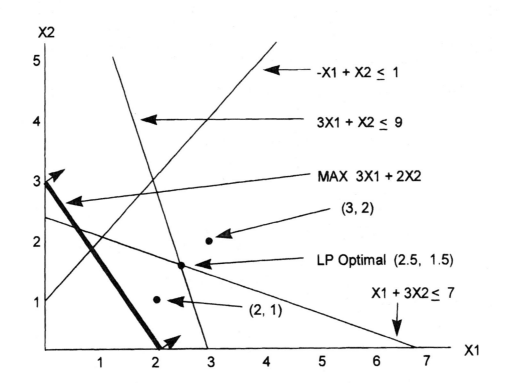

e) Spreadsheet showing data and formulas

	A	B	C	D
1		LHS Coefficients		
2	Constraint	X1	X2	RHS
3	#1	3	1	9
4	#2	1	3	7
5	#3	-1	1	1
6	Obj.Coefficients	3	2	
7				
8			X1	X2
9	Optimal Decision Variable Values			
10	Maximized Objective Function		=B6*C9+C6*D9	
11				
12	Constraints	LHS		RHS
13	#1	=B3*$C$9+C3*$D$9	<=	=D3
14	#2	=B4*$C$9+C4*$D$9	<=	=D4
15	#3	=B5*$C$9+C5*$D$9	<=	=D5

The necessary steps in using Excel to solve an integer LP are as follows:
- Step 1: Select the **Tools** pull-down menu.
- Step 2: Select the **Solver** option.
- Step 3: When the **Solver Parameters** dialog box appears:
    - Enter C10 in the **Set Target Cell** box.
    - Select the **Equal To: Max** option.
    - Enter C9:D9 in the **By Changing Cells** box.
    - Select **Add**.
        - When the **Add Constraint** dialog box appears:
            - Enter B13:B15 in the **Cell Reference** box.
            - Select <=.
            - Enter D13:D15 in the **Constraint** box.
            - Click **OK**.
    - Select **Add**.
        - When the **Add Constraint** dialog box appears:
            - Enter C9:D9 in the **Cell Reference** box.
            - Select **int**.
            - Click **OK**.
- Step 4: When the **Solver Parameters** dialog box appears:
    - Choose **Options**.
- Step 5: When the **Solver Options** dialog box appears:
    - Select **Assume Non-Negative**.
    - Click **OK**.
- Step 6: When the **Solver Parameters** dialog box appears:
    - Choose **Solve**.
- Step 8: When the **Solver Results** dialog box appears:
    - Select **Keep Solver Solution**.
    - Click **OK**.

Spreadsheet showing optimal solution

	A	B	C	D
8			X1	X2
9	Optimal Dec. Var. Values		3.000	2.942E-12
10	Maximized Obj. Function		9.000	
11				
12	Constraints	LHS		RHS
13	#1	9	<=	9
14	#2	3	<=	7
15	#3	-3	<=	1

PROBLEM 2

Given the following problem:

$$\text{MIN} \quad 2x_1 + x_2$$
$$\text{s. t.} \quad x_1 + 3x_2 \geq 5$$
$$8x_1 + 3x_2 \geq 17$$
$$x_1, x_2 \geq 0$$

a) Solve for the optimal solution to the linear program.

b) Suppose only x_2 were restricted to be an integer. What is the optimal solution to this mixed integer linear program?

c) Suppose both x_1 and x_2 were restricted to be integers. Determine the optimal solution to the all-integer linear program.

d) Make a comment about the relative optimal values of the objective functions for the linear program, mixed integer program, and the all-integer program.

SOLUTION 2

a) The optimal solution is $x_1 = 1\ 5/7$, $x_2 = 1\ 2/21$, $z = 4\ 11/21$.

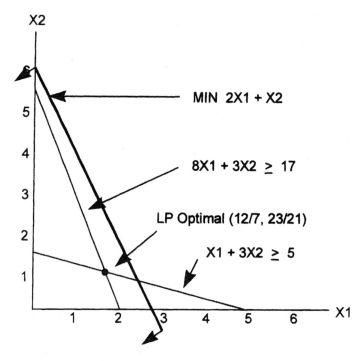

b) Since only x_2 must be integer, the solution must lie on a line of constant integer value of x_2, i.e. $x_2 = 0$, or $x_2 = 1$, etc. The graph below indicates that for each integer value for x_2 the minimum value of z is attained at the minimum value for x_1.

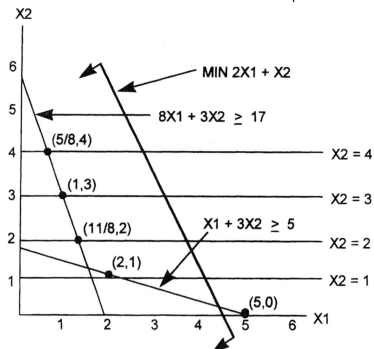

The following table summarizes the corresponding values for x_1 and z for increasing values of x_2.

x_2	Smallest Feasible x_1	z
0	5	10
1	2	5
2	1 3/8	4 3/4
3	1	5
4	5/8	5 1/4

(For higher values of x_2, z continues to increase.)

Hence, the optimal mixed integer solution is $x_1 = 1\ 3/8$, $x_2 = 2$, $z = 4\ 3/4$.

c) The graph below indicates feasible integer points that are closest to the constraint boundaries. It can be observed that the otimal solution to the all-integer problem is attained at $x_1 = 2$, $x_2 = 1$ and at $x_1 = 1$, $x_2 = 3$. Both give $z = 5$.

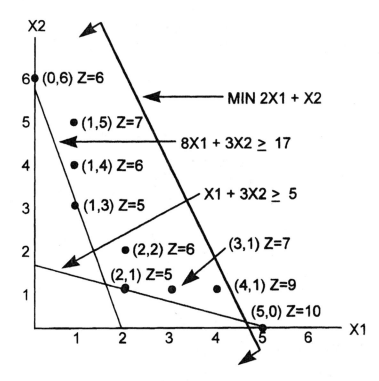

d) The optimal value for the linear program is "better" than for the mixed integer program, which in turn is better than that for the all-integer program.

PROBLEM 3

Metropolitan Microwaves, Inc. is planning to expand its operations into other electronic appliances. The company has identified seven new product lines it can carry. The required initial investment and floor space (sq. ft.) and the expected rate of return on each line.

Product Line	Initial Investment	Floor Space	Expected Rate of Return
1. TV/VCRs	$ 6,000	125	8.1%
2. Color TVs	12,000	150	9.0
3. Projection TVs	20,000	200	11.0
4. VCRs	14,000	40	10.2
5. DVD Players	15,000	40	10.5
6. Video Games	2,000	20	14.1
7. Home Computers	32,000	100	13.2

Metropolitan has decided that they should not stock projection TVs unless they stock either TV/VCRs or color TVs. Also, they will not stock both VCRs and DVD players, and they will stock video games if they stock color TVs. Finally, the company wishes to introduce at least three new product lines. If the company has $45,000 to invest and 420 sq. ft. of floor space available, formulate an integer linear program for Metropolitan to maximize its overall expected return. Use a spreadsheet to solve this binary LP.

SOLUTION 3

Define variables: $x_j = 1$ if product line j is introduced; $= 0$ otherwise.

Define objective function: Maximize total overall expected return:
MAX $.081(6000)x_1 + .09(12000)x_2 + .11(20000)x_3 + .102(14000)x_4 + .105(15000)x_5$
$\qquad + .141(2000)x_6 + .132(32000)x_7$

Define constraints:
1) Money: $6x_1 + 12x_2 + 20x_3 + 14x_4 + 15x_5 + 2x_6 + 32x_7 \leq 45$

2) Space: $125x_1 + 150x_2 + 200x_3 + 40x_4 + 40x_5 + 20x_6 + 100x_7 \leq 420$

3) Stock projection TVs only if stock TV/VCRs or color TVs: $x_1 + x_2 \geq x_3$ or $x_1 + x_2 - x_3 \geq 0$

4) Not stock both VCRs and DVD players: $x_4 + x_5 \leq 1$

5) Stock video games if they stock color TV's: $x_2 - x_6 = 0$.

6) At least 3 new lines: $x_1 + x_2 + x_3 + x_4 + x_5 + x_6 + x_7 \geq 3$

7) Variables are 0 or 1: $x_j = 0$ or 1 for $j = 1,,,7$.

INTEGER LINEAR PROGRAMMING

Spreadsheet showing data and formulas

	A	B	C	D	E	F	G	H	I
1		\multicolumn{7}{c}{LHS Coefficients}								
2	Constraints	X1	X2	X3	X4	X5	X6	X7	RHS
3	#1	6	12	20	14	15	2	32	45
4	#2	125	150	200	40	40	20	100	420
5	#3	1	1	-1					0
6	#4				1	1			1
7	#5		1				-1		0
8	#6	1	1	1	1	1	1	1	3
9	Obj.Func.Coeff.	486	1080	2200	1428	1575	282	4224	
10									
11		X1	X2	X3	X4	X5	X6	X7	
12	Dec.Values								
13	Maximized Total Expected Return				#			
14									
15	Constraints		LHS				RHS	
16	Money		##		<=		=I3	
17	Space		etc.		<=		=I4	
18	TVs		etc.		>=		=I5	
19	VCRs/DVDs		etc.		<=		=I6	
20	Video		etc.		=		=I7	
21	Lines		etc.		>=		=I8	

\# = B9*B12+C9*C12+D9*D12+E9*E12+F9*F12+G9*G12+H9*H12
\## = B3*$B$12+C3*$C$12+D3*$D$12+E3*$E$12+F3*$F$12+G3*$G$12+H3*$H$12

Refer to Problem 1 above for coverage of the steps to follow in using a spreadsheet to solve an LP. The additional/exceptional steps necessary to solve a <u>binary</u> LP using *Excel* are as follows:

    In the **Solver Parameters** dialog box:
        Select **Add**.
        When the **Add Constraint** dialog box appears:
            Enter B12:H12 in the **Cell Reference** box.
            Select **binary**.
            Click **OK**.
        Select **Options**.
        When the **Solver Options** dialog box appears:
            Enter 0 in the **Tolerance** box.
            Click **OK**.

Optimal Solution:
    Metropolitan should stock TV/VCRs, projection TVs, and DVD players for a total expected return of $4,261.

PROBLEM 4

Tom's Tailoring has five idle tailors and four custom garments to make. The estimated time (in hours) it would take each tailor to make each garment is listed below. (An 'X' in the table indicates an unacceptable tailor-garment assignment.)

	Tailor				
Garment	1	2	3	4	5
Wedding gown	19	23	20	21	18
Clown costume	11	14	X	12	10
Admiral's uniform	12	8	11	X	9
Bullfighter's outfit	X	20	20	18	21

Formulate and solve an integer program for determining the tailor-garment assignments that minimize the total estimated time spent making the four garments. No tailor is to be assigned more than one garment and each garment is to be worked on by only one tailor.

Note: This problem is referred to as an assignment problem. An alternative approach to optimally solving an assignment problem is covered in Chapter 10.

SOLUTION 4

This problem can be formulated as a 0-1 integer program. The LP solution to this problem will automatically be integer (0-1).

Define the decision variables:

x_{ij} = 1 if garment i is assigned to tailor j; = 0 otherwise.
Number of decision variables = [(number of garments)(number of tailors)]
- (number of unacceptable assignments) = [4(5)] - 3 = 17.

Define the objective function:

Minimize total time spent making garments:
MIN $19x_{11} + 23x_{12} + 20x_{13} + 21x_{14} + 18x_{15} + 11x_{21} + 14x_{22} + 12x_{24} + 10x_{25}$
$+ 12x_{31} + 8x_{32} + 11x_{33} + 9x_{35} + 20x_{42} + 20x_{43} + 18x_{44} + 21x_{45}$

Define the constraints:

Exactly one tailor per garment:
1) $x_{11} + x_{12} + x_{13} + x_{14} + x_{15} = 1$
2) $x_{21} + x_{22} + x_{24} + x_{25} = 1$
3) $x_{31} + x_{32} + x_{33} + x_{35} = 1$
4) $x_{42} + x_{43} + x_{44} + x_{45} = 1$

No more than one garment per tailor:
5) $x_{11} + x_{21} + x_{31} \leq 1$
6) $x_{12} + x_{22} + x_{32} + x_{42} \leq 1$
7) $x_{13} + x_{33} + x_{43} \leq 1$
8) $x_{14} + x_{24} + x_{44} \leq 1$
9) $x_{15} + x_{25} + x_{35} + x_{45} \leq 1$

Nonnegativity: $x_{ij} \geq 0$ for i = 1,..,4 and j = 1,..,5

Optimal Solution:
  Assign wedding gown to tailor 5
  Assign clown costume to tailor 1
  Assign admiral uniform to tailor 2
  Assign bullfighter outfit to tailor 4

  Total time spent = 55 hours

PROBLEM 5

Market Pulse Research has conducted a study for Lucas Furniture on some designs for a new commercial office desk. Three attributes were found to be most influential in determining which desk is most desirable: number of file drawers, the presence or absence of pullout writing boards, and simulated wood or solid color finish. Listed below are the part-worths for each level of each attribute provided by a sample of 7 potential Lucas customers.

Consumer	File Drawers			Pullout Writing Boards		Finish	
	0	1	2	Present	Absent	Simul. Wood	Solid Color
1	5	26	20	18	11	17	10
2	18	11	5	12	16	15	26
3	4	16	22	7	13	11	19
4	12	8	4	18	9	22	14
5	19	9	3	4	14	30	19
6	6	15	21	8	17	20	11
7	9	6	3	13	5	16	28

Suppose the overall utility (sum of part-worths) of the current favorite commercial office desk is 50 for each customer. What is the product design that will maximize the share of choices for the seven sample participants? Formulate and solve, using Lindo or Excel, this 0 – 1 integer programming problem.

SOLUTION 5

Define the decision variables:

There are 7 I_{ij} decision variables, one for each level of attribute.
I_{ij} = 1 if Lucas chooses level i for attribute j; 0 otherwise.

There are 7 Y_k decision variables, one for each consumer in the sample.
Y_k = 1 if consumer k chooses the Lucas brand, 0 otherwise.

Define the objective function:

Maximize the number of consumers preferring the Lucas brand desk.
MAX $Y_1 + Y_2 + Y_3 + Y_4 + Y_5 + Y_6 + Y_7$

Define the constraints:

There is one constraint for each consumer in the sample.
$5l_{11} + 26l_{21} + 20l_{31} + 18l_{12} + 11l_{22} + 17l_{13} + 10l_{23} - 50Y_1 \geq 1$
$18l_{11} + 11l_{21} + 5l_{31} + 12l_{12} + 16l_{22} + 15l_{13} + 26l_{23} - 50Y_2 \geq 1$
$4l_{11} + 16l_{21} + 22l_{31} + 7l_{12} + 13l_{22} + 11l_{13} + 19l_{23} - 50Y_3 \geq 1$
$12l_{11} + 8l_{21} + 4l_{31} + 18l_{12} + 9l_{22} + 22l_{13} + 14l_{23} - 50Y_4 \geq 1$
$19l_{11} + 9l_{21} + 3l_{31} + 4l_{12} + 14l_{22} + 30l_{13} + 19l_{23} - 50Y_5 \geq 1$
$6l_{11} + 15l_{21} + 21l_{31} + 8l_{12} + 17l_{22} + 20l_{13} + 11l_{23} - 50Y_6 \geq 1$
$9l_{11} + 6l_{21} + 3l_{31} + 13l_{12} + 5l_{22} + 16l_{13} + 28l_{23} - 50Y_7 \geq 1$

There is one constraint for each attribute.
$l_{11} + l_{21} + l_{31} = 1$
$l_{12} + l_{22} = 1$
$l_{13} + l_{23} = 1$

Optimal Solution:

Lucas should choose these product features:
1 file drawer ($l_{21} = 1$)
No pullout writing boards ($l_{22} = 1$)
Simulated wood finish ($l_{13} = 1$)

Three sample participants would choose the Lucas design:
Participant 1 ($Y_1 = 1$)
Participant 5 ($Y_5 = 1$)
Participant 6 ($Y_6 = 1$)

ANSWERED PROBLEMS

PROBLEM 6

Given the following all-integer linear program:

$$\text{MAX} \quad 15x_1 + 2x_2$$
$$\text{s. t.} \quad 7x_1 + x_2 \leq 23$$
$$3x_1 - x_2 \leq 5$$
$$x_1, x_2 \geq 0 \text{ and integer}$$

a) Solve the problem as an LP, ignoring the integer constraints.

b) What solution is obtained by rounding up fractions greater than or equal to 1/2? Is this the optimal integer solution?

c) What solution is obtained by rounding down all fractions? Is this the optimal integer solution? Explain.

d) Show that the optimal z value for the ILP is lower than that for the optimal LP.

e) Why is the optimal z value for the ILP problem always less than or equal to the corresponding LP's optimal z value? When would they be equal? Comment on the MILP's optimal z compared to the corresponding LP & ILP.

PROBLEM 7

Given the following all-integer linear programming problem:

$$\text{MAX} \quad 3x_1 + 10x_2$$
$$\text{s. t.} \quad 2x_1 + x_2 \leq 5$$
$$x_1 + 6x_2 \leq 9$$
$$x_1 - x_2 \geq 2$$
$$x_1, x_2 \geq 0 \text{ and integer}$$

a) Solve the problem graphically as a linear program.

b) Show that there is only one integer point and it is optimal.

c) Suppose the third constraint was changed to $x_1 - x_2 \geq 2.1$. What is the new optimal solution to the LP? ILP?

PROBLEM 8

Given the following all-integer linear programming problem:

$$\text{MAX} \quad 5x_1 + 4x_2$$
$$\text{s. t.} \quad 4x_1 + x_2 \leq 10$$
$$5x_1 + 3x_2 \leq 15$$
$$x_1 + x_2 \leq 4$$
$$x_1, x_2 \geq 0 \text{ and integer}$$

a) Solve the LP relaxation of this problem. Does this provide an upper bound or lower bound for the value of the objective function for the ILP?

b) Solve by inspection for the optimal solution to the ILP.

c) Suppose the objective function was changed to MAX $z = 5x_1 + 3x_2$. What is the new optimal solution to this problem?

PROBLEM 9

Tower Engineering Corporation is considering undertaking several proposed projects for the next fiscal year. The projects, the number of engineers and the number of support personnel required for each project, and the expected profits for each project are summarized in the following table:

	\multicolumn{6}{c}{Project}					
	1	2	3	4	5	6
Engineers Required	20	55	47	38	90	63
Support Personnel Required	15	45	50	40	70	70
Profit ($1,000,000s)	1.0	1.8	2.0	1.5	3.6	2.2

Formulate and solve an integer program that maximizes Tower's profit subject to the following management constraints:

(1) Use no more than 175 engineers
(2) Use no more than 150 support personnel
(3) If either project 6 or project 4 is done, both must be done
(4) Project 2 can be done only if project 1 is done
(5) If project 5 is done, project 3 must not be done and vice versa
(6) No more than three projects are to be done.

PROBLEM 10

Kloos Industries has projected the availability of capital over each of the next three years to be $850,000, $1,000,000, and $1,200,000, respectively. It is considering four options for the disposition of the capital:

(1) Research and development of a promising new product
(2) Plant expansion
(3) Modernization of its current facilities
(4) Investment in a valuable piece of nearby real estate

Monies not invested in these projects in a given year will NOT be available for following year's investment in the projects. The expected benefits three years hence from each of the four projects and the yearly capital outlays of the four options are summarized in the table below in $1,000,000's.

In addition, Kloos has decided to undertake exactly two of the projects, and if plant expansion is selected, it will also modernize its current facilities.

Options	Capital Outlays			Projected Benefits
	Year 1	Year 2	Year 3	
New Product R&D	.35	.55	.75	5.2
Plant Expansion	.50	.50	0	3.6
Modernization	.35	.40	.45	3.2
Real Estate	.50	0	0	2.8

Formulate and solve this problem as a binary programming problem.

PROBLEM 11

Given the following problem:

$$\text{MIN} \quad 5x_1 + 8x_2$$

$$\text{s. t.} \quad x_1 + x_2 \geq 7.2$$

$$x_1 + 4x_2 \geq 20$$

$$5x_1 + 3x_2 \leq 34$$

$$x_1, x_2 \geq 0$$

a) Solve the mixed-integer problem where x_1 must be integer.

b) Solve the problem where both x_1 and x_2 must be integer.

PROBLEM 12

A business manager for a grain distributor is asked to decide how many containers of each of two grains to purchase to fill its 1,600 pound capacity warehouse. The table below summarizes the container size, availability, and expected profit per container upon distribution.

Grain	Container Size	Containers Available	Container Profit
A	500 lbs.	3	$1,200
B	600 lbs.	2	$1,500

a) Formulate as a linear program with the decision variables representing the number of containers purchased of each grain. Solve for the optimal solution.

b) What would be the optimal solution if you were not allowed to purchase fractional containers?

c) There are three possible results from rounding an LP solution to obtain an integer solution:
  (1) the rounded optimal LP solution will be the optimal IP solution;
  (2) the rounded optimal LP solution gives a feasible, but not optimal IP solution;
  (3) the rounded optimal LP solution is an infeasible IP solution.

For this problem (i) round <u>down</u> all fractions; (ii) round <u>up</u> all fractions; (iii) round <u>off</u> (to the nearest integer) all fractions (NOTE: Two of these are equivalent.) Which result above (1, 2, or 3) occurred under each rounding method?

PROBLEM 13

Given the following problem:

$$\text{MIN} \quad 4x_1 + 7x_2$$
$$\text{s. t.} \quad x_1 + 6x_2 \geq 14$$
$$2x_1 + 5x_2 \geq 22$$
$$3x_1 + x_2 \geq 10$$
$$x_1, x_2 \geq 0$$

a) Solve the mixed integer problem where x_1 is required to be integer.

b) Using (a), solve the all-integer linear program.

INTEGER LINEAR PROGRAMMING

TRUE/FALSE

____ 14. The optimal solution to a linear program gave $x_1 = 2.58$ and $x_2 = 1.32$. If x_1 and x_2 were restricted to be integers, then $x_1 = 3$, $x_2 = 1$ will give a feasible solution, but not necessarily an optimal integer solution.

____ 15. The objective function coefficients are all integers for an ILP. The optimal solution to the LP is $x_1 = 2.58$, $x_2 = 1.32$, and $z = 14.28$. The solution was rounded to $x_1 = 3$, $x_2 = 1$ and $z = 14$, and this is a feasible solution. Then this must be the optimal solution to the ILP.

____ 16. Project 5 must be completed before project 6 is started. The constraint would be: $x_5 - x_6 \leq 0$.

____ 17. If the LP relaxation of an integer program has a feasible solution, them the integer program has a feasible solution.

____ 18. A mixed-integer linear program involves both discrete and continuous variables.

____ 19. The optimal solution to an integer linear program still must occur at an extreme point of the feasible region formed by the functional constraints.

____ 20. If at most three of five projects are to be completed, the constraint would be: $x_1 + x_2 + x_3 + x_4 + x_5 \leq 3$.

____ 21. Multiple-choice constraints involve binary variables.

____ 22. The classic assignment problem can be modeled as a 0-1 integer program.

____ 23. An optimal integer solution of $x_1 = 3$, $x_2 = 5$, $z = 150$ has been found for an ILP. For the linear programming formulation, the range of optimality for c_1 was between 15 and 30 with a current $c_1 = 20$. Then if c_1 is increased to 21 in the ILP, the new optimal value of the objective function must be 153.

____ 24. For a minimization ILP, a lower bound is found when any integer solution is determined.

____ 25. For some types of ILP problems their LP relaxation solutions are optimal.

____ 26. The product design and market share optimization problem is sometimes called the share of choices problem.

____ 27. Generally, the optimal solution to an integer linear program is less sensitive to the constraint coefficients than is a linear program.

____ 28. The use of integer variables involves a tradeoff between modeling flexibility and solution difficulty.

CHAPTER 12

Project Scheduling: PERT/CPM

KEY CONCEPTS

CONCEPT	ILLUSTRATED PROBLEMS	ANSWERED PROBLEMS
Construction of PERT Networks	1,2	6,8,12,13
PERT Analysis with Certain Times	1	6,7,9
PERT Analysis with Uncertain Times	2,4	8,10,11
Activity Earliest/Latest Times	1,2,5	7,9,14
Time/Cost Analysis	3,4	9-13
LP Formulation for Crashing	3	12,13

REVIEW

1. <u>PERT</u> (Program Evaluation Review Technique) is used to plan the scheduling of individual activities that make up a project.

2. A <u>PERT network</u> can be constructed to model the <u>precedence</u> of the activities. The <u>nodes</u> of the network represent the activities.

3. PERT can be used to determine the earliest/latest start and finish times for each activity, the entire project completion time and the <u>slack time</u> for each activity. (See algorithm at the end of the Review.)

4. A <u>critical path</u> for the network is a path consisting of activities with <u>zero slack</u>.

5. In the <u>three-time estimate approach</u>, the time to complete an activity is assumed to follow a <u>Beta distribution</u>. Its mean is $t = (a + 4m + b)/6$, and its variance is $\sigma^2 = ((b-a)/6)^2$. Here a = the <u>optimistic</u> completion time estimate, b = the <u>pessimistic</u> completion time estimate, and m = the <u>most likely</u> completion time estimate.

6. In the three-time estimate approach, the <u>critical path</u> is determined as if the mean times for the activities were fixed times. The overall project completion time is assumed to have a normal distribution with mean equal to the sum of the means along the critical path and variance equal to the sum of the variances along the critical path.

7. In the <u>CPM</u> (Critical Path Method) approach to project scheduling, it is assumed that the normal time to complete an activity, t_j, which can be met at a normal cost, c_j, can be <u>crashed</u> to a reduced time, t'_j, under maximum crashing for an increased cost, c'_j.

8. Using CPM, activity j's maximum time reduction, M_j, may be calculated by: $M_j = t_j - t'_j$. It is assumed that its cost per unit reduction, K_j, is linear and can be calculated by:
$K_j = (c'_j - c_j)/M_j$.

9. Linear programming may be used to solve a CPM problem to minimize the crashing costs needed to complete a project within a specified time limit.

PERT/CPM CRITICAL PATH PROCEDURE

Step 1: Develop a list of the activities that make up the project.

Step 2: Determine the immediate predecessors for each activity in the project.

Step 3: Estimate the completion time for each activity.

Step 4: Draw a project network depicting the activities (on nodes) and immediate predecessors listed in Steps 1 and 2.

Step 5: Use the project network and the activity time estimates to determine the earliest start and the earliest finish time for each activity by making a forward pass through the network. The earliest finish time for the last activity in the project identifies the total time required to complete the project.

Step 6: Use the project completion time identified in Step 5 as the latest finish time for the last activity and make a backward pass through the network to identify the latest start and finish time for each activity.

Step 7: Use the difference between the latest start time and the earliest start time for each activity to determine the slack for the activity.

Step 8: Find the activities with zero slack; these are the critical activities.

Step 9: Use the information from Steps 5 and 6 to develop the activity schedule for the project.

FLOW CHART FOR COMPUTING PROJECT COMPLETION TIME PROBABILITY

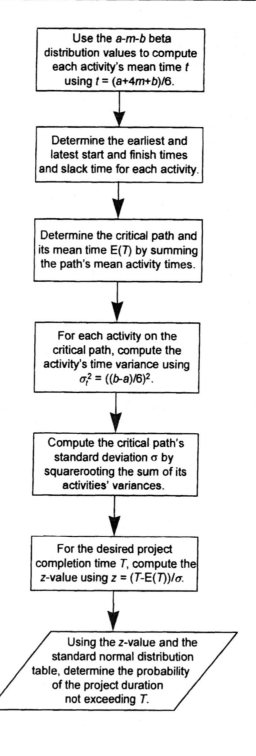

FLOW CHART FOR PROJECT CRASHING

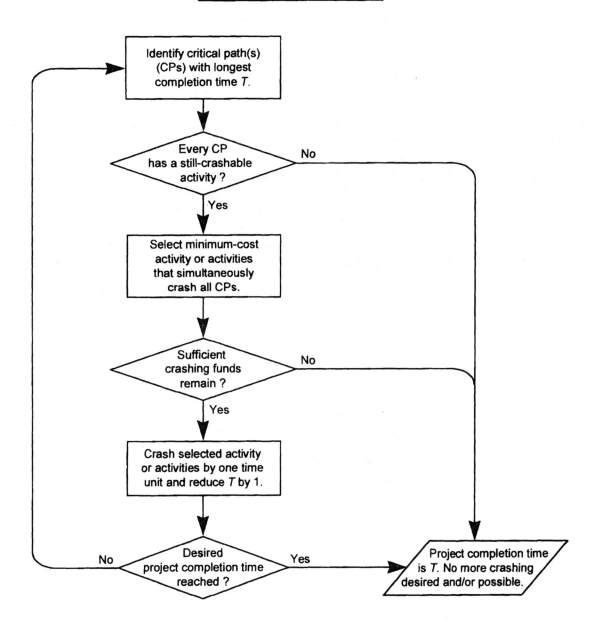

ILLUSTRATED PROBLEMS

PROBLEM 1

Kraft's Kustom Kars is in the business of producing custom automobile assemblies. In particular, Kraft operates a shop that builds and assembles the body and frame of the cars.

Kraft's operations begin with the processing of initial paperwork. This must be done before any other operations are commenced. Once the paperwork has been completed, the body of the car can be built in Room A and the frame of the car can be built in Room B. When the body is built, it is transferred to Room C for finishing work. Similarly, when the frame is built, it is transferred to Room D for finishing work.

When both are built, although not necessarily finished, the final paperwork can be completed. When both the body and frame are finished, they are transported to the assembly Room E, where the body is mounted to the frame.

In the frame building room, Room B, certain chemicals are used which must be completely eliminated by a thorough washdown to prevent gaseous fumes from becoming a health hazard. The project is considered complete when the final paperwork has been completed, the body has been mounted to the frame, and Room B has been completely washed down.

The table below gives the expected completion times in hours for each activity of the project.

Activity	Description	Completion Time
A	Initial Paperwork	3
B	Build Body	3
C	Build Frame	2
D	Finish Body	3
E	Finish Frame	7
F	Final Paperwork	3
G	Mount Body to Frame	6
H	Room B Washdown	2

a) Draw the PERT network that corresponds to this problem.

b) Find the earliest and latest start and finish times for each activity of the project. How long should the project take?

c) Which activities must not be delayed if the project is to be completed in the time calculated in part (b)?

d) Suppose the body finish operation (D) were delayed four hours. By how much would the entire project be delayed?

SOLUTION 1

Before constructing the PERT network, summarize in the following table the immediate predecessor activities for each activity.

Activity	Immediate Predecessors	Completion Times (Hr.)
A	--	3
B	A	3
C	A	2
D	B	3
E	C	7
F	B,C	3
G	D,E	6
H	C	2

To construct a PERT network, there must be a node for each activity. An activity's immediate predecessors are the nodes that immediately precede it.

The result is:

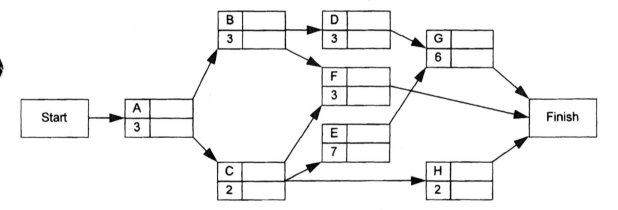

b) The earliest start time (ES) for activity A 0. Then the earliest finish time (EF) for an activity is given by: EF = ES + (completion time). Then ES for an activity = max (EF for all immediately preceding activities).

The forward pass to calculate ES and EF:

Activity	Earliest Start (ES)	Earliest Finish (EF)
A	0	0 + 3 = 3
B	3	3 + 3 = 6
C	3	3 + 2 = 5
D	6	6 + 3 = 9
E	5	5 + 7 = 12
H	5	5 + 2 = 7
F	MAX(5,6) = 6	6 + 3 = 9
G	MAX(9,12) = 12	12 + 6 = 18

The max(EF of the activities immediately preceding the Finish node) = 18, so the completion time of the project is 18.

The backward pass to calculate LF and LS:

Activity	Latest Finish (LF)	Latest Start (LS)
H	18	18 - 2 = 16
G	18	18 - 6 = 12
F	18	18 - 3 = 15
E	12	12 - 7 = 5
D	12	12 - 3 = 9
C	MIN(15,5,16) = 5	5 - 2 = 3
B	MIN(9,15) = 9	9 - 3 = 6
A	MIN(6,3) = 3	3 - 3 = 0

This gives the completed network below.

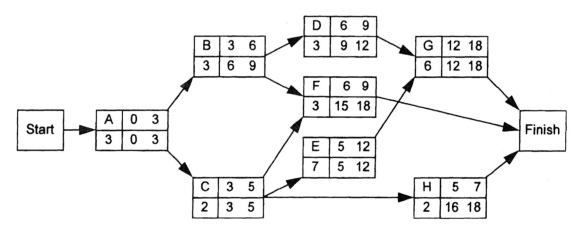

The slack time for each activity (LS - ES) is summarized below:

Activity	ES	EF	LS	LF	Slack
A	0	3	0	3	0
B	3	6	6	9	3
C	3	5	3	5	0
D	6	9	9	12	3
E	5	12	5	12	0
F	6	9	15	18	9
G	12	18	12	18	0
H	5	7	16	18	11

NOTE: One way to partially check the accuracy of your activity earliest/latest time calculations is to compute every activity's slack two ways: Slack = (LS - ES) = (LF - EF) If the preceding is not true, you have an error! Also, the number of different values in your slack column should not exceed (but does not have to equal) the number of paths in the project network.

c) Activities A, C, E, G have 0 slack times -- they form the critical path.

d) Activity D has a slack of 3. Hence a 4 hour delay in D would delay the entire project 4 - 3 = 1 hour.

PROBLEM 2

Consider the following project:

Activity	Immediate Predecessors	Optimistic Time (Hr.)	Most Probable Time (Hr.)	Pessimistic Time (Hr.)
A	--	4	6	8
B	--	1	4.5	5
C	A	3	3	3
D	A	4	5	6
E	A	0.5	1	1.5
F	B,C	3	4	5
G	B,C	1	1.5	5
H	E,F	5	6	7
I	E,F	2	5	8
J	D,H	2.5	2.75	4.5
K	G,I	3	5	7

a) Construct the PERT network for this problem.

b) Solve for the expected earliest and latest start and finish times for each activity.

c) Identify the critical path and give the estimated project completion time.

d) What is the probability the project will be completed within one day (24 hours)?

SOLUTION 2

a) Nodes are needed for the start node, the finish node, and all of the activities.

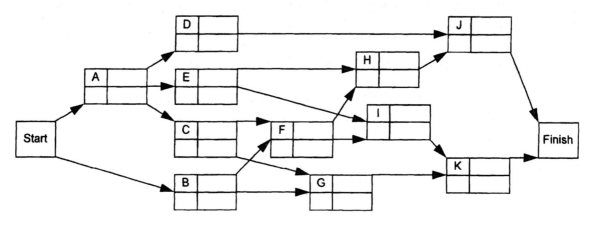

b) Calculate the expected times, t, and variances, σ^2, for each activity.

$$t = (a + 4m + b)/6 \qquad \sigma^2 = ((b-a)/6)^2$$

Activity	Expected Time	Variance
A	6	4/9
B	4	4/9
C	3	0
D	5	1/9
E	1	1/36
F	4	1/9
G	2	4/9
H	6	1/9
I	5	1
J	3	1/9
K	5	4/9

Thus, using the algorithm illustrated in problem 1, we have:

Activity	ES	EF	LS	LF	Slack
A	0	6	0	6	0
B	0	4	5	9	5
C	6	9	6	9	0
D	6	11	15	20	9
E	6	7	12	13	6
F	9	13	9	13	0
G	9	11	16	18	7
H	13	19	14	20	1
I	13	18	13	18	0
J	19	22	20	23	1
K	18	23	18	23	0

PROJECT SCHEDULING   267

c) The critical path is the path of 0 slack = A-C-F-I-K. The estimated project completion time is the Max EF of the activities immediately preceding the Finish node = 23.

d) $z = (24 - 23)/\sigma$. Here, $\sigma^2 = \sigma^2_A + \sigma^2_C + \sigma^2_F + \sigma^2_I + \sigma^2_K$

$$= 4/9 + 0 + 1/9 + 1 + 4/9 = 2.$$

Hence, $\sigma = 1.414$. Thus $z = (24-23)/1.414 = .71$. From Appendix C,

$$P(z < .71) = .5 + .2612 = .7612.$$

PROBLEM 3

National Business Machines (NBM) has just developed a new microcomputer it plans to put into full scale production in a few months. The table and the PERT network below show the precedence relations and give the activity times and costs under normal operations and maximum crashing for a daily (22-hour) production of 1000 microcomputers at its local plant.

NBM wants to know the minimum cost of producing the 1000 microcomputers within the 22-hour period. Formulate and solve a linear program that will yield this information.

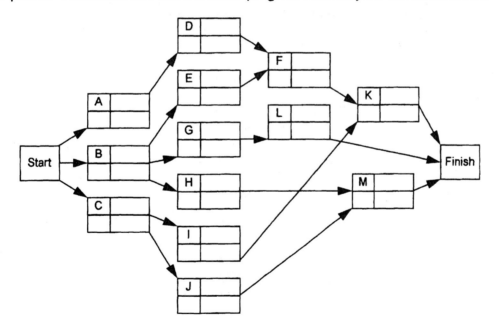

Activity	Normal Time (hr)	Normal Cost	Crash Time (hr)	Crash Cost
A	2	$2,000	1.5	$3,000
B	4	3,000	3	3,500
C	1	1,500	1	1,500
D	4	5,300	2.5	8,000
E	6	5,400	5	7,000
F	10	6,000	8	9,000
G	8	4,800	5	9,900
H	2	2,800	1	2,900
I	5	4,500	4	5,000
J	12	6,000	6	9,600
K	7	7,000	4	9,700
L	11	8,800	9	9,200
M	4	1,000	1	7,000

SOLUTION 3

First prepare a chart giving maximum crashing,

$$M = \text{(Normal Time)} - \text{(Time Under Maximum Crashing)}$$

and the marginal cost per hour for crashing,

$$K = ([\text{Cost Under Maximum Crashing}] - [\text{Normal Cost}])/M$$

Activity	M	K
A	0.5	2000
B	1	500
C	0	0
D	1.5	1800
E	1	1600
F	2	1500
G	3	1700
H	1	100
I	1	500
J	6	600
K	3	900
L	2	200
M	3	2000

Let x_A = earliest finish time for Activity A, and y_A = amount of time Activity A is crashed.

The linear program must minimize the total extra cost:

MIN $2000y_A + 500y_B + 1800y_D + 1600y_E + 1500y_F + 1700y_G$
$+ 100y_H + 500y_I + 600y_J + 900y_K + 200y_L + 2000y_M$

Subject to:

(1) The project must be completed within 22 hours: $x_K \leq 22$, $x_L \leq 22$, $x_M \leq 22$

(2) The amount an activity is crashed cannot exceed its maximum crashing:

$$y_A \leq 0.5$$
$$y_B \leq 1$$
$$y_D \leq 1.5$$
$$y_E \leq 1$$
$$y_F \leq 2$$
$$y_G \leq 3$$
$$y_H \leq 1$$
$$y_I \leq 1$$
$$y_J \leq 6$$
$$y_K \leq 3$$
$$y_L \leq 2$$
$$y_M \leq 3$$

(3) For each activity:
(Earliest Finish Time) $\geq$ (Earliest Finish Time of Preceding Activity)
+ [(Normal Activity Time) - (Amount of Time the Activity is crashed)]

$x_A \geq 0 + (2 - y_A)$	or	$x_A + y_A \geq 2$
$x_B \geq 0 + (4 - y_B)$	or	$x_B + y_B \geq 4$
$x_C \geq 0 + (1 - 0)$	or	$x_C \geq 1$
$x_D \geq x_A + (4 - y_D)$	or	$x_D + y_D - x_A \geq 4$
$x_E \geq x_B + (6 - y_E)$	or	$x_E + y_E - x_B \geq 6$
$x_G \geq x_B + (8 - y_G)$	or	$x_G + y_G - x_B \geq 8$
$x_H \geq x_B + (2 - y_H)$	or	$x_H + y_H - x_B \geq 2$
$x_I \geq x_C + (5 - y_I)$	or	$x_I + y_I - x_C \geq 5$
$x_J \geq x_C + (12 - y_J)$	or	$x_J + y_J - x_C \geq 12$
$x_F \geq x_D + (10 - y_F)$	or	$x_F + y_F - x_D \geq 10$
$x_F \geq x_E + (10 - y_F)$	or	$x_F + y_F - x_E \geq 10$
$x_L \geq x_G + (11 - y_L)$	or	$x_L + y_L - x_G \geq 11$
$x_K \geq x_F + (7 - y_K)$	or	$x_K + y_K - x_F \geq 7$
$x_K \geq x_I + (7 - y_K)$	or	$x_K + y_K - x_I \geq 7$
$x_M \geq x_H + (4 - y_M)$	or	$x_M + y_M - x_H \geq 4$
$x_M \geq x_J + (4 - y_M)$	or	$x_M + y_M - x_J \geq 4$

(4) Non-negativity of the variables:

$$x_i \geq 0 \text{ for } i = A, B, C, ..., M$$
$$y_j \geq 0 \text{ for } j = A, B, D, ..., M$$

The *Management Scientist* provided the following solution output:

OBJECTIVE FUNCTION VALUE =   4700.000

VARIABLE	VALUE	REDUCED COSTS
XA	5.000	0.000
XB	3.000	0.000
XC	1.000	0.000
XD	9.000	0.000
XE	9.000	0.000
XF	18.000	0.000
XG	11.000	0.000
XH	13.000	0.000
XI	18.000	0.000
XJ	13.000	0.000
XK	22.000	0.000
XL	22.000	0.000
XM	17.000	0.000
YA	0.000	2000.000
YB	1.000	0.000
YD	0.000	1800.000
YE	0.000	100.000
YF	1.000	0.000
YG	0.000	1500.000
YH	0.000	100.000
YI	0.000	500.000
YJ	0.000	600.000
YK	3.000	0.000
YL	0.000	0.000
YM	0.000	2000.000

Summary:

Crash activity B one hour, activity F one hour, and activity K three hours. The 1000-unit daily production output will be achieved in 22 hours at a crashing cost of $4700.

PROBLEM 4

Given the following PERT network for Gus's Painters:

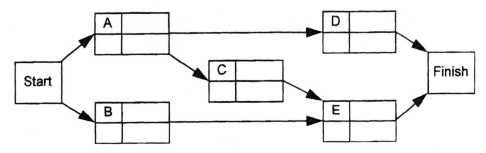

The following means and standard deviations were calculated for the activities:

Activity	t	σ
A	6	2
B	3	1
C	6	1
D	15	2
E	12	2

There is a $100,000 bonus for completing the project in 26 weeks. Currently activity E is assigned to Wilson Brothers. Gus has the option of hiring Jones Inc. for activity E. Their expected completion time for activity E is 8 weeks (with a standard deviation of 2), but they will cost Gus $15,000 more to do E than Wilson Brothers. Should Gus hire Jones?

SOLUTION 4

Analysis using Wilson Brothers to do E:

Activity	ES	EF	LS	LF
A	0	6	0	6
B	0	3	9	12
C	6	12	6	12
D	6	21	9	21
E	12	24	12	24

Hence the critical path is A - C - E and the overall expected project completion time is 24.

The variance of the critical path is:

$$\sigma^2 = \sigma^2_A + \sigma^2_C + \sigma^2_E = (2)^2 + (1)^2 + (2)^2 = 9. \text{ Thus } \sigma = 3.$$

To find the probability of completing the project in 26 weeks, calculate,

$$z = (26 - 24)/3 = .67.$$

Thus the probability of finishing in 26 weeks is $P(z < .67)$. From Appendix C, the table gives the $P(0 < z < .67) = .2486$. Therefore the probability of completing the project within 26 weeks is $.5 + .2486 = .7486$.

Thus the expected bonus using Wilson Brothers to do E is:

$$(.486)(100,000) + (.2514)(0) = \$74,860.$$

Analysis using Jones, Inc. to do E:

Activity	ES	EF	LS	LF
A	0	6	0	6
B	0	3	10	13
C	6	12	7	13
D	6	21	6	21
E	12	20	13	21

Hence the critical path is A - D and the overall expected project completion time is 21. The variance of the critical path is:

$$\sigma^2 = \sigma^2_A + \sigma^2_D = (2)^2 + (2)^2 = 8. \text{ Thus } \sigma = 2.828.$$

To find the probability of completing the project in 26 weeks, calculate,

$$z = (26 - 21)/2.828 = 1.77.$$

Thus the probability of finishing in 26 weeks is $P(z < 1.77)$. From Appendix C, the table gives the $P(0 < z < 1.77) = .4616$. Therefore the probability of completing the project within 26 weeks is $.5 + .4616 = .9616$.

Thus the expected bonus using Jones, Inc. to do E is:

$$(.9616)(100,000) + (.0384)(0) = \$96,160.$$

Decision:

The difference in expected returns between the two firms is: $96,160 - $74,860 = $21,300. Since this is greater than the $15,000 cost to hire Jones, Inc., Gus should hire Jones, Inc.

PROBLEM 5

For the project represented below, determine the earliest and latest start and finish times for each activity as well as the expected overall completion time.

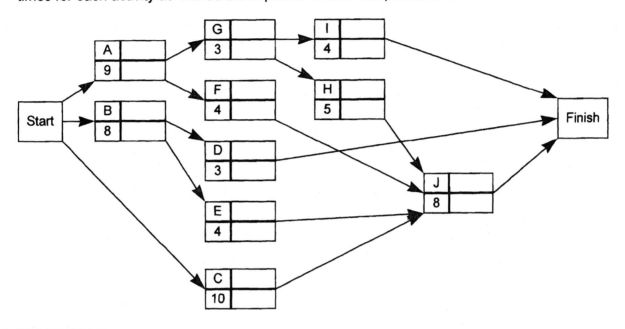

SOLUTION 5

a) Solving the PERT network for the ES, EF, LS, LF, and slack times by the method illustrated in problem 1, we can summarize:

Activity	ES	EF	LS	LF	Slack
A	0	9	0	9	0
B	0	8	5	13	5
C	0	10	7	17	7
D	8	11	22	25	14
E	8	12	13	17	5
F	9	13	13	17	4
G	9	12	9	12	0
H	12	17	12	17	0
I	12	16	21	25	9
J	17	25	17	25	0

The overall project completion time is 25 weeks.

274 CHAPTER 12

ANSWERED PROBLEMS

PROBLEM 6

Consider a project that has been modeled as follows:

Activity	Immediate Predecessors	Completion Time (hr.)
A	---	7
B	---	10
C	A	4
D	A	30
E	A	7
F	B,C	12
G	B,C	15
H	E,F	11
I	E,F	25
J	E,F	6
K	D,H	21
L	G,J	25

a) Draw the PERT network for this project and determine project's expected completion time and its critical path.

> NOTE: If you are only interested in identifying the critical path and expected project completion time, there might be an easier approach than determining every activity's earliest start and finish times. List (if there are not too many) every path in the network and, for each one, sum the expected times of the activities on that path. You are looking for the path with the largest sum.

b) Can activities E and G be performed simultaneously without delaying the minimum project completion time?

c) Can one person perform A, G, and I without delaying the project?

d) By how much can activities G and L be delayed without delaying the entire project?

e) How much would the project be delayed if activity G was delayed by 7 hours and activity L was delayed by 4 hours? Explain.

PROBLEM 7

Given the following PERT network of tasks with completion times in hours for scheduling interns in a hospital:

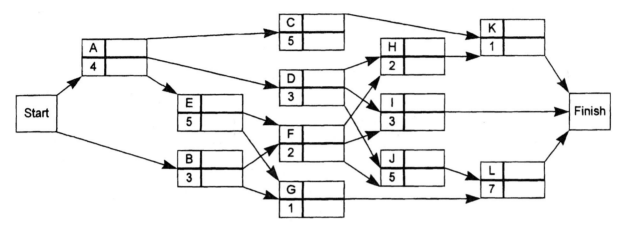

a) If interns start at midnight (00:00), determine each activity's ES, EF, LS, LF, and slack.

b) What is the critical path and project completion time?

c) If an intern can do any job and works a 24-hour shift, show that two interns can complete all the jobs. (Hint: One intern is assigned the critical path jobs.)

PROBLEM 8

A project consists of five activities. Naturally the paint mixing precedes the painting activities. Also, both ceiling painting and floor sanding must be done prior to floor buffing.

Activity	Optimistic Time (hr.)	Most Probable Time (hr.)	Pessimistic Time (hr.)
Floor sanding	3	4	5
Floor buffing	1	2	3
Paint mixing	0.5	1	1.5
Wall painting	1	2	9
Ceiling painting	1	5.5	7

a) Construct a PERT network for this problem.

b) What is the expected completion time of this project?

c) What is the probability that the project can be completed within 9 hr.?

PROBLEM 9

Given the following PERT network modeling new home construction by Bonanza Development:

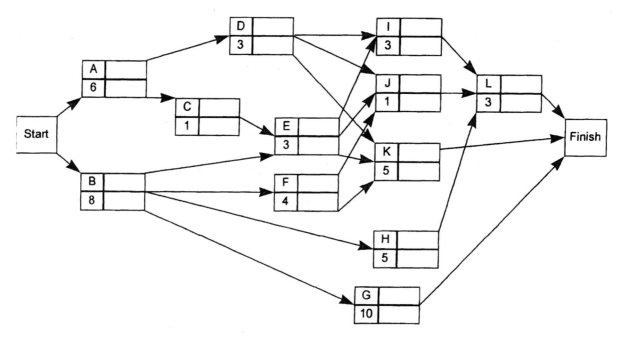

a) Prepare a table of the earliest and latest start and finish times and slack times for each activity in Bonanza's project.

b) What is the critical path and the expected project completion time?

c) Reliable Plumbers, the subcontractor performing activity J, is going to be delayed 5 weeks. If the project is delayed, it will cost Bonanza $2000 per week of delay of the entire project. Reliable is charging $3000 for the plumbing.

Bonanza has three options:

(1) Keep Reliable Plumbers.
(2) Cancel the contract with Reliable Plumbers and hire Local Plumbers, Inc. Local Plumbers, however, will take two weeks to do activity J, and will charge $7000.
(3) Bonanza can train its own employees who are currently performing activity E to do the work. This involves a two-week training period as soon as activity E is done. Then it is expected it will take them three weeks to perform activity J. The cost of the training is $800 per week and the cost of them doing activity J is $1,000 per week.

Which alternative do you recommend to Bonanza? Explain.

PROBLEM 10

Consider the following PERT network with estimated times in weeks. The project is scheduled to begin on May 1.

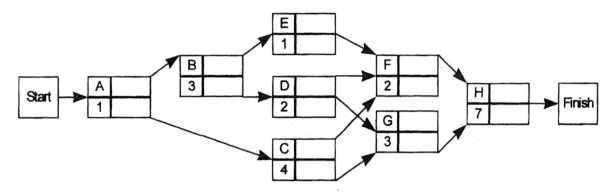

The three-time estimate approach was used to calculate the expected times and the following table gives the variance for each activity:

Activity	Variance	Activity	Variance
A	1.1	E	.3
B	.5	F	.6
C	1.2	G	.6
D	.8	H	1.0

a) Give the expected project completion date and the critical path.

b) By what date are you 99% sure the project will be completed?

c) The project has a target completion date of August 28 (17 weeks). If the project is completed by August 28, the profit on the project will be $10,000. If work is not completed by August 28, a $5,000 penalty will be incurred, reducing the project's profit to $5,000. For $2,000 more than is currently being spent for a firm to do activity H, a more experienced firm can complete the activity in just 5 weeks. Should this offer be accepted? (Assume the variance for H will not change.)

PROBLEM 11

Consider the following PERT network.

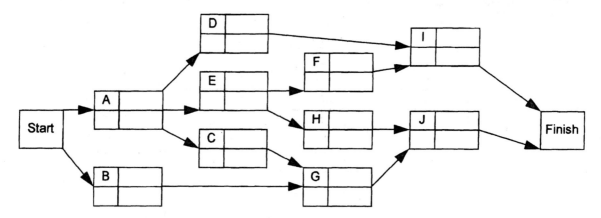

The following chart has been prepared giving the optimistic time (a), the most likely time (m), and the pessimistic time (b), in weeks for each activity.

Activity	a	m	b	Activity	a	m	b
A	2	8	14	F	2	3	10
B	3	12	21	G	3	9	15
C	2	5	8	H	7	8	9
D	4	5	12	I	3	11	13
E	1	3	17	J	7	10	13

a) Determine the critical path, expected project completion time, and standard deviation of the project completion time.

b) Management insists that the project be completed in 36 weeks and will be charged a $100,000 fine for any time overrun. If, at a cost of $3,000, the expected completion time of activity A could be reduced by 2 weeks, should the extra money be spent? (Assume A's variance does not change.)

c) Why is considering only critical path activities for project completion time not always a good assumption in probabilistic cases? (HINT: Consider activity E.)

PROBLEM 12

Plane, Inc. is a manufacturer of heavy equipment and is considering introducing a new line of small steamrollers. Development is to proceed as follows.

A feasibility study will first be performed. Upon receiving a successful feasibility report, a manufacturing building is to be secured and a project leader hired. Once the building is secured, Plane will be committed to the project. Therefore an advertising group will be selected and the raw materials for the manufacturing process will be purchased. When, in addition to securing the building, a project leader has been named, a manufacturing staff will be recruited.

After the manufacturing staff has been selected and the raw materials purchased, a prototype model of the steamroller will be produced. Following the completion of the prototype, work will begin on a production run of 100 steamrollers. When both the prototype model has been built and the advertising staff selected, an intensive advertising campaign will be launched.

The development phase of this project will be complete with the production of the first 100 steamrollers and the initiation of the advertising campaign.

Two separate cost analyses have been prepared. One is effective under current (normal) conditions, while the other is effective if the development phase of the activities is "crashed". These are summarized below. (Times are in weeks.)

	Normal		Crash	
Activity	Time	Cost	Time	Cost
Feasibility Study (A)	6	$ 80,000	5	$100,000
Building Purchased (B)	4	100,000	4	100,000
Project Leader Hired (C)	3	50,000	2	100,000
Advertising Staff Selected (D)	6	150,000	3	300,000
Materials Purchased (E)	3	180,000	2	250,000
Manufacturing Staff Hired (F)	10	300,000	7	480,000
Prototype Manufactured (G)	2	100,000	2	100,000
Production Run of 100 (H)	6	450,000	5	800,000
Advertising Campaign (I)	8	350,000	4	650,000

a) Draw the PERT network for this problem.

b) Formulate and solve a linear program for determining the minimum cost of completing this project in half a year (=26 weeks).

c) What assumptions are made in calculating the "marginal" costs for the activities?

d) Interpret the meaning of the shadow price that would be associated with the constraint that set the maximum completion time to 26 weeks.

PROBLEM 13

Joseph King has ambitions to be mayor of Williston, North Dakota. Joe has determined the breakdown of the steps to the nomination and has estimated normal and crash costs and times for the campaign as follows (times are in weeks):

Activity	Normal Time	Normal Cost	Crash Time	Crash Cost	Immediate Predecessors
A. Solicit Volunteers	6	$5,000	4	$9,000	----
B. Initial "Free" Exposure	3	$4,000	3	$4,000	----
C. Raise Money	9	$4,000	6	$10,000	A
D. Organize and Co-ordinate Schedule	4	$1,000	2	$2,000	A
E. Hire Advertising Firm	2	$1,500	1	$2,000	B
F. Arrange Major TV Interview	3	$4,000	1	$8,000	B
G. Advertising Campaign	5	$7,000	4	$12,000	C,E
H. Personal Campaigning	7	$8,000	5	$20,000	D,F

a) Joe King is not a wealthy man and would like to organize a four month (16 week) campaign at minimum cost. Write and solve a linear program to accomplish this task.

b) Dan Wetzel is an independent who is also trying to make a bid to become mayor of Williston. He has promised a clean campaign, one that he will initially finance on his own. He has $50,000 to invest in his campaign. Being a student of recent successful political campaigns, he knows that his best chance to win is be a "fresh new face" at nomination time. Hence he wishes to keep the entire campaign from beginning to end at a minimum. Write a linear program that, when solved, will minimize the total time of the campaign while keeping expenditures to a maximum of $50,000.

PROBLEM 14

For the project represented below, determine the earliest and latest start and finish times for each activity as well as the expected overall completion time.

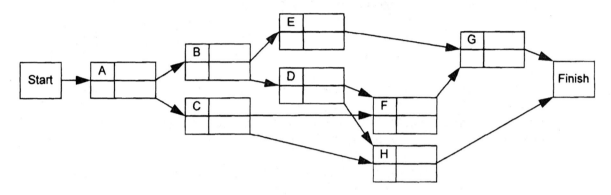

Activity	Time	ES	EF	LS	LF	Slack
A	4					
B	3					
C	4					
D	2					
E	5					
F	2					
G	5					
H	6					

TRUE/FALSE

___ 15. All activities on a critical path have zero slack time.

___ 16. In PERT, it is assumed that the amount of time to complete any one activity is independent of the amount of time to complete any other activity in the project.

___ 17. In PERT, an activity's most probable time is the same as its expected time.

___ 18. In PERT, it is assumed that the underlying distribution for each activity in the three-time estimate approach is a normal distribution.

___ 19. In a given PERT problem, activities F and G are on the critical path. If each is delayed two weeks, then in all cases, the project will be delayed four weeks.

___ 20. The difference between an activity's earliest finish and latest finish equals the difference between its earliest start and latest start.

___ 21. In a given PERT problem, activities F and G are not on the critical path and each has two weeks slack time. If both are delayed by two weeks each, then in all cases, the project will not be delayed.

___ 22. An activity can be started as soon as any one of the activities that immediately precedes it is finished.

___ 23. In a given PERT problem, activity F is on the critical path. If its time is reduced by five weeks, then in all cases, the overall project completion time will be reduced by five weeks.

___ 24. In PERT, the critical path is the path of longest distance through the network.

___ 25. In CPM, the marginal cost per week's saving of an activity's completion time is valid only between its normal time and the time after maximum crashing.

___ 26. It is possible to have more than one critical path at a time.

___ 27. In CPM, if activity B's normal completion time is 8 weeks and normal cost is $10,000, and its completion time after maximum crashing is 5 weeks at a cost of $15,000, then the assumption is that if $13,000 is spent on activity B, its completion time is 6.8 weeks.

___ 28. The project manager should monitor the progress of any activity with a large variance even if the expected times do not identify the activity as a critical activity.

___ 29. A critical activity can be part of a noncritical path.

CHAPTER 13

Inventory Models

KEY CONCEPTS

CONCEPT	ILLUSTRATED PROBLEMS	ANSWERED PROBLEMS
Economic Order Quantity Model	1	9,10,13,18
Economic Production Lot Size Model	2	10,11,18,19
Planned Shortage Model	3	12,13,20
Quantity Discount Model	4	14,24,25
EOQ Model With Stochastic Demand	5	15,26
Reorder Point Based on Service Level	5	15,21,26
Single Period Inventory Model:		
Normal Demand Distribution	6	16,22
Uniform Demand Distribution	8	23
Periodic Review Systems	7	17,27

REVIEW

1. The study of inventory models is concerned with <u>two basic questions</u>: (1) <u>how much</u> should be ordered each time, and (2) <u>when</u> should the reordering occur. The objective is to minimize total variable cost over a specified time period (assumed to be annual in the following review).

2. Potential variable costs include:

    (1) <u>Ordering cost</u> -- salaries and expenses of processing an order, regardless of the order quantity
    (2) <u>Holding cost</u> -- usually a percentage of the value of the item assessed for keeping an item in inventory (including finance costs, insurance, security costs, taxes, warehouse overhead, and other related variable expenses)
    (3) <u>Backorder cost</u> -- costs associated with being out of stock when an item is demanded (including lost goodwill)
    (4) <u>Purchase cost</u> -- the actual price of the items
    (5) Other Costs

3. The simplest inventory models assume demand and the other parameters of the problem to be deterministic and constant. The <u>deterministic models</u> covered in this chapter are: a) economic order quantity (EOQ), b) economic production lot size, c) EOQ with planned shortages, and d) EOQ with quantity discounts.

4. The most basic of the deterministic inventory models is the <u>economic order quantity (EOQ)</u>. The variable costs in this model are annual holding cost and annual ordering cost. For the EOQ, these two costs are equal.

5. The <u>economic production lot size</u> model is a variation of the basic EOQ model. A replenishment order is not received in one lump sum as it is in the basic EOQ model. Instead, inventory is replenished gradually as the order is produced (which requires the production rate to be greater than the demand rate). This model's variable costs are annual holding cost and annual set-up cost (equivalent to ordering cost). For the optimal lot size, these two costs are equal.

6. A <u>stockout</u> (or <u>shortage</u>) is a demand that cannot be immediately satisfied. A <u>backorder</u> is a stockout in which the customer waits until the next replenishment order arrives and then the demand is satisfied.

7. With the <u>EOQ with planned shortages</u> model, a replenishment order does not arrive at or before the inventory position drops to zero. Instead, shortages occur until a predetermined backorder quantity is reached, at which time the replenishment order arrives. The variable costs in this model are annual holding, backorder, and ordering. For the optimal order and backorder quantity combination, the sum of the annual holding and backordering costs equals the annual ordering cost

8. The <u>EOQ with quantity discounts</u> model is applicable where a supplier offers a lower purchase cost when an item is ordered in larger quantities. This model's variable costs are annual holding, ordering and purchase costs.

9. EOQ-based inventory models give results that are rather <u>insensitive to changes in the parameters</u>. Small and sometimes even moderate changes in costs, demands, etc. will have only minor effects on overall total costs.

10. The decision maker must determine if a set of <u>assumptions</u> is appropriate for his particular problem. If the assumptions are approximately correct, employing them will simplify the solution procedure, but the results should only be used as <u>guidelines</u> for an inventory policy.

11. There may be reasons, not built into the model, for <u>modifying the results of a model</u> (such as rounding a reorder time from 12.8 days to 2 weeks to simplify reordering and make bookkeeping control easier).

12. In many cases demand (or some other factor) is not known with a high degree of certainty and a probabilistic inventory model should actually be used. These models tend to be more complex than deterministic models. The <u>probabilistic models</u> covered in this chapter are: a) single-period order quantity, b) reorder-point quantity, and c) periodic-review order quantity.

13. A <u>single-period order quantity</u> model (sometimes called the <u>newsboy problem</u>) deals with a situation in which only one order is placed for the item and the demand is probabilistic. If the period's demand exceeds the order quantity, the demand is not backordered and revenue (profit) will be lost. If demand is less than the order quantity, the surplus stock is sold at the end of the period (usually for less than the original purchase price).

14. A firm's <u>inventory position</u> consists of the <u>on-hand inventory</u> plus <u>on-order inventory</u> (all amounts previously ordered but not yet received). An inventory item is reordered when the item's inventory position reaches a predetermined value, referred to as the <u>reorder point</u>.

15. The <u>reorder point</u> represents the quantity available to meet demand during lead time. <u>Lead time</u> is the time span starting when the replenishment order is placed and ending when the order arrives.

16. Under deterministic conditions, when both demand and lead time are constant, the reorder point associated with EOQ-based models is relatively simple to determine. The reorder point is set equal to <u>lead time demand</u>.

17. Under probabilistic conditions, when demand and/or lead time varies, the reorder point often includes safety stock. <u>Safety stock</u> is the amount by which the reorder point exceeds the expected (average) lead time demand.

18. The amount of safety stock in a reorder point determines the odds (chance) of a stockout during lead time. The complement of this chance is called the service level. <u>Service level</u>, in this context, is defined as the probability of not incurring a stockout during any one lead time. Also, it is the long-run proportion of lead times in which no stockouts occur.

19. A <u>periodic review system</u> is one in which the inventory level is checked and reordering is done only at specified points in time (at fixed intervals usually). Assuming the demand rate varies, the order quantity will vary from one review period to another. This is in contrast to the <u>continuous review system</u> in which inventory is monitored continuously and an order (of a fixed amount) can be placed whenever the reorder point is reached.

20. At the time a <u>periodic-review order quantity</u> is being decided, the concern is that the on-hand inventory and the quantity being ordered is enough to satisfy demand from the time this order is placed until the next order is received (not placed).

DETERMINISTIC INVENTORY MODELS -- ASSUMPTIONS/RESULTS

I. ECONOMIC ORDER QUANTITY (EOQ)

Assumptions

1. Demand is constant throughout the year at D items per year.
2. Ordering cost: $\$C_o$ per order.
3. Holding cost: $\$C_h$ per item in inventory per year.
4. Purchase cost per unit is constant (no quantity discount).
5. Delivery time (lead time) is constant.
6. Planned shortages are not permitted.

Results

1. Optimal order quantity: $Q^* = \sqrt{2DC_o/C_h}$
2. Number of orders per year: D/Q^*
3. Time between orders (cycle time): Q^*/D years
4. Total annual cost: [holding + ordering] = $[(1/2)Q^*C_h] + [DC_o/Q^*]$

II. ECONOMIC PRODUCTION LOT SIZE

Assumptions

1. Demand occurs at a constant rate of D items per year.
2. Production rate is P items per year (and $P > D$).
3. Set-up cost: $\$C_o$ per run.
4. Holding cost: $\$C_h$ per item in inventory per year.
5. Purchase cost per unit is constant (no quantity discount).
6. Set-up time (lead time) is constant.
7. Planned shortages are not permitted.

Results

1. Optimal production lot-size: $Q^* = \sqrt{2DC_o/[(1-D/P)C_h]}$
2. Number of production runs per year: D/Q^*
3. Time between set-ups (cycle time): Q^*/D years
4. Total annual cost: [holding + ordering] = $[(1/2)(1-D/P)Q^*C_h] + [DC_o/Q^*]$

III. PLANNED-SHORTAGE ORDER QUANTITY

Assumptions

1. Demand occurs at a constant rate of D items per year.
2. Ordering cost: $\$C_o$ per order.
3. Holding cost: $\$C_h$ per item in inventory per year.
4. Backorder cost: $\$C_b$ per item backordered per year.
5. Purchase cost per unit is constant (no quantity discount).
6. Set-up time (lead time) is constant.
7. Planned shortages are permitted (backordered demand units are withdrawn from a replenishment order when it is delivered).

Results

1. Optimal order quantity: $Q^* = \sqrt{2DC_o/C_h} \ \sqrt{(C_h+C_b)/C_b}$
2. Maximum number of backorders: $S^* = Q^*(C_h/(C_h+C_b))$
3. Number of orders per year: D/Q^*
4. Time between orders (cycle time): Q^*/D years
5. Total annual cost: $[C_h(Q^*-S^*)^2/2Q^*] + [DC_o/Q^*] + [S^{*2}C_b/2Q^*]$
   [holding + ordering + backordering]

IV. QUANTITY-DISCOUNT ORDER QUANTITY

Assumptions

1. Demand occurs at a constant rate of D items per year.
2. Ordering Cost: $\$C_o$ per order.
3. Holding Cost: $\$C_h = \$C_i I$ per item in inventory per year (note: holding cost is based on the cost of the item, C_i).
4. Purchase Cost: $\$C_1$ per item if the quantity ordered is between 0 and x_1, $\$C_2$ if the order quantity is between x_1 and x_2, etc.
5. Delivery time (lead time) is constant.
6. Planned shortages are not permitted.

Results

1. Optimal order quantity: use procedure on the next page to determine Q^*
2. Number of orders per year: D/Q^*
3. Time between orders (cycle time): Q^*/D years
4. Total annual cost: $[(1/2)Q^*C_h] + [DC_o/Q^*] + DC$
   [holding + ordering + purchase]

INVENTORY MODELS

FLOW CHART OF QUANTITY DISCOUNT PROCEDURE

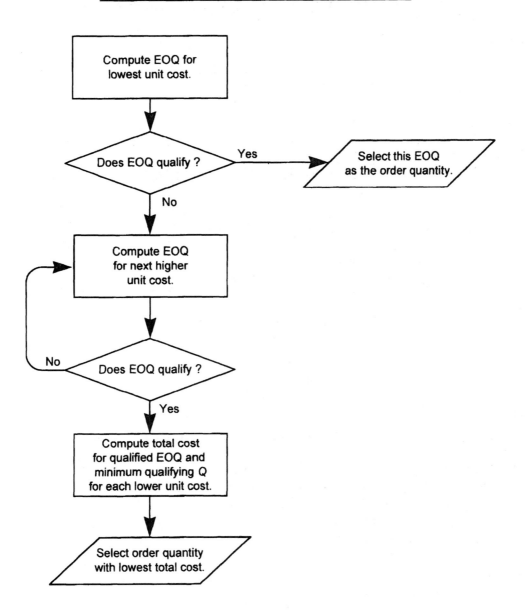

PROBABILISTIC INVENTORY MODELS -- ASSUMPTIONS/RESULTS

I. SINGLE-PERIOD ORDER QUANTITY

Assumptions

1. Period demand follows a known probability distribution.
    a. normal: mean is μ, standard deviation is σ
    b. uniform: minimum is a, maximum is b
2. Cost of overestimating demand: $\$c_o$
3. Cost of underestimating demand: $\$c_u$
4. Shortages are not backordered.
5. Period-end stock is sold for salvage (not held in inventory).

Results

1. Optimal probability of no shortage: $P(\text{demand} \leq Q^*) = c_u/(c_u+c_o)$

2. Optimal probability of shortage: $P(\text{demand} > Q^*) = 1 - c_u/(c_u+c_o)$

3. Optimal order quantity, based on demand distribution --
    a. normal: $Q^* = \mu + z\sigma$
    b. uniform: $Q^* = a + P(\text{demand} \leq Q^*)(b-a)$

II. REORDER POINT

Assumptions

1. Lead-time demand is normally distributed with mean μ and standard deviation σ.
2. Approximate optimal order quantity: EOQ
3. Service level is defined in terms of the probability of no stockouts during lead time and is reflected in z.
4. Shortages are not backordered.
5. Inventory position is reviewed continuously.

Results

1. Reorder point: $r = \mu + z\sigma$

2. Safety stock: $z\sigma$

3. Average inventory: $1/2(Q) + z\sigma$

4. Total annual cost: $[(1/2)Q^*C_h] + [z\sigma C_h] + [DC_o/Q^*]$
    [holding(normal) + holding(safety) + ordering]

III. PERIODIC-REVIEW ORDER QUANTITY

Assumptions

1. Inventory position is reviewed at constant intervals (periods).
2. Demand during review period plus lead time period is normally distributed with mean μ and standard deviation σ.
3. Service level is defined in terms of the probability of no stockouts during a review period and is reflected in z.
4. On-hand inventory at ordering time: I
5. Shortages are not backordered.
6. Lead time is less than the length of the review period.

Results

1. Replenishment level: $M = \mu + z\sigma$

2. Order quantity: $Q = M - I$

ILLUSTRATED PROBLEMS

PROBLEM 1

Bart's Barometer Business (BBB) is a retail outlet that deals exclusively with weather equipment. Currently BBB is trying to decide on an inventory and reorder policy for home barometers.

These cost BBB $50 each and demand is about 500 per year distributed fairly evenly throughout the year. Reordering costs are $80 per order and holding costs are figured at 20% of the cost of the item. BBB is open 300 days a year (6 days a week and closed two weeks in August). Lead time is 60 working days.

a) Develop a total variable cost model for this system.

b) What is the optimal reorder quantity and reorder point?

c) How many times per year would BBB reorder?

d) What total annual variable cost does the model give?

e) Given your answer to (b) and (c), choose a more convenient order quantity. Compare the resulting total cost with (d) and comment.

SOLUTION 1

a) Total Costs = (Holding Cost) + (Ordering Cost) = $[C_h(Q/2)] + [C_o(D/Q)]$
   TC = $[.2(50)(Q/2)] + [80(500/Q)] = 5Q + (40,000/Q)$

b) $Q^* = \sqrt{2DC_o/C_h} = \sqrt{2(500)(80)/10} = 89.44 \approx 90$

   Lead time is $m = 60$ days, and daily demand is $d = 500/300$ or 1.667. Thus the reorder point $r = (1.667)(60) = 100$. Bart should reorder 90 barometers when his inventory position reaches 100, i.e. 10 on hand and one outstanding order.

c) Number of reorder times per year = (500/90) = 5.56 or once every (300/5.56) = 54 working days -- about every 9 weeks.

d) TC = 5(90) + (40,000/90) = 450 + 444 = $894.

e) It might be more convenient to order 100 at a time and order 5 times per year (every 10 weeks). This total cost is TC = 5(100) + (40,000/100) = 500 + 400 = $900. This $6 difference represents only a 0.6% change in total cost.

PROBLEM 2

Non-Slip Tile Company (NST) has been using production runs of 100,000 tiles, 10 times per year to meet the demand of 1,000,000 tiles annually. The set-up cost is $5,000 per run and holding cost is estimated at 10% of the manufacturing cost of $1 per tile. The production capacity of the machine is 500,000 tiles per month. The factory is open 365 days per year.

a) Develop a model for the total annual variable cost for this problem.

b) What production schedule do you recommend?

c) How much is NST losing annually with their present production schedule?

d) How long is the machine idle between production runs?

e) What is the maximum number of tiles in inventory under the current policy? under the optimal policy?

f) What fraction of time is the machine producing tiles?

SOLUTION 2

This is an economic production lot size problem with
$$D = 1,000,000, \ P = 6,000,000, \ C_h = .10, \ C_o = 5,000.$$

a) $TC = [C_h(Q/2)(1 - D/P)] + [DC_o/Q] = .04167Q + 5,000,000,000/Q$

b) $Q^* = \sqrt{2DC_o/[C_h(1-D/P)]} = \sqrt{2(1,000,000)(5,000)/[.1(1-1/6)]} = 346,410$
The number of runs per year = $D/Q^* = 2.89$ times per year

c) Optimal TC = .04167(346,410) + 5,000,000,000/346,410 = $28,868
Current TC = .04167(100,000) + 5,000,000,000/100,000 = $54,167
Difference = 54,167 - 28,868 = $25,299

d) There are 2.89 cycles per year, so each cycle lasts (365/2.89) = 126.3 days.
The time to produce 346,410 per run = 346,410/6,000,000)365 = 21.1 days.
The machine is idle 126.3 - 21.1 = 105.2 days between runs.

e) Current maximum inventory = $(1-D/P)Q^* = (1-1/6)100,000 = 83,333$.
Optimal maximum inventory = $(1-1/6)346,410 = 288,675$.

f) The machine is producing tiles $D/P = 1/6$ of the time.

PROBLEM 3

Hervis Rent-a-Car has a fleet of 2,500 Rockets serving the Los Angeles area. All Rockets are maintained at a central garage. On the average, eight Rockets per month require a new engine. Engines cost $850 each. There is also a $120 order cost (independent of the number of engines ordered).

Hervis has an annual holding cost rate of 30% on engines. It takes two weeks to obtain the engines after they are ordered. For each week a car is out of service, Hervis loses $40 profit.

a) Determine Hervis' optimal order policy for engines.

b) How many days after receiving an order does Hervis run out of engines? How long is Hervis without any engines per cycle?

SOLUTION 3

This can be modeled as a planned shortage model with the following annual data:
$$D = 8 \times 12 = 96; \ C_o = \$120; \ C_h = .30(850) = \$255; \ C_b = 40 \times 52 = 2080.$$

a) $Q^* = \sqrt{2DC_o/C_h} \sqrt{(C_h+C_b)/C_b} = \sqrt{2(96)(120)/255} \sqrt{(255+2080)/2080} = 10.07 \approx 10$

$S^* = Q^*(C_h/(C_h+C_b)) = 10(255/(255+2080)) = 1.09 \approx 1$

Demand is 8 per month or 2 per week. Since lead time is 2 weeks, lead time demand is 4. Thus, since the optimal policy is to order 10 to arrive when there is one backorder, the order should be placed when there are 3 engines remaining in inventory.

b) Inventory exists for $C_b/(C_b+C_h) = 2080/(255+2080) = .8908$ of the order cycle. (Note, $(Q^*-S^*)/Q^* = .8908$ also, before Q^* and S^* are rounded.) An order cycle is $Q^*/D = .1049$ years = 38.3 days. Thus, Hervis runs out of engines .8908(38.3) = 34 days after receiving an order. Hervis is out of stock for approximately 38 - 34 = 4 days.

PROBLEM 4

Nick's Camera Shop carries Zodiac instant print film. The film normally costs Nick $3.20 per roll, and he sells it for $5.25. Zodiac film has a shelf life of 18 months. Nick's average sales are 21 rolls per week. His annual inventory holding cost rate is 25% and it costs Nick $20 to place an order with Zodiac.

If Zodiac offers a 7% discount on orders of 400 rolls or more, a 10% discount for 900 rolls or more, and a 15% discount for 2000 rolls or more, determine Nick's optimal order quantity.

SOLUTION 4

This can be modeled as a quantity discount problem with the following annual data:

$$D = 21(52) = 1092;\ C_h = .25(C_i);\ C_o = 20.$$

For each unit-price, starting with the lowest and working up, determine the most economical, <u>feasible</u> order quantity.

For $C_4 = .85(3.20) = \$2.72$:

To receive a 15% discount Nick must order at least 2,000 rolls. Unfortunately, the film's shelf life is 18 months. The demand in 18 months (78 weeks) is 78 X 21 = 1638 rolls of film, if he ordered 2,000 rolls he would have to scrap 372 of them. This would cost more than the 15% discount would save.

For $C_3 = .90(3.20) = \$2.88$:

$$Q_3^* = \sqrt{2DC_o/C_h} = \sqrt{2(1092)(20)/[.25(2.88)]} = 246.31\ \text{(not feasible)}$$

The most economical, feasible quantity for C_3 is $Q_3^* = 900$.

For $C_2 = .93(3.20) = \$2.976$:

$$Q_2^* = \sqrt{2DC_o/C_h} = \sqrt{2(1092)(20)/[.25(2.976)]} = 242.30\ \text{(not feasible)}$$

The most economical, feasible quantity for C_2 is $Q_2^* = 400$.

For $C_1 = 1.00(3.20) = \$3.20$ (no discount):

$$Q_1^* = \sqrt{2DC_o/C_h} = \sqrt{2(1092)(20)} = 233.67 \approx 234 \text{ (feasible, so we stop computing Qs.}$$
In this problem we have no more C_is anyway.)

Compute the total cost for the most economical, feasible order quantity in each price category for which a Q^* was computed.

$TC_i = (1/2)(Q_i^* C_h) + (DC_o/Q_i^*) + DC_i$
$TC_3 = (1/2)(900)(.720) + ((1092)(20)/900) + (1092)(2.880) = \$3493$
$TC_2 = (1/2)(400)(.744) + ((1092)(20)/400) + (1092)(2.976) = \$3453$
$TC_1 = (1/2)(234)(.800) + ((1092)(20)/234) + (1092)(3.200) = \$3681$

Comparing the total costs for 234, 400 and 900, the lowest total annual cost is $3453. Nick should order 400 rolls at a time.

PROBLEM 5

Robert's Drugs is a drug wholesaler supplying 55 independent drug stores. Roberts wishes to determine an optimal inventory policy for Comfort brand headache remedy. Sales of Comfort are relatively constant as the past 10 weeks of data indicate:

Week	Sales (cases)	Week	Sales (cases)
1	110	6	120
2	115	7	130
3	125	8	115
4	120	9	110
5	125	10	130

a) Each case of Comfort costs Roberts $10 and Roberts uses a 14% annual holding cost rate for its inventory. If the cost to prepare a purchase order for Comfort is $12, determine the optimal inventory ordering quantity for Comfort.

b) The lead time for a delivery of Comfort has averaged four working days. Lead time has therefore been estimated as having a normal distribution with a mean of 80 cases and a standard deviation of 10 cases. Roberts wants at most a 2% probability of selling out of Comfort during this lead time. What should be Roberts' reorder point?

c) On the basis of parts (a) and (b) determine the total annual inventory cost for Comfort.

SOLUTION 5

a) The average sales over the 10 week period is 120 cases. Hence

$$D = 120 \times 52 = 6{,}240 \text{ cases per year}; \quad C_h = (.14)(10) = 1.40; \quad C_o = 12.$$

$$Q^* = \sqrt{2DC_o/C_h} = \sqrt{2(6240)(12)/1.40} = 327$$

b) Lead time demand is normally distributed with $\mu = 80$, $\sigma = 10$. Since Roberts wants at most a 2% probability of selling out of Comfort, the corresponding z value (see Appendix C) is 2.06. That is, $P(z > 2.06) = .0197$ (about .02). Roberts should reorder Comfort when supply reaches $\mu + z\sigma = 80 + 2.06(10) = 101$ cases. The safety stock is 21 cases.

c) The total annual cost of this solution is:

Ordering: $(DC_o/Q^*) = ((6240)(12)/327)$ = $229
Holding--Normal: $(1/2)Q^*C_o = (1/2)(327)(1.40)$ = $229
Holding--Safety Stock: $C_h(21) = (1.40)(21)$ = $ 29
Total = $487

PROBLEM 6

The publishers of the *Fast Food Restaurant Menu Book* wish to determine how many copies to print. There is a fixed cost of $5,000 to produce the book and the incremental profit per copy is $.45. Sales for this edition are estimated to be normally distributed. The most likely sales volume is 12,000 copies and they believe there is a 5% chance that sales will exceed 20,000.

a) If any unsold copies of the book can be sold at salvage at a $.55 loss, how many copies should be printed?

b) If any unsold copies of the book can be sold at salvage at a $.65 loss, how many copies should be printed? Comment.

SOLUTION 6

a) $\mu = 12{,}000$. To find σ note that $z = 1.65$ corresponds to a 5% tail probability. Therefore,

$$(20{,}000 - 12{,}000) = 1.65\sigma \text{ or } \sigma = 4848.$$

Using incremental analysis with $c_o = .55$ and $c_u = .45$,

$$(c_u/(c_u+c_o)) = .45/(.45+.55) = .45.$$

Find Q^* such that $P(D \leq Q^*) = .45$. From Appendix C, $z = -.12$ gives this probability. Thus,

$$Q^* = 12{,}000 - .12(4848) = 11{,}418 \text{ books.}$$

b) Using incremental analysis as above but with $c_o = .65$,

$$(c_u/(c_u+c_o)) = .45/(.45 + .65) = .4091.$$

Find Q^* such that $P(D \leq Q^*) = .4091$. From Appendix C, $z = -.23$ gives this probability. Thus,

$$Q^* = 12{,}000 - .23(4848) = 10{,}885 \text{ books.}$$

However, since this is less than the breakeven volume of 11,111 books (= 5000/.45), <u>no copies should be printed</u> because if the company produced only 10,885 copies it will not recoup its $5,000 fixed cost of producing the book.

> NOTE: A common mistake is thinking that the probability of a stockout in some period is equivalent to the portion of the period's demand that will be backordered or lost, when demand is stochastic. A .10 stockout probability does NOT mean that ninety percent of demand will be satisfied. This point applies to all of the stochastic models in this chapter.

PROBLEM 7

Joe Walsh is a salesman for the Ace Brush Company. Every three weeks he contacts Dollar Department Store so that they may place an order to replenish their stock.

Weekly demand for Ace brushes at Dollar approximately follows a normal distribution with a mean of 60 brushes and a standard deviation of 9 brushes. Once Joe submits an order, the lead time until Dollar receives the brushes is one week.

Dollar would like at most a 2% chance of running out of stock during any replenishment period. If Dollar has 75 brushes in stock when Joe contacts them, how many should they order?

> NOTE: When ordering with a periodic review system, we are concerned about having enough inventory NOT until we order again, but rather until the next order is received.

SOLUTION 7

This can be modeled as a periodic review problem with probabilistic demand. The review period plus the following lead time totals 4 weeks. This is the amount of time that will elapse before the next brush shipment arrives.

Weekly demand is normally distributed with:
  Mean weekly demand, $\mu = 60$
  Weekly standard deviation, $\sigma = 9$
  Weekly variance, $\sigma^2 = 81$

Thus the distribution of demand for 4 weeks is normal with:
  Mean demand over 4 weeks, $\mu = 4 \times 60 = 240$
  Variance of demand over 4 weeks, $\sigma^2 = 4 \times 81 = 324$
  Standard deviation over 4 weeks, $\sigma = (324)^{1/2} = 18$

> **NOTE:** The standard deviation of the "whole" does NOT equal the sum of the standard deviations of the "parts". For example, if the standard deviation of daily demand equals 10, the standard deviation of two-day demand does <u>not</u> equal 20.
> However, the variance of the "whole" does equal the sum of the variances of the "parts". So, for this example, the variance of two-day demand is $2(10)^2 = 200$ and the standard deviation of two-day demand equals $\sqrt{200} = 14.14$ or $\sqrt{2}$ days$(10) = 14.14$.

The replenishment level, M, is given by the formula $M = \mu + z\sigma$, where z is determined by the desired stockout probability. For a 2% stockout probability (2% tail area in Appendix C), $z = 2.05$. Thus, $M = 240 + 05(18) = 277$.

As the store currently has 75 brushes in stock, Dollar should order: $277 - 75 = \underline{202}$ brushes from Joe. The safety stock is $z\sigma = (2.05)(18) = 37$ brushes.

PROBLEM 8

Mark Hall manages a small greeting card shop that sells artificial Christmas trees during the six weeks prior to Christmas. Based on past experience and current circumstances, Mark estimates that he will sell somewhere between 20 and 80 trees and that the actual number is equally likely to fall anywhere in that range.

A tree costs Mark $28.00 and he sells it for $46.00. Due to very limited storage space, Mark is forced to sell any trees that remain after Christmas at half price ($23.00).

a) What is the optimal probability of stocking out (having more demand than supply)?

b) How many trees should Mark order?

c) Based on your order quantity in (b), what is the probability that Mark will have ten or more trees to sell at a discount after Christmas?

SOLUTION 8

This can be modeled as a single-period problem with uniform demand. The demand limits are $a = 20$ and $b = 80$. $c_u = 46 - 28 = \$18$; $c_o = 28 - 23 = \$5$.

a) Using incremental analysis, the optimal probability of NOT running out of stock is:

$$P(D \leq Q^*) = c_u/(c_u+c_o) = 18/(18+5) = .7826$$

Thus, the optimal probability of stocking out is:

$$P(D > Q^*) = 1 - P(D \leq Q^*) = 1 - .7826 = .2174$$

b) The optimal order quantity is:

$$Q^* = a + P(D \leq Q^*)(b-a) = 20 + .7826(80-20) = 20 + 46.956 = 67 \text{ trees.}$$

c) If Mark orders 67 trees and later is left with ten or more of them, demand would have to be less than or equal to 57 trees. The probability of this is:

$$P(D \leq 57) = (57-a)/(b-a) = (57-20)/(80-20) = 37/60 = .6167 \approx .62$$

INVENTORY MODELS

ANSWERED PROBLEMS

PROBLEM 9

Terri's Tie Shop (TTS) is the exclusive retail outlet for Trophy Ties. Although Trophy Tie demand was slightly higher in December (Christmas) and June (Father's Day), it was relatively constant throughout last year:

Month	Demand	Month	Demand	Month	Demand	Month	Demand
Jan	75	Apr	76	Jul	68	Oct	70
Feb	70	May	69	Aug	75	Nov	76
Mar	72	Jun	85	Sep	74	Dec	90

The average cost of a Trophy Tie is $4 to TTS. TTS figures inventory costs at 15% yearly and reorder costs are $25 per order. There is no reason to assume demand will change much this year.

a) What is the average monthly demand?

b) Assume demand is constant throughout the year and the lead time is two months. Determine an optimal inventory policy for the model.

c) What is the total annual variable cost of the model?

d) Make some suggestions to modify the inventory policy of the model to fit the "real" problem. Comment.

PROBLEM 10

One decision faced by many manufacturing firms is whether to make or buy a particular component of the manufacturing process. Harrison Sound Corporation manufactures stereo systems. The company has a choice of either manufacturing the digital display unit for the Model 243 receiver themselves or purchasing the unit from Allied Electronics.

Allied will charge Harrison $7.50 per unit and Harrison estimates the cost of placing an order with Allied at $48. On the other hand, if Harrison manufactures the units themselves, there will be a set-up cost for production of $1,600, an annual production rate of 50,000 units a year is possible, and the per unit production cost will be $7.00.

a) If Harrison expects the annual demand for these display units to be about 10,000, and the holding cost rate is 20%, determine the total annual variable costs of both policies.

b) Which policy would you recommend to management? Why?

300 CHAPTER 13

PROBLEM 11

National Business Machines (NBM) is trying to develop the effective use of one of its production lines which produces transistors for circuits of computers. In general, allowing for defectives, NBM needs 1,000,000 transistors per year for their NBM 470 series.

A production rate of 3,000,000 per year is possible if the production line were in continuous operation 24 hours a day, 365 days a year. The cost of storing a transistor is $1 per year. Production set-up costs $4,000 and takes two weeks.

a) What is the optimal number of transistors NBM should make per production run?

b) What is the duration of a production run?

c) If workers have (and must take) vacation between the end of one production run and the start-up of another, how much vacation time do they get per year? (Note they participate in the set-up).

PROBLEM 12

Andy's Auto Parts has been stocking an unusually fine grade of racing oil on which it makes a $.10 per quart profit. Demand has been 2,000 quarts per week. Storage costs are $.01 per quart per week. Reorder costs are $10 per order. If Andy allows backorders, he figures his demand will drop to 1,900 quarts per week. Andy will, however, give a $.03 per week discount per can backordered.

a) Derive an optimal inventory policy for Andy.

b) Based on your answer in (a), what is the maximum time a backordered customer would have to wait?

PROBLEM 13

Wiley's TV Town sells Apex large screen TV's. Weekly demand has averaged 20 Apex TV's per week. Wiley makes a gross profit of $50 per TV sold (not including inventory costs). Holding costs are $260 per TV per year and reorder costs are $32 per order. Lead time is 1 week.

a) Determine: (1) the optimal number of TV's Wiley should order; (2) his reorder point; and, (3) his yearly net profit.

b) Wiley is considering allowing backorders. Wiley intends to offer customers a discount of $20 per week for each week the customer must wait for his TV. Wiley estimates that this policy will result in a drop in demand to 19 TV's per week. Order and holding costs will remain the same. Should Wiley adopt this policy? Why or why not?

PROBLEM 14

Rosato's Pizza Parlor uses tomato sauce at a fairly constant rate of 3600 cans per year. It costs Rosato $40 to place an order for tomato sauce. The holding cost rate is 30 percent per can per year.

Rosato must pay shipping cost, based on weight, and a can of tomato paste weighs 15 pounds. Both the shipping cost and the purchase cost depend on the order quantity. Using the information below, determine the most economical number of cans to order at one time. (Hint: Start by converting the shipping cost discount schedule from pounds to cans.)

Shipping Cost		Purchase Cost	
Pounds	Cost/Pound	Quantity	Cost/Can
1 - 1799	$0.26	1 - 149	$17.80
1800 - 4499	0.23	150 - 349	17.50
4500 or more	0.20	350 or more	17.40

NOTE: A common mistake is to think the EOQ equation (square-root equation) always determines the optimal order quantity. This is not true for quantity discount problems. Recall that an optimal solution must be a feasible solution. With quantity discounts, an EOQ-derived order quantity must be checked to see if it is feasible (falls in the required quantity range for the price being used).

PROBLEM 15

Kelly's Service Station does a large business in tune-ups. Demand has been averaging 210 spark plugs per week. Holding costs are $.01 per plug per week and reorder costs are estimated at $10 per order.

Kelly does not want to be out of stock on more than 1% of his orders. There is a one day delivery time. The standard deviation of demand is five plugs per day. Assume a normal distribution of demand during lead time and a 7-day work week.

a) What inventory policy do you suggest for Kelly's station?

b) What is the average amount of safety stock for the reorder point in (a)?

c) What are the total variable weekly costs including safety stock costs?

PROBLEM 16

Winkies Donuts is a small chain of donut shops in Lemon County. Winkies' success is built largely around its jelly donut. Recently, Winkies management has received a number of complaints concerning store #17 running out of jelly donuts late in the afternoon. Thus, Winkies has undertaken a study of the store's operations. The study has indicated the following:

(1) Afternoon demand for jelly donuts is approximately normally distributed with mean 150 and standard deviation 30 donuts.
(2) The cost to manufacture a jelly donut is $.09.
(3) The selling price is $.20.
(4) Donuts unsold at the end of the day are given to a charity which gives Winkies a tax savings of $.03 per donut.

Winkies management feels there is a goodwill loss of $.75 for each sale lost when it is out of stock of jelly donuts. Based on this information, how many jelly donuts should the baker prepare for the afternoon?

PROBLEM 17

A lawn and garden shop that is open for business seven days a week orders bags of grass seed every OTHER Monday. Lead time for seed orders is 5 days. On Monday, at ordering time, a clerk found 112 bags of seed in stock, and so he ordered 198 bags. Daily demand for grass seed is normally distributed with a mean of 15 bags and a standard deviation of four bags.

The manager would like to know what the probability is that a grass seed stockout will occur before the NEXT order arrives.

PROBLEM 18

Zak's Zippers is contemplating manufacturing their own zippers rather than distributing the zippers it receives from ZZZ, Inc. Zak's figures it must sell the zipper at the same price or else the yearly demand of 4,000 dozen zippers will be greatly affected.

Presently the purchase cost per dozen zippers is $10, whereas the proposed manufacturing cost for labor and raw materials is estimated at $8 per dozen. In any event, the holding costs are estimated at 20% of the purchase or manufacturing cost of the item. Reorder costs are currently $40 per order. Set-up costs for a production run are estimated at $400.

If Zak's can lease a machine capable of making 8,000 dozen zippers per year at an annual cost of $5,000, should Zak's make their own zippers?

PROBLEM 19

Rancher Jim's Luncheon Meat Company produces fresh luncheon meats. Demand is for 300,000 pounds of meat annually. Rancher Jim has his choice of two machines to process the meats. The annual lease costs and annual processing capacities are given below:

	Annual Lease Cost	Annual Processing Capacity
Machine I	$10,000	250,000 lb.
Machine II	$12,000	1,000,000 lb.

Set-up costs are $1,000 per run and holding costs are $.10 per pound per year. Rancher Jim's gross profit is $.20 per pound (not including annual lease, set-up or holding costs).

a) Show that Machine II gives the maximum profit for processing.

b) Why would you likely recommend machine I for Rancher Jim?

PROBLEM 20

Honest Archie's Appliance Co. has a policy of giving loaner TV's to customers who purchase a set from Archie when he is out of stock of they model they want. As soon as Archie gets the set in stock, he delivers it and picks up the loaner from the customer.

Demand at Archie's for the Apex 19-inch remote control TV is 5 units per week. The sets have an annual holding cost of $55. The cost to place an order for the sets is $80, and it typically takes three weeks for the sets to arrive after the order is placed.

a) If the loaner TV's cost Archie $3 per week to rent, determine his optimal order policy.

b) What is the total annual variable cost of this policy?

PROBLEM 21

Clearview Optical gives a customer a complimentary carrying case with each pair of eye glasses bought. Lead time demand for these cases is normally distributed with a mean of 230 cases and a standard deviation of 42 cases.

a) If Clearview reorders cases when inventory reaches 285 cases, what is the probability that there will be a stockout during lead time?

b) If Clearview desires a .05 probability of a stockout during lead time, what should the reorder point be for glass cases?

PROBLEM 22

Amazing Bakers sells bread to 40 supermarkets. It costs Amazing $1,250 per day to operate its plant. The profit per loaf of bread sold in the supermarket is $.025. Any unsold bread is returned to be sold at the Amazing Thrift store at a loss of $.015.

a) If sales follow a normal distribution with $\mu = 70,000$ and $\sigma = 5,000$ per day, how many loaves should Amazing bake daily?

b) Amazing is considering a different sales plan for which the profit per loaf of bread sold in the supermarket is $.03 and the loss per loaf bread returned is $.018. If $\mu = 60,000$ and $\sigma = 4,000$ per day, how many loaves should Amazing bake daily?

PROBLEM 23

Every year in early October Steven King buys pumpkins of one size from a farmer in Maine and then hires an artist to carve bewitching faces in them. He then tries to sell them at his produce stand in a public market in Boston.

The farmer charges Steven $2.50 per pumpkin and the artist is paid $2.00 per carved pumpkin. Steven sells a carved pumpkin for $8.00. Any pumpkins not sold by 5:00 p.m. on Halloween are donated to Steven's favorite children's hospital. Steven pays the artist $0.75 per pumpkin to rush the pumpkins to the hospital for the youngsters to enjoy.

Steven estimates the demand for his pumpkins this season to be uniformly distributed within a range of 30 to 70.

a) How many pumpkins should Steven have available for sale?

b) Based on your answer to (a), what is the probability that Steven will be short five or more pumpkins?

PROBLEM 24

Bank Drugs sells Jami Michelle lipstick. The Jami Michelle Company offers a 6% discount on orders of at least 500 tubes, a 10% discount on orders of at least 1,000 tubes, a 12% discount on orders of at least 1,800 tubes and a 15% discount on orders at least 2,500 tubes.

Bank sells an average of 40 tubes of Jami Michelle lipstick weekly. The normal price paid by Bank drugs is $1 per tube. If it costs Bank $30 to place an order, and Bank's annual holding cost rate is 27%, determine the optimal order policy for Bank Drugs.

PROBLEM 25

A company has the following choices for purchasing a product for which demand is 100 per week.

Option	Purchase Cost/Item	Quantity
I	$10.00	0 - 599
II	$ 9.80	600 or more
III	$ 9.90	exactly 100

Under option III, the ordering company will have no paperwork as the 100 items will be delivered every week automatically. Thus under option III, the only work associated with an order is in filing an invoice which is assumed to have zero cost. Otherwise reorder costs are $75 per order.

If holding costs are .5% per week, what is the optimal order quantity?

PROBLEM 26

Demand for the Kansas Systems Model 402 printer at the Computer Town chain of computer stores has averaged 35 units per week with a standard deviation of 10 units per week.

The units cost Computer Town $910 each and there is a $500 order cost. Computer Town's an annual holding cost rate is 20%.

The lead time for these printers is approximately a month with lead time demand being normally distributed with a mean of 140 units and a standard deviation of 20 units.

a) If Computer Town wants to experience an average of at most one stockout per year on these printers, determine an optimal inventory policy for the store.

b) Determine the annual cost of the policy for Computer Town.

c) If Computer Town was just starting to stock printers, use the data above to determine how many printers they should order.

PROBLEM 27

Chez Paul Restaurant orders special styrofoam "doggy bags" for its customers once a month and lead time is one week. Weekly demand for doggy bags is approximately normally distributed with an average of 120 bags and a standard deviation of 25.

Chez Paul wants at most a 3% chance of running out of doggy bags during the replenishment period. If he has 150 bags in stock when he places an order, how many additional bags should he order? What is the safety stock in this case?

TRUE/FALSE

___ 28. At the optimal order quantity, Q^*, in the EOQ model, annual order costs equal annual holding costs.

___ 29. If an item's per-unit backorder cost is greater than its per-unit holding cost, no intentional shortage should be planned.

___ 30. At the optimal order quantity for the quantity discount model, the sum of the annual holding and ordering costs is minimized.

___ 31. As lead time for an item increases, the cycle time increases.

___ 32. If an item's per-unit backorder cost equals one-half of its per-unit holding cost, it is optimal to plan to incur the first shortage one-third of the way through the order cycle.

___ 33. If the annual production rate for an item increases, then the optimal production lot size, Q^*, will also increase.

___ 34. When demand increases by 100 percent, the economic order quantity increases by less than 50 percent.

___ 35. If the cost of underestimating demand, c_u, is greater than the cost of overestimating demand, c_o, the optimal single-period order quantity is greater than expected demand.

___ 36. An assumption in the economic production lot size model is that there is storage capacity to hold the entire production lot.

___ 37. In the EOQ model, an item's optimal order quantity, Q^*, cannot be greater than its annual demand, D.

___ 38. The single-period inventory model is most applicable to items that are perishable or have seasonal demand.

___ 39. In the single-period inventory model, an increase in the item's salvage value will cause a decrease in the optimal quantity.

___ 40. To avoid a stockout of a periodic-review item, the item's order quantity plus inventory on hand at ordering time must last until the time the item can be ordered again.

___ 41. If the optimal production lot size decreases, average inventory increases.

___ 42. In the EOQ model, the doubling of both the ordering and holding costs would result in no change in the optimal order quantity.

CHAPTER 14

Waiting Line Models

CHAPTER 14

KEY CONCEPTS

CONCEPT	ILLUSTRATED PROBLEMS	ANSWERED PROBLEMS
Poisson Arrival Process	1	10
Exponential Service Time Distribution	1	10
Queuing Systems		
M/M/1	1,4,5	10-15,17
M/M/k	2,3	11,15,16,18
M/M/1 w/Finite Calling Population	6	23
M/G/1	7	19
M/G/k w/Blocked Customers Cleared	8	20,21
M/D/1	9	22
Economic Analysis of Queuing Systems	3,4,5	11,14,16,17
Spreadsheet Example	1,2	

* Note: <u>Unless otherwise stated</u>, all problems in this chapter assume a Poisson arrival process, exponential service time distribution, first-come-first-served queue discipline and unlimited potential queue length.

WAITING LINE MODELS

REVIEW

1. <u>Queuing theory</u> is the study of waiting lines. Four characteristics of a queuing system are: (1) the manner in which customers arrive; (2) the time required for service; (3) the priority determining the order of service; and (4) the number and configuration of servers in the system.

2. In general, the arrival of customers into the system is a random event. Frequently the <u>arrival pattern</u> is modeled as a <u>Poisson process</u>. The Poisson distribution defines the probability of x arrivals during a specified time period as:

$$P(x) = \lambda^x e^{-\lambda}/(x!)$$

Here λ is the mean number of arrivals during the specified period and $e = 2.71828...$ Appendix D provides a table of the quantity $e^{-\lambda}$ for specified values of λ. Sample data should be collected to determine the appropriateness of using the Poisson distribution as well as estimating the value for λ.

3. <u>Service time</u> is also usually a random variable. A distribution commonly used to describe this time is the <u>exponential distribution</u>. The exponential distribution has a probability density function:

$$f(t) = \mu e^{-\mu t}$$

Here μ is the mean number of customers that can be served in a specified time period.

4. For the exponential distribution, the cumulative probability:

$$P(t \leq T) = 1 - e^{-\mu t}$$

Thus the probability of a service taking longer than T is $e^{-\mu t}$.

5. The most common <u>queue discipline</u> is first come, first served (FCFS). An elevator is an example of last come, first served (LCFS) queue discipline.

6. A <u>three part code</u> of the form A/D/k is used to describe various queuing systems. Here, A identifies the arrival distribution, D the service (departure) distribution and <u>k</u> the number of servers for the system.

    Frequently used symbols for the arrival and service processes are: <u>M</u> - Markov distributions (Poisson/exponential), <u>D</u> - Deterministic (constant) and <u>G</u> - General distribution (with a known mean and variance).

    Thus the notation, <u>M/M/k</u> refers to a queuing situation in which arrivals occur according to a Poisson distribution, service times follow an exponential distribution and there are k servers each working at an identical service rate.

7. In order for an M/M/k system not to have an infinitely large queue, λ must be less than $k\mu$, where λ is the arrival rate and μ is the service rate for each server.

8. For a single server system, one defines the ratio λ/μ as the utilization factor for the queue. This can be thought of as the long run proportion of time the server is busy, the probability there is someone in the system, or the probability that an arriving customer must wait for service.

9. In determining the most economical queuing system configuration or evaluating service parameters for the system, one is frequently interested in long run or steady state results. For steady state results to exist, $\lambda < \mu$ for an M/M/1 or an M/G/1 queue, and $\lambda < k\mu$ for an M/M/k queue.

10. Notation used for various queue measures are as follows:

    P_0 = probability of no units in the system
    P_n = probability of n units in the system
    P_w = probability an arriving unit must wait for service
    L_q = average number of units in the waiting line
    L = average number of units in the system
    W_q = average time a unit spends in the waiting line
    W = average time a unit spends in the system
    λ = mean number of arrivals per time period (the arrival rate)
    μ = mean number of services per time period (the service rate)
    $1/\lambda$ = average time between arrivals
    $1/\mu$ = average service time
    σ = standard deviation of the service time

11. For nearly all queuing systems, there is a relationship between the average time a unit spends in the system or queue and the average number of units in the system or queue. These relationships, known as Little's flow equations are:

    $$L = \lambda W \quad \text{and} \quad L_q = \lambda W_q$$

12. When the queue discipline is FCFS, analytical formulas have been derived for several different queuing models including the following: M/M/1, M/M/k, M/G/1, M/G/k with blocked customers cleared, and M/M/1 with a finite calling population. These formulas are presented on the next few pages.

13. Analytical formulas are not available for all possible queuing systems. In this event, insights may be gained through a simulation of the system.

STEADY STATE RESULTS

Quantity	M/M/1 Queues	M/M/k Queues	M/G/1 Queues*
P_0	$1 - \dfrac{\lambda}{\mu}$	$\dfrac{1}{\sum_{n=0}^{k-1} \dfrac{(\lambda/\mu)^n}{n!} + \dfrac{(\lambda/\mu)^k}{k!}\left(\dfrac{k\mu}{k\mu - \lambda}\right)}$	$1 - \dfrac{\lambda}{\mu}$
P_n	$\left(\dfrac{\lambda}{\mu}\right)^n P_0$	$\dfrac{(\lambda/\mu)^n}{n!} P_0$ for $n \leq k$ $\dfrac{(\lambda/\mu)^n}{k!\,k^{(n-k)}} P_0$ for $n > k$	No formula
L_q	$\dfrac{\lambda^2}{\mu(\mu - \lambda)}$	$\dfrac{\lambda\mu(\lambda/\mu)^k}{(k-1)!(k\mu - \lambda)^2} P_0$	$\dfrac{\lambda^2 \sigma^2 + (\lambda/\mu)^2}{2(1 - \lambda/\mu)}$
L	$L_q + \dfrac{\lambda}{\mu}$	$L_q + \dfrac{\lambda}{\mu}$	$L_q + \dfrac{\lambda}{\mu}$
W_q	$\dfrac{L_q}{\lambda}$	$\dfrac{L_q}{\lambda}$	$\dfrac{L_q}{\lambda}$
W	$W_q + \dfrac{1}{\mu}$	$W_q + \dfrac{1}{\mu}$	$W_q + \dfrac{1}{\mu}$
P_w	$\dfrac{\lambda}{\mu}$	$\dfrac{1}{k!}\left(\dfrac{\lambda}{\mu}\right)^k \left(\dfrac{k\mu}{k\mu - \lambda}\right) P_0$	$\dfrac{\lambda}{\mu}$

* Note: In the M/G/1 queue, if G is the exponential distribution, then $\sigma = 1/\mu$ and the above formulas reduce to those given for the M/M/1 queue.

STEADY STATE RESULTS

Quantity	M/D/1 Queues	M/G/k Queues with Blocked Customers Cleared	M/M/1 Queues* with Finite Calling Population (Size N)
P_0	$1 - \dfrac{\lambda}{\mu}$	$\dfrac{1}{\sum_{i=0}^{k}(\lambda/\mu)^i / i!}$	$\dfrac{1}{\sum_{n=0}^{N}\dfrac{N!}{(N-n)!}(\dfrac{\lambda}{\mu})^n}$
P_n	No formula	$\dfrac{(\lambda/\mu)^n / n!}{\sum_{i=0}^{k}(\lambda/\mu)^i / i!}$	$\dfrac{N!}{(N-n)!}(\dfrac{\lambda}{\mu})^n P_0$
L_q	$\dfrac{(\lambda/\mu)^2}{2(1-\lambda/\mu)}$	No queue	$N - \dfrac{\lambda+\mu}{\lambda}(1-P_0)$
L	$L_q + \dfrac{\lambda}{\mu}$	$\dfrac{\lambda}{\mu}(1-P_k)$	$L_q + (1-P_0)$
W_q	$\dfrac{L_q}{\lambda}$	No queue	$\dfrac{L_q}{(N-L)\lambda}$
W	$W_q + \dfrac{1}{\mu}$	$\dfrac{1-P_k}{\mu}$	$W_q + \dfrac{1}{\mu}$
P_w	$\dfrac{\lambda}{\mu}$	No waiting	$1 - P_0$

* Note: In the finite calling population model, λ represents the arrival rate for each unit.

ILLUSTRATED PROBLEMS

NOTE: Students frequently confuse <u>rates</u> and <u>times</u> in this chapter. They are not the same; they have an inverse relationship. The equations throughout the chapter assume λ and μ are <u>rates</u>. If you are given average interarrival time or average service time, use the inverse.

Also, λ and μ should be stated in the same unit of time (per hour, for example). Finally, be careful converting the standard deviation of service times σ from one unit of time to another. (It might be easier to convert the arrival rate's time basis to that of the service rate.)

PROBLEM 1

Joe Ferris is a stock trader on the floor of the New York Stock Exchange for the firm of Smith, Jones, Johnson, and Thomas, Inc. Stock transactions arrive at a rate of 20 per hour. Each order received by Joe requires an average of two minutes to process.

a) What is the probability that no orders are received within a 15-minute period?

b) What is the probability that exactly 3 orders are received within a 15-minute period?

c) What is the probability that more than 6 orders arrive within a 15-minute period?

d) What is the service rate per hour?

e) What percentage of the orders will take less than one minute to process?

f) What percentage of the orders will be completed in exactly 3 minutes?

g) What percentage of the orders will take more than 3 minutes to process?

h) What is the average time an order must wait from the time Joe receives the order until its processing is finished (i.e. its turnaround time)?

I) What is the average number of orders Joe has waiting to be processed?

j) What percentage of the time is Joe processing orders?

k) Develop a spreadsheet to find the values for the operating characteristics of this M/M/1 queuing system.

SOLUTION 1

Orders arrive at a rate of 20 per hour or one order every 3 minutes. Therefore, in a 15 minute interval the average number of orders arriving will be $\lambda = 15/3 = 5$. From Appendix D note that $e^{-5} = .0067$.

a) $P(x = 0) = (5^0 e^{-5})/0! = e^{-5} = .0067$.

b) $P(x = 3) = (5^3 e^{-5})/3! = 125(.0067)/6 = .1396$.

c) $P(x > 6) = 1 - P(x = 0) - P(x = 1) - P(x = 2) - P(x = 3) - P(x = 4) - P(x = 5) - P(x = 6)$
$= 1 - .762 = .238$.

d) Since Joe Ferris' average processing time is 2 minutes (= 2/60 hr.), the service rate, μ, is $\mu = 1/(\text{mean service time})$, or $60/2 = 30/\text{hr}$.

e) Since the units are expressed in hours, $P(T \leq 1 \text{ minute}) = P(T \leq 1/60 \text{ hour})$. Using the exponential distribution, $P(T \leq t) = 1 - e^{-\mu t}$. Hence,

$$P(T \leq 1/60) = 1 - e^{-30(1/60)} = 1 - e^{-.5} = 1 - .6065 = .3935.$$

f) Since the exponential distribution is a continuous distribution, the probability a service time exactly equals any specific quantity is 0.

g) The percentage of orders requiring more than 3 minutes to process is:

$$P(T > 3/60) = e^{-30(3/60)} = e^{-1.5} = .2231.$$

h) This is an M/M/1 queue with $\lambda = 20$ per hour and $\mu = 30$ per hour. The average time an order waits in the system is:

$$W = 1/(\mu - \lambda) = 1/(30 - 20) = 1/10 \text{ hour or 6 minutes}.$$

I) The average number of orders waiting in the queue is:

$$L_q = \lambda^2/[\mu(\mu - \lambda)] = (20)^2/[(30)(30-20)] = 400/300 = 4/3.$$

j) The percentage of time Joe is processing orders is equivalent to the utilization factor, λ/μ. Thus, the percentage of time he is processing orders is: $\lambda/\mu = 20/30 = 2/3$ or 66 2/3%.

WAITING LINE MODELS   315

k) Spreadsheet showing data and formulas

	A	B	C	D	E	F	G	H
1	Poisson Arrival Rate						λ	20
2	Exponential Service Rate						μ	30
3	Operating Characteristics							
4	Probability of no orders in system						Po	=1-H1/H2
5	Average number of orders waiting						Lq	=H1^2/(H2*(H2-H1))
6	Average number of orders in system						L	=H5+H1/H2
7	Average time an order waits						Wq	=H5/H1
8	Average time an order is in system						W	=H7+1/H2
9	Probability an order must wait						Pw	=H1/H2

Spreadsheet showing computed values for operating characteristics

	A	B	C	D	E	F	G	H
1	Poisson Arrival Rate						λ	20
2	Exponential Service Rate						μ	30
3	Operating Characteristics							
4	Probability of no orders in system						Po	0.333
5	Average number of orders waiting						Lq	1.333
6	Average number of orders in system						L	2.000
7	Average time an order waits						Wq	0.067
8	Average time an order is in system						W	0.100
9	Probability an order must wait						Pw	0.667

PROBLEM 2

Smith, Jones, Johnson, and Thomas, Inc. (see problem 1) has begun a major advertising campaign which it believes will increase its business 50%. To handle the increased volume, the company has hired an additional floor trader, Fred Hanson, who works at the same speed as Joe Ferris.

a) Why will Joe Ferris alone not be able to handle the increase in orders?

b) What is the probability that neither Joe nor Fred will be working on an order at any point in time?

c) What is the average turnaround time for an order with both Joe and Fred working?

d) What is the average number of orders waiting to be filled with both Joe and Fred working?

SOLUTION 2

We first note that the new arrival rate of orders, λ, is 50% higher than that of problem 1. Thus, $\lambda = 1.5(20) = 30$ per hour.

a) Since Joe Ferris processes orders at a rate of $\mu = 30$ per hour, then $\lambda = \mu = 30$ and the utilization factor is 1. This implies the queue of orders will grow infinitely large. Hence, Joe alone cannot handle this increase in demand.

b) This is now an M/M/2 queuing system with $\lambda = 30$, $\mu = 30$, $k = 2$ and $\lambda/\mu = 1$. The probability that neither Joe nor Fred will be working is:

$$P_0 = \frac{1}{\sum_{n=0}^{k-1}\frac{(\lambda/\mu)^n}{n!} + \frac{(\lambda/\mu)^k}{k!}\left(\frac{k\mu}{k\mu - \lambda}\right)}$$

$$= \frac{1}{[(1 + (1/1!)(30/30)^1] + [(1/2!)(1)^2][2(30)/(2(30)-30)]}$$

$$= 1/(1 + 1 + 1)$$

$$= 1/3$$

(Note: Table 14.4 in the textbook shows $P_0 = .3333$ for $k = 2$ and $\lambda/\mu = 1$.)

c) The average turnaround time is the average waiting time in the system, W. $W = L/\lambda$ and $L = L_q + (\lambda/\mu)$. Now,

$$L_q = \frac{\lambda\mu(\lambda/\mu)^k}{(k-1)!(k\mu - \lambda)^2} P_0$$

$$= \frac{(30)(30)(30/30)^2}{(1!)((2)(30)-30)^2}(1/3)$$

$$= 1/3$$

Hence, $L = 1/3 + (30/30) = 4/3$. $W = (4/3)/30 = 4/90$ hr. $= 2.67$ min.

d) The average number of orders waiting to be filled is L_q. This was calculated above in part (c) as 1/3.

PROBLEM 3

The advertising campaign of Smith, Jones, Johnson and Thomas, Inc. (see problems 1 and 2) was so successful that business actually doubled. The rate of stock orders arriving at the exchange is now 40 per hour and the company must decide how many floor traders to employ. Each floor trader hired can process an order, on average, in 2 minutes.

Based on a number of factors the brokerage firm has determined the average waiting cost per minute for an order to be $.50. Floor traders hired will earn $20 per hour in wages and benefits. Using this information compare the total hourly cost of hiring 2 traders with that of hiring 3 traders.

> NOTE: Economic analysis of queuing systems usually involves a tradeoff between the cost of service and the <u>cost of waiting</u>. To get the latter, you might be inclined to multiply the waiting cost per unit per time period, c_W, by the average wait time in the system, W. If you take this approach you are not finished until you also multiply by the average number of units entering the system per hour, λ. You get the same results by simply multiplying c_W by L, the average number of units in the system.

SOLUTION 3

Hourly Cost = (Total Hourly Salary Cost) + (Total Hourly Cost for Orders in the System)
= ($20/trader/hour) x (Number of Traders)
  + ($30 waiting cost/hour) x (average number of orders in the system)
= $20k + 30L$.

Thus, L must be determined for $k = 2$ traders and for $k = 3$ traders with $\lambda = 40$/hour and $\mu = 30$/hour (since the average service time is 2 minutes).

$k = 2$

$$P_0 = \frac{1}{\sum_{n=0}^{k-1} \frac{(\lambda/\mu)^n}{n!} + \frac{(\lambda/\mu)^k}{k!}\left(\frac{k\mu}{k\mu - \lambda}\right)}$$

$$= \frac{1}{[1+[(1/1!)(40/30)]+[(1/2!)(40/30)^2(60/(60-40))]} = \frac{1}{1+(4/3)+(8/3)} = 1/5$$

Thus,

$$L_q = \frac{\lambda\mu(\lambda/\mu)^k}{(k-1)!(k\mu - \lambda)^2} P_0$$

$$= \frac{40(30)(40/30)^2}{1!(2(30)-40)^2} = 16/15$$

$L = L_q + (\lambda/\mu) = 16/15 + 4/3 = 12/5.$

Total Cost = $(20)(2) + 30(12/5) = \$112.00/hr.$

$k = 3$

$$P_0 = \frac{1}{[1+(1/1!)(40/30)+(1/2!)(40/30)^2] + [(1/3!)(40/30)^3(90/(90-40))]}$$

$$= \frac{1}{[1+ 4/3 + 8/9] + [32/45]} = 15/59$$

$$L_q = \frac{(30)(40)(40/30)^3}{(2!)(3(30)-40)^2} (15/59) = 128/885 \ (=.1446)$$

Thus, $L = 128/885 + 40/30 = 1308/885 \ (= 1.4780).$

Total Cost = $(20)(3) + 30(1308/885) = \$104.35$ per hour

Thus, the cost of having 3 traders is less than that of 2 traders.

PROBLEM 4

Frederick's Auto Company currently receives an average of 22 letters a day and has a typist who can type a letter in an average time of 20 minutes. It is considering taking some relief action to ease the typist's workload. It can either hire an additional typist (who also works at a rate of one letter per 20 minutes) at a cost of $40 per day or it can lease one of three models of word processing systems listed below:

Model	Increase in Cost Per Day	Typist's Efficiency
I	$37	50%
II	$39	75%
III	$43	150%

Frederick's has determined that the cost of a letter waiting to be mailed is $.80/hour. If typists work 8 hours/day, what action should be taken?

SOLUTION 4

For each of the 5 alternatives, determine the average number of letters in the system and the total daily cost. The total daily cost = (extra typist cost or lease cost) + 6.40L, where $6.40 is $.80/hour times 8 hours/day.

Another Typist
This is an M/M/2 system with $\lambda = 22$, $\mu = 24$ and $k = 2$. Using the formulas from the review section we compute $P_0 = .3714$ and $L = 1.16$.
Hence total cost = 40 + (6.40)(1.16) = $47.42.

Lease Machine 1
This is an M/M/1 system with $\lambda = 22$ and $\mu = (1.5)(24) = 36$. Using the formulas from the review section for an M/M/1 system, $L = 22/(36-22) = 1.57$.
Total Cost = 37 + (6.40)(1.57) = $47.05.

Lease Machine 2
This is an M/M/1 system with $\lambda = 22$ and $\mu = (1.75)(24) = 42$. $L = 22/(42-22) = 1.1$.
Total cost = 39 + (6.40)(1.1) = 46.04.

Lease Machine 3
This is an M/M/1 system with $\lambda = 22$ and $\mu = (2.5)(24) = 60$. $L = 22/(60-22) = .58$.
Total cost = 43 + (6.40)(.58) = $46.71.

No Action
This is an M/M/1 system with $\lambda = 22$ and $\mu = 24$. $L = 22/(24-22) = 11$.
Total cost = 0 + (6.40)(11) = $70.40.

Conclusion   Thus, the best course of action is to lease Machine II.

PROBLEM 5

Jerry's Jewelry Store is seeking a salesman for its evening shift. Three applicants with former experience have applied for the position, each demanding different salaries. Jerry has contacted the former supervisor of each who has supplied him with information on average service times for each applicant. The applicant's salary demands and average service times are as follows:

Applicant	Hourly Wage	Average Service Time
Martha Miller	$ 6	6 min.
Ken Weeks	$10	5 min.
Eddie Smith	$14	4 min.

Customers arrive to the store at the rate of 8 per hour and you have estimated the cost of having a customer in the store to be $4 per customer per hour (for security, customer relations, etc.) Which applicant should Jerry hire?

SOLUTION 5

For each applicant calculate the total hourly cost = (Hourly Wage) + 4L, where L is the average number of customers in the system. Each case is an M/M/1 system with $\lambda = 8$.

Martha Miller
Since the average service time is 6 min., the service rate, $\mu = 1/6$ per minute or 10 per hour.

$L = 8/(10-8) = 4.$ Total hourly cost = 6 + (4)(4) = $22.00.

Ken Weeks
Since the average service time is 5 min., the service rate, $\mu = 1/5$ per minute or 12 per hour.

$L = 8/(12-8) = 2.$ Total hourly cost = 10 + (4)(2) = $18.00.

Eddie Smith
Since the average service time is 4 min., the service rate, $\mu = 1/4$ per minute or 15 per hour.

$L = 8/(15-8) = 8/7.$ Total hourly cost = 14 + (4)(8/7) = $18.56.

Conclusion   Based on this study, hire Ken Weeks.

PROBLEM 6

Biff Smith is in charge of maintenance for four of the rides at the Algorithmland Amusement Park: the Pivot, the Traveling Salesman's Adventure, Minimax Regret, and the CPM Crash. On the average, each ride operates four hours before needing repair. When repair is needed, the average repair time is 10 minutes.

Assuming that the time between machine repairs and the service times follow exponential distributions, determine the following:

a) the proportion of time Biff is idle

b) the average time a ride is "down" for repairs

> NOTE: For an M/M/1 system with a finite calling population, the arrival rate is the average number of times <u>one</u> unit from the finite population arrives per time period, not the average number of total arrivals per time period.

SOLUTION 6

This problem can be modeled as an M/M/1 queue with a finite calling population of size $N = 4$ (rides). Here

$\lambda = 1/(4 \text{ hours}) = .25$ per hour. $\mu = 60/(10 \text{ minutes}) = 6$ per hour. $\lambda/\mu = .25/6 = 1/24.$

a) The proportion of time Biff is idle is P_0:

$$P_0 = \frac{1}{\sum_{n=0}^{N} \frac{N!}{(N-n)!}(\frac{\lambda}{\mu})^n}$$

$$P_0 = \frac{1}{\frac{4!}{4!}(1/24)^0 + \frac{4!}{3!}(1/24)^1 + \frac{4!}{2!}(1/24)^2 + \frac{4!}{1!}(1/24)^3 + \frac{4!}{0!}(1/24)^4}$$

$$= \frac{1}{1 + 4(1/24)^1 + 12(1/24)^2 + 24(1/24)^3 + 24(1/24)^4} = .840825$$

Hence, Biff is idle approximately 84% of the time.

b) To find the average time a ride is "down" for repairs we need to calculate W. This is given by the formula:

$$W = W_q + \frac{1}{\mu} \text{ where } W_q = \frac{L_q}{(N-L)\lambda}, \quad L_q = N - \frac{\lambda+\mu}{\lambda}(1-P_0), \text{ and } L = L_q + (1-P_0)$$

Substituting the appropriate values gives:

L_q = 4 - ((.25+6)/.25)(1-.840825) = .020625

L = .020625 + (1-.840825) = .1798

W_q = .020625/((4-.1798).25) = .02159573

W = .02159573 + 1/6 = .18826 hours or 11.3 minutes

PROBLEM 7

The Bowmar University Student Union has one self service copying machine. On the average, 12 customers per hour arrive to make copies. (The arrival process follows a Poisson distribution.) The time to copy documents follows approximately a <u>normal</u> distribution with a mean of two and one-half minutes and a standard deviation of 30 seconds.

The Union president has received several complaints from students regarding the long lines at the copying machine. On the basis of this information determine:

a) the average number of customers waiting or using the copy machine.

b) the probability an arriving customer must wait in line.

c) the proportion of time the copy machine is idle.

d) the average time a customer must wait in line before using the copy machine.

SOLUTION 7

This situation can be modeled as an M/G/1 system with

λ = 12 per hour = 12/60 = .2/minute
$1/\mu$ = 2.5 minutes
μ = 1/(2.5) = .4 per minute
σ = 30 seconds = .5 minutes

a) The average number of customers waiting or using the copy machine is:

$$L = L_q + \frac{\lambda}{\mu} \quad \text{where } L_q = \frac{\lambda^2 \sigma^2 + (\lambda/\mu)^2}{2(1 - \lambda/\mu)}$$

Substituting the appropriate values gives:

$$L_q = \frac{(.2)^2(.5)^2 + (.2/.4)^2}{2(1-(.2/.4))} = .26 \quad \text{Thus } L = .26 + (.2/.4) = .76$$

b) The probability an arriving customer must wait in line is:
   $P_W = \lambda/\mu = .2/.4 = .5$

c) The proportion of time the copy machine is idle is:
   $P_0 = 1 - \lambda/\mu = 1 - .2/.4 = .5$

d) The average time a customer must wait in line before using the copy machine is:
   $W_q = L_q/\lambda = .26/.2 = 1.3$ minutes

PROBLEM 8

Several firms are over the counter (OTC) market makers of Probabillistics stock. A broker wishing to trade this stock for a client will call on these firms to execute the order. If the market maker's phone line is busy, a broker will immediately try calling another market maker to transact the order.

Richardson and Company is one such OTC market maker. It estimates that on the average, a broker will try to call to execute a stock transaction every two minutes. The time required to complete the transaction averages 75 seconds. The firm has four traders staffing its phones. Assume calls arrive according to a Poisson distribution.

a) What percentage of its potential business will be lost by Richardson?

b) What percentage of its potential business would be lost if only three traders staffed its phones?

SOLUTION 8

This problem can be modeled as an M/G/k system with block customers cleared with:
$1/\lambda$ = 2 minutes = 2/60 hour
λ = 60/2 = 30 per hour
$1/\mu$ = 75 seconds = 75/60 minutes = 75/3600 hours
μ = 3600/75 = 48 per hour

a) As there are four traders staffing the phones, $k = 4$, and the system will be blocked when there are four customers in the system. Hence, the answer is P_4. To find P_4, first find P_0 by:

$$P_0 = \frac{1}{\sum_{i=0}^{k}(\lambda/\mu)^i/i!} \quad \text{where } k = 4$$

$$= \frac{1}{1 + (30/48) + (30/48)^2/2! + (30/48)^3/3! + (30/48)^4/4!}$$

$$= \frac{1}{1 + (.625) + (.625)^2/2 + (.625)^3/6 + (.625)^4/24}$$

$$= .536$$

Now, $P_4 = \frac{(\lambda/\mu)^4}{4!} P_0 = \frac{(30/48)^4}{24}(.536) = .003$

Thus with four traders 0.3% of the potential customers are lost.

b) In this case $k = 3$, and the answer is P_3. Recalculate P_0:

$$P_0 = \frac{1}{1 + (30/48) + (30/48)^2/2! + (30/48)^3/3!} = .537$$

Now, $P_3 = \frac{(\lambda/\mu)^3}{3!} P_0 = \frac{(30/48)^3}{6}(.537) = .022$

Thus with three traders 2.2% of the potential customers are lost.

PROBLEM 9

The long distance viewing scope at the scenic rest stop on Interstate 20 provides two minutes of viewing for $.25. People wanting to use the scope arrive according to a Poisson distribution with a rate of 15 per hour.

a) What fraction of time is the scope idle?

b) What is the average number of sightseers waiting to use the scope?

c) What is the average time a sightseer waits to use the scope?

SOLUTION 9

This situation can modeled as an M/D/1 system with

$\lambda = 15$ per hour
$1/\mu = 2$ minutes $= 2/60$ hour
$\mu = 60/2 = 30$ per hour
$\sigma = 0$

a) The fraction of time the scope is idle is the complement of the utilization factor:

$1 - \lambda/\mu = 1 - 15/30 = .5$

b) The average number of customers waiting in line is:

$$L_q = \frac{(\lambda/\mu)^2}{2(1 - \lambda/\mu)} = \frac{(15/30)^2}{2(1 - 15/30)} = .25 \text{ people}$$

c) The average time a customer must wait to use the scope is:

$W_q = L_q/\lambda = .25/15 = .01667$ hours or 1.0 minutes.

ANSWERED PROBLEMS

PROBLEM 10

Customers arrive at the Roney Tax Preparation office at a rate of one per hour. The average time it takes Ms. Roney to prepare a customer's income tax form is 45 minutes.

a) What is the probability of no customers arriving in 2 hours?

b) What is the probability that an income tax form is finished within 45 minutes from the time it is started?

c) What is the average time a customer spends waiting to see Ms. Roney?

d) What is the probability Ms. Roney has 3 customers in the office (i.e. 2 waiting customers plus the one being served)?

PROBLEM 11

Ms. Roney is contemplating a computer system to help her with her income tax preparation (see problem 10). She estimates that such a system will reduce the average time to prepare a return from 45 minutes to 30 minutes. The computer leases for $40 per day.

Ms. Roney estimates the average cost to her of having a waiting customer (due to good will, lost sales, etc.) is $3 per hour. She is open 10 hours per day.

a) Should she lease the computer?

b) If Ms. Roney has the option of hiring another tax preparer for $60 per day who works at the same speed she does, should she do this rather than lease the computer?

PROBLEM 12

The postmaster at the Oak Hill Post Office expects the arrival rate of people to her customer counter will soon increase by fifty percent due to a large apartment complex being built.

Currently, the arrival rate is 15 people per hour. The postmaster can serve an average of 25 people per hour.

By what percentage must the postmaster's service rate increase when the apartment complex is completed in order that the average time spent at the post office remains at its current value?

PROBLEM 13

Cars travel down Main Street at the rate of 20,000 per hour. The probability of any of these cars stopping at the drive-in window of the Burger Prince Restaurant is .002. Cars at Burger Prince are serviced at the rate of 60 per hour.

a) What is the arrival rate to the drive-in window?

b) What is the average number of cars waiting to be served?

c) What is the average number of cars both being served and waiting to be served?

d) What is the probability an arriving car to the Burger Prince Restaurant must wait for service?

e) Management believes that if the number of cars waiting to be served was reduced to less than 1, the probability a car would stop at the drive-in window would increase to .003 and is contemplating implementing changes to speed up service. Given the increased arrival rate, what is the minimum value of the service rate that will meet management's objective?

PROBLEM 14

Cabinet Crafters manufactures specially designed kitchen cabinets and bathroom vanities. One piece of machinery used on most projects is a stationary router. The average number of times carpenters use the router daily is 20 and the average time required for each use is 15 minutes.

The company is planning to purchase a digital router which should reduce the average time required for each use to 12 minutes. The digital router will cost an additional $60 per day. If carpenters earn an average of $20 per hour and work 9 hours per day, determine whether or not Cabinet Crafters should purchase the digital router.

PROBLEM 15

Shear's Department Store has 2 catalog order desks, one at each entrance to the store. On the average, a customer arrives at each order desk every 12 minutes. The service rate at each order desk is an average of 8 customers per hour.

Shear's is considering consolidating its two desks into one location staffed by two order clerks. They would continue to serve customers at the rate of 8 customers per hour. Shear's figures it would not lose customers under this arrangement and hence arrivals to this single desk would occur every 6 minutes on the average.

For each configuration of the catalog order desks, determine:

a) the average number of customers waiting to be served

b) the probability no customers are present in the entire system

c) the average time a customer spends at the order desk (waiting time plus service time)

d) the probability both clerks are busy

e) should Shear's consolidate its catalog order desks? Explain.

PROBLEM 16

The insurance department at Shear's has two agents, each working at a mean speed of 8 customers per hour. Customers arrive at the insurance desk at a rate of one every six minutes and form a single queue.

Management feels that some customers are going to find the wait at the desk too long and take their business to Word's, Shear's competitor. In order to reduce the time required by an agent to serve a customer Shear's is contemplating installing one of two minicomputer systems: System A which leases for $18 per day and will increase an agent's efficiency by 25%; or, System B which leases for $23 per day and will increase an agent's efficiency by 50%. Agents work 8-hour days.

If Shear's estimates its cost of having a customer in the system at $3 per hour, determine if Shear's should install a new minicomputer system, and if so, which one.

PROBLEM 17

A company has tool cribs where workmen draw parts. Two men have applied for the position of distributing parts to the workmen. George Fuller is fresh out of trade school and expects a $6 per hour salary. His average service time is 4 minutes. John Cox is a veteran who expects $12 per hour. His average service time is 2 minutes. A workman's time is figured at $10 per hour. Workmen arrive to draw parts at a rate of 12 per hour.

a) What is the average waiting time a workman would spend in the system under each applicant?

b) Which applicant should be hired?

PROBLEM 18

Dollar Supermarkets currently has 5 checkout positions. On the average, one customer per minute enters the store and spends an average of 55 minutes choosing his items. Each checker can check out a customer in an average of 4 minutes.

a) What is the average time a customer will spend in the store?

b) Management is considering reducing the number of checkout positions to 3 and hiring baggers. This would reduce the average time required to check out a customer to 2.5 minutes. Would this reduce the average waiting time?

PROBLEM 19

Ted "Tank" Fuller operates a small Texxon gas station that has one gas pump. The arrival rate of vehicles to the pump follows a Poisson probability distribution having a mean of 10 per hour. The time it takes Tank to service a vehicle follows a normal distribution with a mean of 4 minutes and a standard deviation of 1.5 minutes.

a) What is the utilization factor for the gas pump?

b) What is the average length of time a gas customer spends waiting for service to begin?

c) What is the average number of full-service vehicles at the gas station?

d) What is the probability that the servicing of a vehicle will take no more than 7 minutes (excluding the queue wait)?

PROBLEM 20

Tom's Towing Service operates three tow trucks and relies solely on towing requests from the city police department. If Tom has a truck available when the police need one, Tom is always their first choice. However, if he does not have a truck available, the police will select an alternative company rather than wait.

On the average, the police request a tow truck once every 50 minutes. Tom estimates his average tow job takes 90 minutes.

a) Is Tom achieving at least 50 percent utilization of his fleet of tow trucks? (HINT: start by determining the average number of his trucks being used.)

b) What percentage of the police department's towing business is Tom losing because he has only three trucks?

WAITING LINE MODELS

PROBLEM 21

Quick Clean Rooter cleans out clogged drains. Due to the competitive nature of the drain cleaning business, if a customer calls Quick Clean and finds the line busy, they immediately try another company and Quick Clean loses the business.

Quick Clean management estimates that on the average, a customer tries to call Quick Clean every three minutes and the average time to take a service order is 200 seconds. The company wishes to hire enough operators so that at most 4% of its potential customers get the busy signal.

a) How many operators should be hired to meet this objective?

b) Given your answer to a), what is the probability that all the operators are idle?

PROBLEM 22

The Quick Snap photo machine at the Lemon County bus station takes four snapshots in <u>exactly</u> 75 seconds. Customers arrive at the machine according to a Poisson distribution at the rate of 20 per hour.

On the basis of this information, determine the following:

a) the average number of customers waiting to use the photo machine

b) the average time a customer spends in the system

c) the probability an arriving customer must wait for service.

PROBLEM 23

Andy Archer, Ph.D., is a training consultant for six mid-sized manufacturing firms. On the average, each of his six clients calls him for consulting assistance once every 25 days. Andy typically spends an average of five days at the client's firm during each consultation.

Assuming that the time between client calls follows an exponential distribution, determine the following:

a) the average number of clients Andy has on backlog

b) the average time a client must wait before Andy arrives to it

c) the proportion of the time Andy is busy.

TRUE/FALSE

___ 24. For an M/M/1 queue, the average time a customer spends in the system is equal to the average time a customer spends in the waiting line plus the average service time.

___ 25. In order to obtain analytical results for a queuing problem one must assume an exponential service time.

___ 26. For a single server queuing system, the average number of customers in the waiting line is one less than the average number in the system.

___ 27. In queuing notation L_q is the average time a customer spends in the waiting line.

___ 28. Little's flow equations apply to any queuing system regardless of the arrival distribution, service time distribution, and number of channels.

___ 29. For an M/M/1 queue, the sum of the utilization factor plus P_0 equals 1.

___ 30. Queue discipline refers to the assumption that a customer has the patience to remain in a slow moving queue.

___ 31. For an M/M/2 system, the probability that the system is empty plus the probability an arriving customer must wait for service equals 1.

___ 32. If some maximum number of customers are allowed in a queuing system at one time, the system has a finite calling population.

___ 33. Assuming the same arrival rates, the average number of customers in the system for an M/M/1 system is the same as that for an M/M/2 system if the average service time in the M/M/2 system is twice as long as for the M/M/1 system.

___ 34. For an M/M/k system, the average number of customers in the system equals the customer arrival rate times the average time a customer spends waiting in the system.

___ 35. Simulation may be used to obtain results for queuing systems.

___ 36. For an M/M/1 queuing system, if the service rate, μ, is doubled, the average wait in the system, W, is cut in half.

___ 37. A multiple-channel system has more than one waiting line.

___ 38. Even when the service rate of the queuing system cannot be increased, a reduction in the service time variation will reduce the average length of the waiting line.

CHAPTER 15

Simulation

CHAPTER 15

KEY CONCEPTS

CONCEPT	ILLUSTRATED PROBLEMS	ANSWERED PROBLEMS
Simulation of:		
One Event	1,2	9
Two Events	3	7,8,10,11,14
Multiple Events	4	13,15
Comparison of Expected and Simulated Outcomes	1,6	9,10
Simulation of Decision Alternatives	2	7,10,11,14
Simulation Applications:		
Inventory Systems	2	10,11,13
Waiting Lines	3,4	8,12,15
Markov Processes	5	16
Project Times	6	9
Spreadsheet Example	1,5	

REVIEW

1. <u>Computer simulation</u> is one of the most frequently employed management science techniques. It is typically used to model random processes that are too complex to be solved by analytical methods.

2. One begins a computer simulation by developing a <u>mathematical statement</u> of the problem. The model should be realistic yet solvable within the speed and storage constraints of the computer system being used. Input values for the model as well as probability estimates for the random variables must then be determined.

3. Random variable values are utilized in the model through a technique known as <u>Monte Carlo simulation</u>. Here each random variable is mapped to a set of numbers so that each time one number in that set is generated, the corresponding value of the random variable is given as an input to the model. The mapping is done in such a way that the likelihood that a particular number is chosen is the same as the probability that the corresponding value of the random variable occurs.

4. Because a computer program generates random numbers for the mapping according to some formula, the numbers are not truly generated in a random fashion. However, using standard statistical tests, the numbers can be shown to appear to be drawn from a random process. Appendix E is a table of <u>pseudo-random digits</u>.

5. In a <u>fixed-time simulation</u> model, time periods are incremented by a fixed amount. For each time period a different set of data from the input sequence is used to calculate the effects on the model.

6. In a <u>next-event simulation</u> model, time periods are not fixed but are determined by the data values from the input sequence.

7. The computer program that performs the simulation is called a <u>simulator</u>. While this program can be written in any general purpose language (e.g. BASIC, FORTRAN, C++, etc.), special languages that reduce the amount of code which must be written to perform the simulation have been developed such as SIMSCRIPT, GPSS, SLAM, and Arena.

8. <u>Validation/verification</u> of both the model and the method used by the computer to carry out the calculations is extremely important. Models which do not accurately reflect real world behavior cannot be expected to generate meaningful results. Likewise, errors in programming can result in nonsensical results.

9. <u>Validation/verification</u> is generally done by having an expert review the model and the computer code for errors. If possible, the simulation should be run using actual past data. Predictions from the simulation model should be compared with historical results.

10. <u>Experimental design</u> is an important consideration in the simulation process. Issues such as the length of time of the simulation and the treatment of initial data outputs from the model must be addressed prior to collecting and analyzing output data.

11. Normally one is interested in results for the <u>steady state</u> (long run) operation of the system being modeled. Hence, the initial data inputs to the simulation generally represent a start-up period for the process and it may be important that the data outputs for this start-up period be ignored for predicting this long run behavior.

12. For each policy under consideration by the decision maker, the simulation is run by considering a long sequence of input data values (given by a pseudo-random number generator). Whenever possible, different policies should be compared by using the <u>same sequence of input data</u>.

13. Among the advantages of computer simulation is the ability to gain insights into the model solution which may be impossible to attain through other techniques. Also, once the simulation has been developed, it provides a convenient experimental laboratory to perform "<u>what if</u>" and <u>sensitivity analysis.</u>

14. Two major disadvantages of simulation are: (1) a large amount of time may be required to develop the simulation; and, (2) there is no guarantee that the solution obtained will actually be optimal. Simulation is, in effect, a <u>trial-and-error method</u> of comparing different policy inputs. It does not determine if some input which was not considered could have provided a better solution for the model.

FLOW CHART OF COMPUTER SIMULATION PROCESS

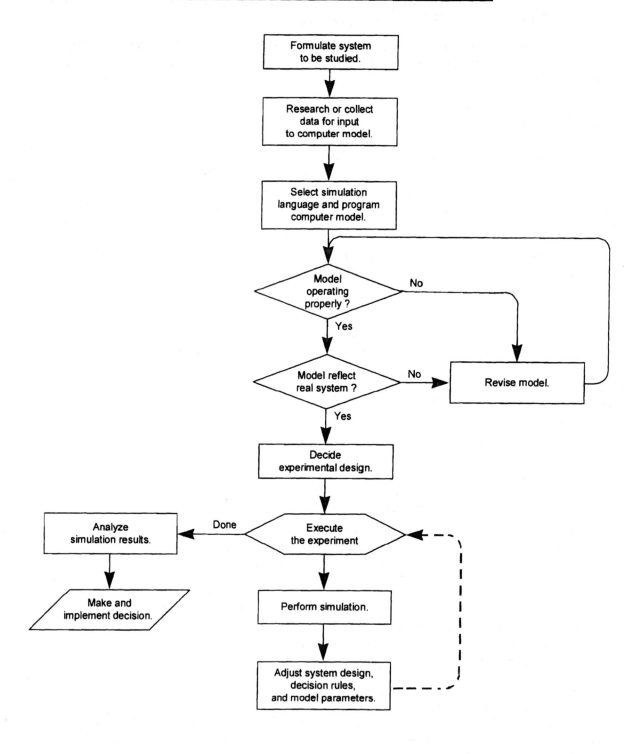

SIMULATION

ILLUSTRATED PROBLEMS

NOTE: Several of the problems in this chapter relate to topics of other chapters such as inventory, waiting lines, markov processes, and projects. Where applicable, you are encouraged to compare your simulation results with those of analytic procedures covered in other chapters. However, you might find differences in the results that are due most likely to the small number of simulation trials performed.

PROBLEM 1

The price change of shares of Probablistics, Inc. has been observed over the past 50 trades. The frequency distribution is as follows:

Price Change	Frequency (Number of Trades)
-3/8	4
-1/4	2
-1/8	8
0	20
+1/8	10
+1/4	3
+3/8	2
+1/2	1
	Total = 50

a) Develop a relative frequency distribution for this data.

b) If the current price per share of Probablistics is 23, use random numbers to simulate the price per share over the next 20 trades. (For random numbers, use the first two numbers at the bottom of column 1 of Appendix E and move up.)

c) Compare the simulated price after 20 trades (from part b) with the expected price one would obtain based on the probability distribution.

d) Using a spreadsheet, simulate 10 trades.

SOLUTION 1

a) To develop a relative frequency distribution for this data, divide the frequency of each price by the total number of trades. The results are shown in the table below.

b) To develop a simulation for the future prices, assign a random number to each price change so that the probability of seeing a certain price corresponds to its probability (relative frequency).

One such assignment of numbers is:

Price Change	Relative Frequency	Random Numbers
-3/8	.08	00 - 07
-1/4	.04	08 - 11
-1/8	.16	12 - 27
0	.40	28 - 67
+1/8	.20	68 - 87
+1/4	.06	88 - 93
+3/8	.04	94 - 97
+1/2	.02	98 - 99

According to the instructions, the first random number will be 62, the second 32, etc. The simulated results are:

Trade Number	Rand. Number	Price Change	Stock Price
1	62	0	23
2	32	0	23
3	71	+1/8	23 1/8
4	94	+3/8	23 1/2
5	04	-3/8	23 1/8
6	97	+3/8	23 1/2
7	58	0	23 1/2
8	67	0	23 1/2
9	78	+1/8	23 5/8
10	14	-1/8	23 1/2
11	54	0	23 1/2
12	48	0	23 1/2
13	84	+1/8	23 5/8
14	66	0	23 5/8
15	73	+1/8	23 3/4
16	13	-1/8	23 5/8
17	91	+1/4	23 7/8
18	10	-1/4	23 5/8
19	62	0	23 5/8
20	32	0	23 5/8

c) Based on the probability distribution, the expected price change per trade can be calculated by: $(.08)(-3/8) + (.04)(-1/4) + (.16)(-1/8) + (.40)(0) + (.20)(1/8) + (.06)(1/4) + (.04)(3/8) + (.02)(1/2) = .005$

The expected price change for 20 trades is $(20)(.005) = .10$. Hence, the expected stock price after 20 trades is $23 + .10 = 23.10$. This is lower than the simulated price of 23.625 (or 23 5/8).

d) Spreadsheet for the stock price simulation

	A	B	C	D	E	F
1	Lower	Upper		Trade	Price	Stock
2	Random	Random	Price	Number	Change	Price
3	Number	Number	Change	1	0.000	23.000
4	0.00	0.08	-0.375	2	0.000	23.000
5	0.08	0.12	-0.250	3	-0.125	22.875
6	0.12	0.28	-0.125	4	-0.375	22.500
7	0.28	0.68	0.000	5	0.125	22.625
8	0.68	0.88	0.125	6	0.000	22.625
9	0.88	0.94	0.250	7	0.000	22.625
10	0.94	0.98	0.375	8	0.125	22.750
11	0.98	1.00	0.500	9	0.125	22.875
12				10	0.125	23.000

=VLOOKUP(RAND(),$A$4:$C$11,3) ⟶

PROBLEM 2

Shelly's Supermarket has just installed a postage stamp vending machine. Based on one month of operation, Shelly's estimates the number of postage stamps sold per day can be approximated by the following distribution:

Number Sold Per Day	Probability
20	.10
30	.15
40	.20
50	.25
60	.20
70	.10

Shelly's makes a $.02 profit per postage stamp. The vending machine holds 230 stamps and it costs Shelly's $2.00 in labor to fill the machine.

a) Determine the mean number of stamps sold per day.

b) Determine the mean time until the machine is empty.

c) Assume that Shelly's adopts the following policy. Shelly's will fill the machine at the beginning of every n-th day, where n is the answer found in part (b). Conduct a 20-day simulation and determine the expected profit per day. Assume the machine must be filled on the first day. (Use the first two numbers of column 3 of Appendix E, beginning at the top for the random numbers.)

d) Suppose Shelly's fills the machine every (n-1)-th day. Repeat the 20-day simulation and compare the answer with that of part (c). Which policy would you recommend?

SOLUTION 2

The flow chart below can assist in setting up the simulations.

a) The mean number of stamps sold daily =
   $(.10)(20)+(.15)(30)+(.20)(40)+(.25)(50)+(.20)(60)+(.10)(70) = 46.$

b) The mean time until the machine is empty is:
   (Machine capacity)/(Mean number of stamps sold per day) = 230/46 = 5 days.

Flow Chart of Vending Machine Simulation

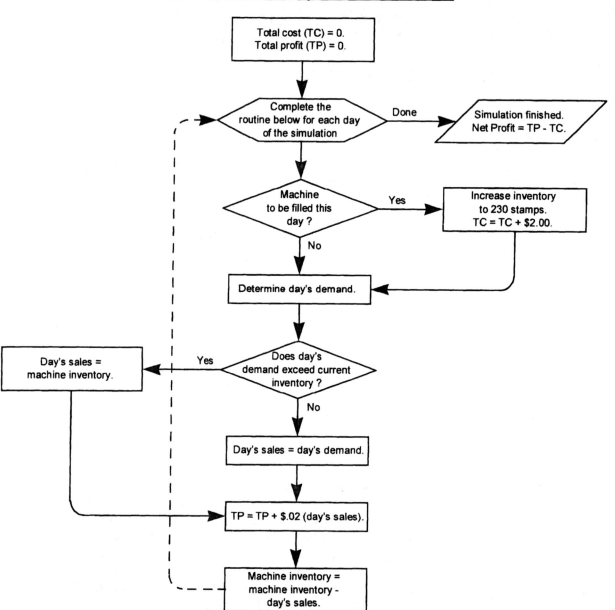

Assuming Shelly's fills the machine every fifth day, determine a set of random numbers corresponding to each sales level.

Number Sold Per Day	Range Of Numbers
20	00 - 09
30	10 - 24
40	25 - 44
50	45 - 69
60	70 - 89
70	90 - 99

Following the instructions of using the first two numbers in column 3 of Appendix E generates the following simulation:

Day	Random Number	Demand	Number Of Stamps Left In Machine	Profit From Sale Of Stamps	Cost Of Refilling Machine	Daily Profit
1	71	60	170	1.20	2.00	- .80
2	95	70	100	1.40	--	1.40
3	83	60	40	1.20	--	1.20
4	44	40	0	.80	--	.80
5	34	40	0	0*	--	0
6	49	50	180	1.00	2.00	-1.00
7	88	60	120	1.20	--	1.20
8	56	50	70	1.00	--	1.00
9	05	20	50	.40	--	.40
10	39	40	10	.80	--	.80
11	75	60	170	1.20	2.00	- .80
12	12	30	140	.60	--	.60
13	03	20	120	.40	--	.40
14	59	50	70	1.00	--	1.00
15	29	40	30	.80	--	.80
16	77	60	170	1.20	2.00	- .80
17	76	60	110	1.20	--	1.20
18	57	50	60	1.00	--	1.00
19	15	30	30	.60	--	.60
20	53	50	0	.60**	--	.60

Total Profit = $9.60

* 0 since the machine was empty
** .60 since there were only 30 stamps left in the machine

Expected Profit per Day = (9.60)/20 = $.48.

d) In order to compare the two policies, the same set of input data should be generated. Thus the simulation below has the same input for daily demand as that of part (c). In this part, the machine is to be filled every fourth day.

Day	Random Number	Demand	Number Of Stamps Left In Machine	Profit From Sale Of Stamps	Cost Of Refilling Machine	Daily Profit
1	71	60	170	1.20	2.00	-.80
2	95	70	100	1.40	--	1.40
3	83	60	40	1.20	--	1.20
4	44	40	0	.80	--	.80
5	34	40	190	0	2.00	-1.20
6	49	50	140	1.00	--	1.00
7	88	60	80	1.20	--	1.20
8	56	50	30	1.00	--	1.00
9	05	20	210	.40	2.00	-1.60
10	39	40	170	.80	--	.80
11	75	60	110	1.20	--	1.20
12	12	30	80	.60	--	.60
13	03	20	210	.40	2.00	-1.60
14	59	50	160	1.00	--	1.00
15	29	40	120	.80	--	.80
16	77	60	60	1.20	--	1.20
17	76	60	170	1.20	2.00	-.80
18	57	50	120	1.00	--	1.00
19	15	30	90	.60	--	.60
20	53	50	40	.60	--	1.00

Total Profit = $8.80

Expected Daily Profit = $ .44

Based on the results of this simulation, Shelly's should fill the machine every fifth day rather than every fourth day.

PROBLEM 3

Wayne International Airport primarily serves domestic air traffic. Occasionally, however, a chartered plane from abroad will arrive with passengers bound for Wayne's two great amusement parks, Algorithmland and Giffith's Cherry Preserve.

Whenever an international plane arrives at the airport the two customs inspectors on duty set up operations to process the passengers.

Incoming passengers must first have their passports and visas checked. This is handled by one inspector. The time required to check a passenger's passports and visas can be described by the following probability distribution:

Time Required to Check a Passenger's Passport and Visa	Probability
20 seconds	.20
40 seconds	.40
60 seconds	.30
80 seconds	.10

After having their passports and visas checked, the passengers next proceed to the second customs official who does baggage inspections. Passengers form a single waiting line with the official inspecting baggage on a first come, first served basis.

The time required for baggage inspection has the following probability distribution:

Time Required For Baggage Inspection	Probability
No Time	.25
1 minute	.60
2 minutes	.10
3 minutes	.05

a) If a chartered plane from abroad lands at Wayne Airport with 80 passengers, use simulation to determine how long it will take for the first 20 passengers to clear customs. (From Appendix E, use the first two digits in column 8 for passport control and the first two digits in column 9 for baggage inspection.)

b) What is the average length of time a customer waits before having his bags inspected after he clears passport control? How is this estimate biased?

NOTE: The value of a decision variable is an input to a simulation model, whereas it is an output to an optimization model (like a linear programming model). A simulation model does not decide the value of a decision variable, it evaluates the result, in terms of some objective, of a given value of a decision variable.

SOLUTION 3

The problem is easiest to set up as a next-event simulation model. The random number mappings are:

Time Required to Check a Passenger's Passport and Visa	Probability	Random Numbers
20 seconds	.20	00 - 19
40 seconds	.40	20 - 59
60 seconds	.30	60 - 89
80 seconds	.10	90 - 99

Time Required For Baggage Inspection	Probability	Random Numbers
No Time	.25	00 - 24
1 minute	.60	25 - 84
2 minutes	.10	85 - 94
3 minutes	.05	95 - 99

For each passenger the following information must be recorded:
  (1) When his service begins at the passport control inspection
  (2) The length of time of this service
  (3) When his service begins at the baggage inspection
  (4) The length of time of this service

Note the following relationships:
  (1) Time a passenger begins service by the passport inspector =
      (Time the previous passenger started passport service)
        + (Time of previous passenger's passport service)

  (2) Time a passenger begins being served by the baggage inspector depends on whether or not the passenger must wait in line for this service.
      If passenger does not wait in line for baggage inspection:
        Time a passenger begins being served by the baggage inspector =
        (Time passenger completes service with the passport control inspector)
      If the passenger does wait in line for baggage inspection:
        Time a passenger begins being served by the baggage inspector =
        (Time previous passenger completes service with the baggage inspector)

  (3) Time a customer completes service at the baggage inspector =
      (Time customer begins service with baggage inspector) +
      (Time required for baggage inspection).

SIMULATION

The following table describes the simulation: (service times in minutes)

Passenger Number	Passport Control				Baggage Inspections			
	Time Begin	Random Number	Service Time	Time End	Time Begin	Random Number	Service Time	Time End
1	0:00	93	1:20	1:20	1:20	13	0:00	1:20
2	1:20	63	1:00	2:20	2:20	08	0:00	2:20
3	2:20	26	:40	3:00	3:00	60	1:00	4:00
4	3:00	16	:20	3:20	4:00	13	0:00	4:00
5	3:20	21	:40	4:00	4:00	68	1:00	5:00
6	4:00	26	:40	4:40	5:00	40	1:00	6:00
7	4:40	70	1:00	5:40	6:00	40	1:00	7:00
8	5:40	55	:40	6:20	7:00	27	1:00	8:00
9	6:20	72	1:00	7:20	8:00	23	0:00	8:00
10	7:20	89	1:00	8:20	8:20	64	1:00	9:20
11	8:20	49	:40	9:00	9:20	36	1:00	10:20
12	9:00	64	1:00	10:00	10:20	56	1:00	11:20
13	10:00	91	1:20	11:20	11:20	25	1:00	12:20
14	11:20	02	:20	11:40	12:20	88	2:00	14:20
15	11:40	52	:40	12:20	14:20	18	0:00	14:20
16	12:20	69	1:00	13:20	14:20	74	1:00	15:20
17	13:20	29	:40	14:00	15:20	75	1:00	16:20
18	14:00	96	1:20	15:20	16:20	29	1:00	17:20
19	15:20	95	1:20	16:40	17:20	80	1:00	18:20
20	16:40	84	1:00	17:40	18:20	25	1:00	19:20

For example, passenger 1 begins being served by the passport control inspector immediately. His service time is 1:20 (80 seconds) at which time he goes to the baggage inspector who waves him through without inspection.

Passenger 2 begins service with passport inspector 1:20 minutes (80 seconds) after arriving there (as this is when passenger 1 is finished) and requires 1:00 minute (60 seconds) for passport inspection. He is waved through baggage inspection as well. The process continues in this manner.

a) Passenger 20 clears customs after 19 minutes 20 seconds.

b) For each passenger calculate his waiting time (in seconds):

(Baggage Inspection Begins) - (Passport Control Ends) =
0+0+0+40+0+20+20+40+40+0+20+20+0+40+120+60+80+60+40+40 = 640 seconds.

This gives an average of 640/20 = 32 seconds per passenger. This is a biased estimate because we assume that the simulation began with the system empty, resulting in underestimated average waiting time.

344　CHAPTER 15

PROBLEM 4

Attendees at the National Management Science Society (NMSS) Conference register by first standing in line to pay their fees. They then proceed to a designated line based on the first letter of their last name to collect their conference materials.

At the conference, it is planned to have three different parallel lines for the collection of materials: one each for people whose last names begin with A-H, I-Q, and R-Z respectively.

During each minute of the morning registration period it is anticipated that attendees will arrive to pay their fees according to the following distribution:

Number of Arrivals	Probability
0	.30
1	.30
2	.30
3	.10

The time to pay one's fees is either one minute or two minutes depending upon whether one uses a check or credit card. The probability of a one-minute time is .60.

After paying his fees, an attendee then goes to the correct line for the conference materials. At this year's conference 35% of the attendees have last names beginning with A-H, 36% with last names beginning with I-Q, and 29% with last names beginning with R-Z. The time required to pick up conference materials is fixed at 2 minutes.

a) Simulate the waiting line for the first 30 attendees during the morning registration. From Appendix E, use column 3 to generate the number of arrivals in any given minute, column 4 to generate registration fee service time, and column 5 to generate the first letter of the last name. Assume registration begins at 8:00 AM.

b) What is the average size of the waiting line to pay fees (not including the person being served), and the average customer waiting time to pay fees based on this simulation?

c) What is the percentage of time each of the three individuals who distribute conference materials is working?

SOLUTION 4

a) For each minute, record the number of arrivals at the fee desk. Then for each arrival determine how long he must wait in line to pay fees, his service time to pay fees, which conference material line is joined, and his waiting and service time in that material line. Random numbers are needed for the number of arrivals in a given minute to the fee line, the length of service time in the fee line, and first letter of the last name to determine to which conference material line the attendee will proceed. The random number mappings are:

Number of Arrivals	Probability	Random Numbers
0	.30	00 - 29
1	.30	30 - 59
2	.30	60 - 89
3	.10	90 - 99

Service Time To Pay Fees	Probability	Random Numbers
1 minute	.60	00 - 59
2 minutes	.40	60 - 99

First Letter Of Last Name	Probability	Random Numbers
A-H	.35	00 - 34
I-Q	.36	35 - 70
R-Z	.29	71 - 99

The simulation then flows chronologically through the events listed as headings on the simulation chart on the next page:
1. Time period begins with Q customers waiting to pay fees. (Q can be determined from the number of previous customers with end times in the fee pay line greater than this time period.)
2. Determine the number of arrivals in the period, #.
3. For each distinct arrival, I, determine wait, begin and end times in the fee pay line.
4. For each distinct arrival, determine the last name and the wait, begin and end times in the conference material line.
5. Repeat until 30 customers have arrived.

b) The average length of the waiting line to pay fees is the average of the entries in the column Q over the 26 time intervals observed, including the 10 at the beginning of 8:26. This is 181/26 = 6.96. The average time a customer waits to pay his fees is the average of the entries in the Fee Pay Wait column over the 30 arrivals = 248/30 = 8.267 minutes.

c) To determine the percentage of time each of the three people who are distributing conference materials are working, take the total amount of time each works and divide by the total length of the simulation (= 42 minutes since 8:41 is the earliest the next arrival could want material service.)

346 CHAPTER 15

ARRIVALS				FEE PAY						CONFERENCE MATERIAL				
Time	RN	#	Q	I	Wait	Begin	RN	Time	End	RN	Name	Wait	Begin	End
8:00	71	2	0	1	0	8:00	51	1	8:01	15	A-H	0	8:01	8:03
				2	1	8:01	79	2	8:03	08	A-H	0	8:03	8:05
8:01	95	3	1	3	2	8:03	09	1	8:04	19	A-H	1	8:05	8:07
				4	3	8:04	67	2	8:06	45	I-Q	0	8:06	8:08
				5	5	8:06	15	1	8:07	76	R-Z	0	8:07	8:09
8:02	83	2	3	6	5	8:07	58	1	8:08	42	I-Q	0	8:08	8:10
				7	6	8:08	04	1	8:09	38	I-Q	1	8:10	8:12
8:03	44	1	5	8	6	8:09	78	2	8:11	47	I-Q	1	8:12	8:14
8:04	34	1	5	9	7	8:11	30	1	8:12	82	R-Z	0	8:12	8:14
8:05	49	1	5	10	7	8:12	56	1	8:13	37	I-Q	1	8:14	8:16
8:06	88	2	6	11	7	8:13	75	2	8:15	49	I-Q	1	8:16	8:18
				12	9	8:15	75	2	8:17	43	I-Q	1	8:18	8:20
8:07	56	1	7	13	10	8:17	05	1	8:18	37	I-Q	2	8:20	8:22
8:08	05	0	7											
8:09	39	1	6	14	9	8:18	49	1	8:19	11	A-H	0	8:19	8:21
8:10	75	2	6	15	9	8:19	70	2	8:21	45	I-Q	1	8:22	8:24
				16	11	8:21	25	1	8:22	55	I-Q	2	8:24	8:26
8:11	12	0	8											
8:12	03	0	7											
8:13	59	1	6	17	9	8:22	26	1	8:23	89	R-Z	0	8:23	8:25
8:14	29	0	6											
8:15	77	2	6	18	8	8:23	10	1	8:24	09	A-H	0	8:24	8:26
				19	9	8:24	46	1	8:25	67	I-Q	1	8:26	8:28
8:16	76	2	7	20	9	8:25	16	1	8:26	84	R-Z	0	8:26	8:28
				21	10	8:26	64	2	8:28	51	I-Q	0	8:28	8:30
8:17	57	1	9	22	11	8:28	72	2	8:30	67	I-Q	0	8:30	8:32
8:18	15	0	9											
8:19	53	1	8	23	11	8:30	50	1	8:31	14	A-H	0	8:31	8:33
8:20	37	1	8	24	11	8:31	15	1	8:32	10	A-H	1	8:33	8:35
8:21	46	1	9	25	11	8:32	79	2	8:34	52	I-Q	0	8:34	8:36
8:22	85	2	9	26	12	8:34	22	1	8:35	03	A-H	0	8:35	8:37
				27	13	8:35	51	1	8:36	02	A-H	1	8:37	8:39
8:23	24	0	10											
8:24	53	1	9	28	12	8:36	01	1	8:37	09	A-H	2	8:39	8:41
8:25	72	2	9	29	12	8:37	47	1	8:38	13	A-H	3	8:41	**8:43**
				30	13	8:38	88	2	8:40	42	I-Q	0	8:40	8:42
8:26			10	---------- Simulation Over ----------										
TOTAL 26		30	181	248				40		21				

In 43 minutes (8:00-8:43) the number of minutes busy were A-H: 20, I-Q: 29, R-Z: 8

A-H worker = 20/43 = 46.5%, I-Q worker = 29/43 = 67.4%, R-Z worker = 8/43 = 18.6%.

PROBLEM 5

Mark is a specialist at repairing large metal-cutting machines that use laser technology. His repair territory consists of the cities of Austin, San Antonio, and Houston. His day-to-day repair assignment locations can be modeled as a Markov process (covered in Chapter 16). The transition matrix is as follows:

		Next Day's Location		
		Austin	San Antonio	Houston
This Day's Location	Austin	.60	.15	.25
	San Antonio	.20	.75	.05
	Houston	.15	.05	.80

a) Show the random number assignments that can be used to simulate Mark's next day location when his current location is Austin, San Antonio, and Houston.

b) Assume Mark is currently in Houston. Simulate where Mark will be over the next 25 days. What percentage of time will Mark be in each of the three cities? (Use column 8 of Appendix E.)

c) Repeat the simulation in part b with Mark currently in Austin. (Use column 9 of Appendix E.) Compare the percentages with those found in part (b).

d) Using a spreadsheet, simulate where Mark will be over the next 20 days.

SOLUTION 5

a)

Currently in Austin		Currently in San Antonio		Currently in Houston	
Next-Day Location	Random Numbers	Next-Day Location	Random Numbers	Next-Day Location	Random Numbers
Austin	00 - 59	Austin	00 - 19	Austin	00 - 14
San Ant.	60 - 74	San Ant.	20 - 94	San Ant.	15 - 19
Houston	75 - 99	Houston	95 - 99	Houston	20 - 99

b) Starting in Houston

Day	Random Number	Day's Location
1	93	Houston
2	63	Houston
3	26	Houston
4	16	San Ant.
5	21	San Ant.
6	26	San Ant.
7	70	San Ant.
8	55	San Ant.
9	72	San Ant.
10	89	San Ant.
11	49	San Ant.
12	64	San Ant.
13	91	San Ant.
14	02	Austin
15	52	Austin
16	69	San Ant.
17	29	San Ant.
18	96	Houston
19	95	Houston
20	84	Houston
21	61	Houston
22	09	Austin
23	06	Austin
24	00	Austin
25	63	San Ant.

Austin = 5/25 = 20%
San Antonio = 13/25 = 52%
Houston = 7/25 = 28%

c) Starting in Austin

Day	Random Number	Day's Location
1	13	Austin
2	08	Austin
3	60	San Ant.
4	13	Austin
5	68	San Ant.
6	40	San Ant.
7	40	San Ant.
8	27	San Ant.
9	23	San Ant.
10	64	San Ant.
11	36	San Ant.
12	56	San Ant.
13	25	San Ant.
14	88	San. Ant.
15	18	Austin
16	74	San Ant.
17	75	San Ant.
18	29	San Ant.
19	80	San Ant.
20	25	San Ant.
21	05	Austin
22	64	San Ant.
23	71	San Ant.
24	83	San Ant.
25	74	San Ant.

Austin = 5/25 = 20%
San Antonio = 20/25 = 80%
Houston = 0/25 = 0%

d) Partial Spreadsheet with Variable Look-up Table

	A	B	C	D	E	F	G	H	I
1	Current Location								
2	Aus.			S.A.			Hou.		
3	LRN	URN	NDL	LRN	URN	NDL	LRN	URN	NDL
6	0.00	0.60	Aus.	0.00	0.20	Aus.	0.00	0.15	Aus.
7	0.60	0.75	S.A.	0.20	0.95	S.A.	0.15	0.20	S.A.
8	0.75	1.00	Hou.	0.95	1.00	Hou.	0.20	1.00	Hou.
9									

LRN = Lower Random Number
URN = Upper Random Number
NDL = Next-Day Location

Partial Spreadsheet with Simulation Table

	A B	C D	E F
11	Current	Random	Next-Day
12	Location	Number	Location
13	Aus.	0.89	Hou.
14	Hou.	0.12	Aus.
15	Aus.	0.45	Aus.
16	Aus.	0.57	Aus.
17	Aus.	0.53	Aus.
18	Aus.	0.89	Hou.
19	Hou.	0.56	Hou.
20	Hou.	0.84	Hou.
21	Hou.	0.93	Hou.
22	Hou.	0.16	S.A.
23	S.A.	0.64	S.A.
24	S.A.	0.30	S.A.
25	S.A.	0.67	S.A.
26	S.A.	0.46	S.A.
27	S.A.	0.37	S.A.
28	S.A.	0.75	S.A.
29	S.A.	0.45	S.A.
30	S.A.	0.37	S.A.
31	S.A.	0.28	S.A.
32	S.A.	0.22	S.A.

IF(A13=$A$2,VLOOKUP(C13,$A$6:$C$8,3),
   IF(A13=$D$2,VLOOKUP(C13,$D$6:$F$8,3),
      VLOOKUP(C13,$G$6:$I$8,3)))

PROBLEM 6

Consider the following house renovation project.

Job and Description	Expected Completion Time (in days)
A - Remove old roof shingles	2
B - Plaster walls and ceilings	4
C - Lay new roof shingles	3
D - Paint exterior	5
E - Paint walls and ceilings	4
F - Hang new gutters	2
G - Install exterior lights	2
H - Lay wall-to-wall carpet	2

350 CHAPTER 15

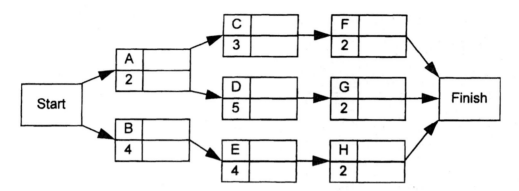

For each job the probabilities of being completed 1 day earlier than expected, on time, 1 day later than expected, and 2 days later than expected are .2, .5, .2, and .1, respectively.

a) Determine the critical path(s) and the project's expected completion time.

b) Do a five trial simulation using the last digit of column 4 of Appendix E for the random numbers to determine the average completion time of the project. Compare this result with your answer to part (a).

SOLUTION 6

a) A path in this context is a set of successive jobs that lead from the project network's starting node (node 1) to the finish node (node 7). There are three different paths in the network. They are A-C-F, A-D-G, and B-E-H. The critical (longest) path consists of jobs B-E-H. The project's expected completion time is the sum of the expected completion times of the jobs on the critical path. Thus, the expected project completion time is 4 + 4 + 2 = 10 days.

Note: Project Scheduling is covered in Chapter 12. The critical path method (CPM) and other techniques are demonstrated.

b) The random number mappings are as follows. Note only single digit random numbers need be generated.

Job Completion Time	Probability	Random Numbers
Expected time minus 1 day	.2	0,1
Expected time	.5	2,3,4,5,6
Expected time plus 1 day	.2	7,8
Expected time plus 2 days	.1	9

The table on the following page describes the simulation. Note that due to the randomness of job times, a path other than B-E-H was critical in four out of the five trials.

SIMULATION

Job	RN	Job's Exp. Time	Time Adjustment	Job's Act. Time	Critical Path and Proj. Compl. Time
A	2	2	0	2	
B	5	4	0	4	
C	1	3	-1	2	
D	2	5	0	5	
E	0	4	-1	3	
F	7	2	+1	3	
G	8	2	+1	3	
H	1	2	-1	1	A-D-G = 10
A	7	2	+1	3	
B	5	4	0	4	
C	8	3	+1	4	
D	5	5	0	5	
E	5	4	0	4	
F	1	2	-1	1	
G	5	2	0	2	
H	0	2	-1	1	A D-G = 10
A	8	2	+1	3	
B	8	4	+1	5	
C	2	3	0	3	
D	5	5	0	5	
E	6	4	0	4	
F	7	2	+1	3	
G	4	2	0	2	
H	4	2	0	2	B-E-H = 11
A	7	2	+1	3	
B	2	4	0	4	
C	3	3	0	3	
D	6	5	0	5	
E	7	4	+1	5	
F	6	2	0	2	
G	3	2	0	2	
H	4	2	0	2	A-D-G = 12
A	8	2	+1	3	
B	5	4	0	4	
C	1	3	-1	2	
D	2	5	0	5	
E	2	4	0	4	
F	5	2	0	2	
G	9	2	+2	4	
H	6	2	0	2	A-D-G = 12

Average project completion time is (10+10+11+12+12)/5 = 11 days.

ANSWERED PROBLEMS

PROBLEM 7

Susan Winslow has two alternative routes to travel from her home in Olport to her office in Lewisburg. She can travel on Freeway 5 to Freeway 57 or on Freeway 55 to Freeway 91. The time distributions are as follows:

Freeway 5		Freeway 57		Freeway 55		Freeway 91	
Time	Relative Frequency	Time	Relative Frequency	Time	Relative Frequency	Time	Relative Frequency
5	.30	4	.10	6	.20	3	.30
6	.20	5	.20	7	.20	4	.35
7	.40	6	.35	8	.40	5	.20
8	.10	7	.20	9	.20	6	.15
		8	.15				

Do a five-day simulation of each of the two combinations of routes using columns 1, 2, 3, and 4 of Appendix E for the random numbers. Based on this simulation, which routes should Susan take if her objective is to minimize her total travel time?

PROBLEM 8

The Rumson Post Office serves a small rural town. In any one minute interval during a Saturday morning either 0, 1, 2, or 3 customers arrive with the following probabilities: $P(0) = .40$; $P(1) = .30$; $P(2) = .20$; $P(3) = .10$.

The only clerk working on Saturday at the post office is Mrs. Smith. Her service time per customer is also a random variable with the following distribution of 1, 2, or 3 minutes per service: $P(1) = .80$; $P(2) = .15$; $P(3) = .05$.

If the system starts empty, simulate the waiting line over a 10 minute interval. Use Appendix E, column 1 to generate the number of arrivals and column 2 to generate service times. What is the average number of customers waiting in line for service?

PROBLEM 9

Consider the following PERT problem.

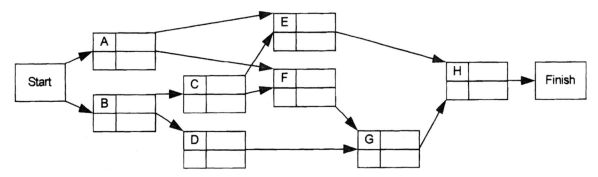

The expected completion time for each job is given below:

Job	Expected Completion Time in Weeks
A	7
B	4
C	3
D	6
E	5
F	3
G	3
H	4

a) Determine the critical path(s) and expected completion time of this project.

b) Suppose that each job has a 25% chance of being completed in a week less than its expected completion time and a 25% chance of being completed in a week more than its expected completion time. Do a five-trial simulation using the last two digits of column 6 of Appendix E for the random numbers to determine the average completion time of this project. Compare this answer to part (a).

PROBLEM 10

Demand for mopeds at the Easy Rider Bike Shop is either 0, 1, 2, or 3 per day with the following probability distribution:

Demand	Probability
0	.15
1	.25
2	.35
3	.25

Each working day a moped remains in inventory costs Easy Rider $2 per moped. It costs Easy Rider $30 to process an order.

The delivery time required to obtain new mopeds from the wholesaler is also random with the following distribution:

Delivery Time (in Working Days)	Probability
1	.20
2	.60
3	.20

Easy Rider has adopted the policy of offering customers an $8 per day discount for each working day they must wait for a moped if Easy Rider is out of stock.

a) Assume Easy Rider is open 200 days per year and delivery time is exactly two working days. Further, assume that although demand varies, it is actually constant at the level of average demand. What would be the optimal order policy?

b) Do a 20-day simulation using the answer you obtained in part (a) assuming a current inventory of 6 mopeds and that delivery time and demand follow the distributions given above. Use the first two digits in column 2 of Appendix E to generate demands and the first two digits of column 3 to generate lead times. What is the average inventory cost per day, where inventory holding costs are assessed on ending inventory?

PROBLEM 11

Miller's Carpets runs a store that carries a certain style of carpet. Over the weeks Miller's has collected data concerning the demand for the number of rolls of this brand of carpet:

Weekly Demand	Probability
0	.10
1	.20
2	.20
3	.20
4	.20
5	.10

Miller's policy has been to reorder 8 rolls whenever its inventory reached 4 rolls or less at the end of the week. Current inventory is 5 rolls. Reordering costs are $40 per order and holding costs are $2 per roll per week. Stockout costs are $10 per occurrence and these sales are lost.

Lead time for an order has been observed to be:

Weeks	Probability
1	.40
2	.50
3	.10

a) Conduct an 8-week simulation of Miller's situation and determine the total cost for this period. Use column 4 of Appendix E for demand and column 6 for lead times.

b) Miller's has been offered a new policy. The wholesaler will automatically deliver 3 rolls to Miller's at the end of each week for a weekly service charge of $12. This eliminates reorder costs and lead times. Should Miller's accept this new policy or keep its current policy?

PROBLEM 12

Customers at Winkies Donuts can either eat their doughnut purchase on the premises or carry it out. Winkies estimates that 35% of its morning business customers enjoy their doughnuts on the premises.

Customers eating their doughnuts on the premises either buy one or two doughnuts and nearly always order a beverage. Winkies estimates the following profits and probabilities hold for service times of the on-the-premises customers:

Service Times	Probability	Expected Profit
25 seconds	.15	$ .08
50 seconds	.60	$ .23
75 seconds	.25	$ .34

Carry-out customers typically purchase a couple of doughnuts or a dozen. Winkies estimates the following profits and probabilities hold for the service times of carry-out customers:

Service Times	Probability	Expected Profit
50 seconds	.30	$ .15
100 seconds	.70	$ .90

During any 25 second interval, either 0, 1, or 2 customers will arrive at Winkies with the following probability distribution:

Customers	Probability
0	.75
1	.16
2	.09

Winkies has only one morning sales clerk on duty.

a) Conduct a 10-minute simulation of sales at Winkies. Using Appendix E, use column 1 for number of arrivals in 25 seconds, column 2 for the eat-in/carry-out decision, column 3 for on-the-premises and carry-out service times.

b) What is the expected profit during this period?

c) What percentage of the time is the server busy?

d) What is the average number of customers at Winkies?

PROBLEM 13

Honest Archie's Appliance Company has decided to cease selling microwave ovens. Archie has slashed prices on the three models he has in stock -- Amana, Litton, and Tappan. He currently has 3 Amanas, 2 Littons, and 4 Tappans left in stock.

Because of his low prices, Archie expects to have from one to four customers come into his store each day inquiring about the ovens. Archie believes the following daily probabilities hold:

Customers	Probability
1	.50
2	.30
3	.10
4	.10

For each inquiry, Archie believes the following probabilities hold: P(not interested in any purchase) = .30; P(desires Amana) = .30; P(desires Litton) = .25; P(desires Tappan) = .15.

If a customer desires a particular brand of oven that Archie is sold out of, Archie will try to sell the customer the brand of oven in which his current stock is the largest. (If there is a tie, he chooses the one that had the largest initial stock.) Archie believes there is a 25% chance he will be successful in convincing the person to switch to that brand and a 75% chance of losing the sale.

Do a simulation to determine the number of days it will take Archie to sell out his stock of microwave ovens. Use column 7 of Appendix E for arriving customers, column 8 for oven preference, and column 9 for brand switching.

PROBLEM 14

As the owner of a rent-a-car agency you have determined the following statistics:

Potential Rentals Daily	Probability	Rental Duration	Probability
0	.10	1 day	.50
1	.15	2 days	.30
2	.20	3 days	.15
3	.30	4 days	.05
4	.25		

The gross profit is $40 per car per day rented. When there is demand for a car when none is available there is a goodwill loss of $80 and the rental is lost. Each day a car is unused costs you $5 per car. Your firm initially has 4 cars.

a) Conduct a 10-day simulation of this business using column 1 of Appendix E for demand and column 2 for rental length.

b) If your firm can obtain another car for $200 for 10 days should you take the extra car?

PROBLEM 15

Scooper Dooper is a small ice cream parlor located next to the Lemonville exit of the Metropolitan Subway Line. Arriving customers to Scooper Dooper purchase either ice cream cones, malts, or sundaes. The following table gives the approximate service time required, expected profit, and the probability of each type of purchase:

Purchase	Probability	Service Time Required	Expected Profit
Ice Cream Cone	.75	30 seconds	$.12
Malt	.15	60 seconds	$.22
Sundae	.10	90 seconds	$.30

The store is small, holding a maximum of four customers. Customers who find the store full go elsewhere for their ice cream. Management estimates that during the lunch hour in each 30 second interval there will be either 0, 1, 2, or 3 arrivals with the following probability distribution:

Number of Arrivals	Probability
0	.30
1	.40
2	.20
3	.10

Simulate the operation of the Scooper Dooper ice cream parlor for a 15-minute period during the lunch hour. From Appendix E, use column 1 to generate the number of arrivals and column 2 to determine the type of purchase.

a) What is the expected profit earned during this period?

b) What percentage of the customers are lost due to a full store?

c) What is the average size of the waiting line?

PROBLEM 16

Three airlines compete on the route between New York and Los Angeles. Stanton Marketing has performed an analysis of first class business travelers to determine their airline choice. Stanton has modeled this choice as a Markov process and has determined the following transition probabilities.

		Next Airline A	B	C
Last Airline	A	.50	.30	.20
	B	.30	.45	.25
	C	.10	.35	.55

a) Show the random number assignments that can be used to simulate the first class business traveler's next airline when her last airline is A, B, and C.

b) Assume the traveler used airline C last. Simulate which airline the traveler will be using over her next 25 flights. What percentage of her flights are on each of the three airlines? (Use column 3 of Appendix E.)

TRUE/FALSE

___ 17. Given the accuracy of computers, it is not necessary to validate a simulation program.

___ 18. Using the next event simulation approach, the time between system updates is variable.

___ 19. A random number mapping always maps a set of occurrences to numbers between 00 and 99.

___ 20. Computer simulation requires a special purpose simulation language.

___ 21. In comparing different policies using simulation, one should use the same set of random numbers whenever possible.

___ 22. Flowcharts are useful in designing a simulation program.

___ 23. In Monte Carlo simulation, outcomes are determined by choosing a random number and selecting the outcome corresponding to that number.

___ 24. A computer simulator is a device which acts like a computer but is not a computer.

___ 25. One is guaranteed an optimal solution to a problem using simulation.

___ 26. A typical way to avoid start-up problems in simulation is to run the program for a specified time without recording any data corresponding to the simulation.

___ 27. Probabilistic inputs to a Monte Carlo simulation must follow a discrete probability distribution.

___ 28. If there are no probabilistic components in a problem, there is no reason to use computer simulation to solve it.

___ 29. Simulation is a trial-and-error approach to problem solving.

___ 30. If a computer simulator is correctly programmed, there will be no difference between the simulated and real distributions for a probabilistic component.

___ 31. If a computer simulator is properly programmed, different random number sequences will not cause different output results.

CHAPTER 16

Markov Processes

KEY CONCEPTS

CONCEPT	ILLUSTRATED PROBLEMS	ANSWERED PROBLEMS
Transition Probabilities	1-4	5-13
Probabilities at Stage n	1,2,3	8,10,11,12,13
Steady State Probabilities	1,2	6,7,8,9,12
Fundamental Matrix	3,4	5,10,11,13
Markov Chain Applications	1-4	5-13

REVIEW

1. <u>Markov process</u> models are useful in studying the evolution of systems over <u>repeated trials</u> or <u>sequential time periods</u> or <u>stages</u>.

2. <u>Transition probabilities</u> govern the manner in which the <u>state</u> of the system changes from one stage to the next. These are often represented in a <u>transition matrix</u>, P.

3. A system has a finite <u>Markov chain</u> with <u>stationary transition probabilities</u> if:
   1) there are a finite number of states,
   2) the transition probabilities remain constant from stage to stage, and
   3) the probability of the process being in a particular state at stage $n+1$ is completely determined by the state of the process at stage n (and not the state at stage n-1). This is referred to as the <u>memory-less property</u>.

4. The <u>state probabilities</u> at any stage of the process can be recursively calculated by multiplying the initial state probabilities by the state of the process at stage n.

5. The probability of the system being in a particular state after a large number of stages is called a <u>steady-state probability</u>. Steady-state probabilities are independent of the initial state of the system.

6. <u>Steady state probabilities</u> can be found by solving the system of equations $\Pi P = \Pi$ together with the condition for probabilities that $\Sigma \pi_i = 1$. Here the matrix P is the transition probability matrix and the vector Π is the vector of steady state probabilities.

MARKOV PROCESSES 361

7. An <u>absorbing state</u> is one in which the probability that the process remains in that state once it enters the state is 1.

8. If a Markov chain has both absorbing and non-absorbing states, the states may be rearranged so that the transition matrix can be written as the following composition of <u>four</u> <u>submatrices</u>: *I*, *0*, *R*, and *Q*:

$$\begin{bmatrix} I & O \\ \hline R & Q \end{bmatrix}$$

where:
- *I* = an identity matrix indicating one always remains in an absorbing state once it is reached,
- *0* = a zero matrix representing 0 probability of transitioning from the absorbing states to the non-absorbing states,
- *R* = the transition probabilities from the non-absorbing states to the absorbing states, and
- *Q* = the transition probabilities between the non-absorbing states.

9. The <u>fundamental matrix</u>, *N*, is the inverse of the difference between the identity matrix and the *Q* matrix, i.e.

$$N = (I - Q)^{-1}$$

10. The <u>*NR* matrix</u>, the product of the fundamental matrix and the *R* matrix, gives the probabilities of eventually moving from each non-absorbing state to each absorbing state. Multiplying any vector of initial non-absorbing state probabilities by *NR* gives the vector of probabilities for the process eventually reaching each of the absorbing states. Such computations enable economic analyses of systems and policies.

362 CHAPTER 16

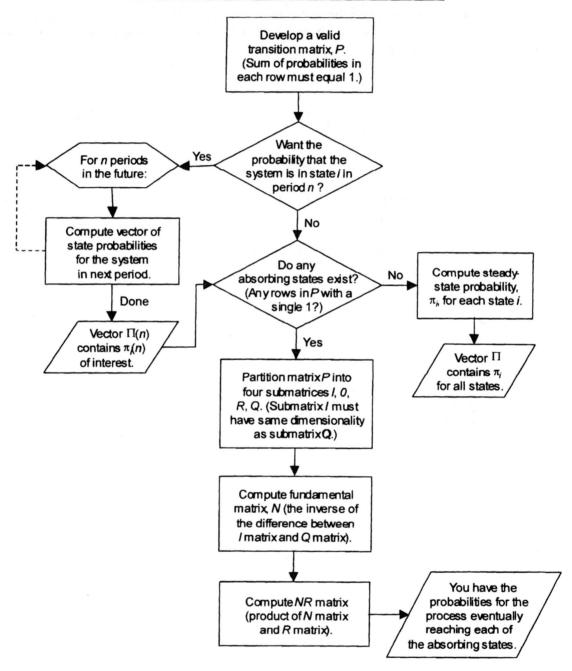

MARKOV PROCESSES 363

ILLUSTRATED PROBLEMS

PROBLEM 1

Henry, a persistent salesman, calls North's Hardware Store once a week hoping to speak with the store's buying agent, Shirley. If Shirley does not accept Henry's call this week, the probability she will do the same next week is .35. On the other hand, if she accepts Henry's call this week, the probability she will not do so next week is .20.

a) Construct the transition matrix for this problem.

b) How many times per year can Henry expect to talk to Shirley?

c) What is the probability Shirley will accept Henry's next two calls if she does not accept his call this week?

d) What is the probability of Shirley accepting exactly one of Henry's next two calls if she accepts his call this week?

SOLUTION 1

a) The transition matrix is:

		Next Week's Call	
		Refuses	Accepts
This Week's Call	Refuses	.35	.65
	Accepts	.20	.80

b) To find the expected number of accepted calls per year, find the long-run probability of a call being accepted and multiply it by 52 weeks.

Let π_1 = the long run proportion of refused calls
π_2 = the long run proportion of accepted calls

Then,

$$[\pi_1 \ \pi_2] \begin{bmatrix} .35 & .65 \\ .20 & .80 \end{bmatrix} = [\pi_1 \ \pi_2]$$

Thus,

$$.35\pi_1 + .20\pi_2 = \pi_1 \quad (1)$$
$$.65\pi_1 + .80\pi_2 = \pi_2 \quad (2)$$

and,

$$\pi_1 + \pi_2 = 1 \quad (3)$$

Solving using equations (2) and (3), (equation 1 is redundant), substitute $\pi_1 = 1 - \pi_2$ into (2) to give:

$$.65(1 - \pi_2) + .80\pi_2 = \pi_2$$

This gives $\pi_2 = .76471$. Substituting back into (3) gives $\pi_1 = .23529$.

Thus the expected number of accepted calls per year is $(.76471)(52) = 39.76$ or about 40.

c) The tree diagram for this problem is:

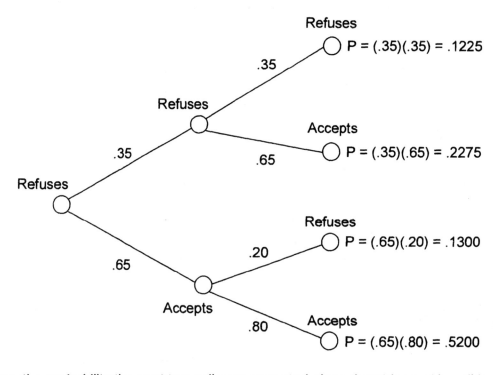

Hence, the probability the next two calls are accepted given that this week's call is refused is $.65(.80) = .52$.

d) The probability of exactly one of the next two calls being accepted if this week's call is accepted can be found by adding the probabilities of (accept next week and refuse the following week) and (refuse next week and accept the following week) = .13 + .16 = .29.

PROBLEM 2

Joe Ferris, a stock trader at the brokerage firm of Smith, Jones, Johnson, and Thomas, Inc. has noticed that price changes in the shares of Dollar Department Stores at each trade are dependent upon the previous trade's price change. His observations can be summarized by the following transition matrix.

		Next Price Change		
		+1/8	0	-1/8
Current Price Change	+1/8	.7	.2	.1
	0	.3	.4	.3
	-1/8	.2	.1	.7

a) What is the long-run average change in the value of a share of Dollar Department Stores' stock per trade?

b) If the shares of Dollar Department Stores are currently traded at $18 and the last trade was at 17 7/8, what is the probability the shares will sell at 18 in two trades?

SOLUTION 2

a) Let, p_1 = the long-run probability of a +1/8 stock price change
p_2 = the long-run probability of a 0 stock price change
p_3 = the long-run probability of a -1/8 stock price change

To determine the long-run (steady state) probabilities, solve:

$$[\pi_1 \; \pi_2 \; \pi_3] \begin{bmatrix} .7 & .2 & .1 \\ .3 & .4 & .3 \\ .2 & .1 & .7 \end{bmatrix} = [\pi_1 \; \pi_2 \; \pi_3]$$

This gives the following set of equations

$$.7\pi_1 + .3\pi_2 + .2\pi_3 = \pi_1 \quad (1)$$
$$.2\pi_1 + .4\pi_2 + .1\pi_3 = \pi_2 \quad (2)$$
$$.1\pi_1 + .3\pi_2 + .7\pi_3 = \pi_3 \quad (3)$$
$$\pi_1 + \pi_2 + \pi_3 = 1 \quad (4)$$

Note, that any one of (1), (2) or (3) can be considered redundant. Delete (1) and solve (2), (3), and (4). From (4) substitute $\pi_1 = 1 - \pi_2 - \pi_3$. This leaves the following two equations in two unknowns:

$$.2(1 - \pi_2 - \pi_3) + .4\pi_2 + .1\pi_3 = \pi_2$$
and $$.1(1 - \pi_2 - \pi_3) + .3\pi_2 + .7\pi_3 = \pi_3$$
or $$.8\pi_2 + .1\pi_3 = .2$$
and $$-.2\pi_2 + .4\pi_3 = .1$$

Solving these two gives $\pi_2 = 7/34$ and $\pi_3 = 12/34$. Substituting into (4) gives $\pi_1 = 15/34$. Hence, the long run average change per trade is:

$$(15/34)(1/8) + (7/34)(0) + (12/34)(-1/8) = \$.011.$$

b) The fact that the last trade was 17 7/8 means that the last price change was +1/8. The tree diagram for the problem is:

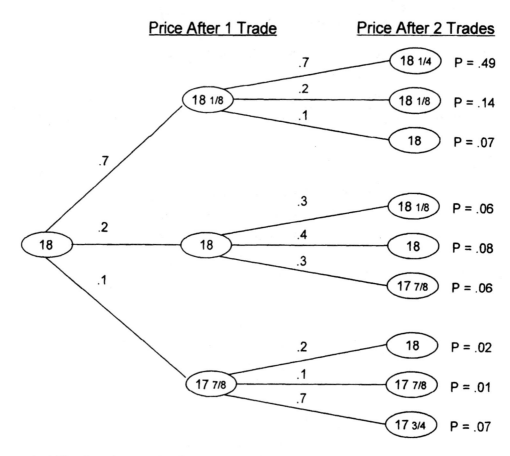

The probability that the stock will sell at 18 after 2 trades is .07 + .08 + .02 = .17.

PROBLEM 3

Joe Isley, the owner of Big I HiFi, believes that the store's inventory can be modeled as a Markov process. If items are either classified as in stock, out of stock, discontinued from stock or put on clearance sale, then the following transition matrix has been estimated:

		Next Month			
		In Stock	Out of Stock	Discontinued from Stock	Put On Clearance Sale
This Month	In Stock	.67	.20	.05	.08
	Out of Stock	.48	.42	.10	0
	Discontinued	0	0	1	0
	Clearance Sale	0	0	0	1

a) Rewrite the transition matrix for the problem in the form of I, O, R, and Q submatrices.

b) Compute the fundamental matrix for this problem.

c) What is the probability of an item currently in stock being out of stock in two months?

d) What is the probability of an item currently out of stock eventually being discontinued from stock?

SOLUTION 3

a) Rearranging the states gives:

		Next Month			
		Discontinued	Clearance	In Stock	Out Of Stock
This Month	Discontinued	1	0	0	0
	Clearance	0	1	0	0
	In Stock	.05	.08	.67	.20
	Out of Stock	.10	0	.48	.42

Note, $R = \begin{bmatrix} .05 & .08 \\ .10 & 0 \end{bmatrix}$ and $Q = \begin{bmatrix} .67 & .20 \\ .48 & .42 \end{bmatrix}$

b)
$$N = (I - Q)^{-1} = \begin{bmatrix} .33 & -.20 \\ -.48 & .58 \end{bmatrix}^{-1}$$

To compute $(I - Q)^{-1}$, first calculate its determinant,

$$d = a_{11}a_{22} - a_{21}a_{12} = (.33)(.58) - (-.48)(-.20) = .0954$$

Then,
$$N = \begin{bmatrix} (.58/.0954) & (.20/.0954) \\ (.48/.0954) & (.33/.0954) \end{bmatrix} = \begin{bmatrix} 6.08 & 2.10 \\ 5.03 & 3.46 \end{bmatrix}$$

c) The tree diagram is:

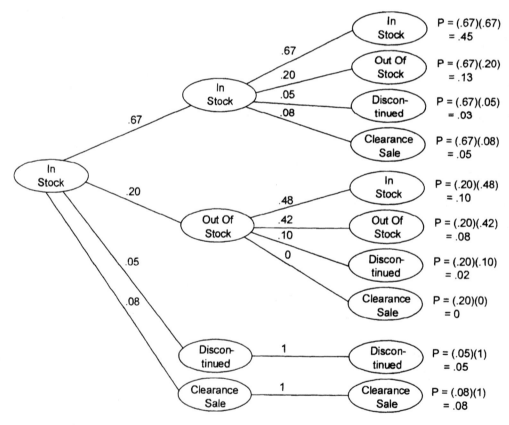

Therefore, the probability of an item currently in stock being out of stock in two months is .13 + .08 = .21.

d) The probability of eventually moving into each of the absorbing states from the nonabsorbing states is given by:

$$NR = \begin{bmatrix} 6.08 & 2.10 \\ 5.03 & 3.46 \end{bmatrix} \times \begin{bmatrix} .05 & .08 \\ .10 & 0 \end{bmatrix}$$

$$= \begin{array}{c} \\ \text{In Stock} \\ \text{Out Of Stock} \end{array} \begin{array}{cc} \text{Discontinued} & \text{Clearance Sale} \\ \begin{bmatrix} .51 & .49 \\ .60 & .40 \end{bmatrix} \end{array}$$

Hence, the probability of an item currently out of stock eventually being discontinued is .60.

> NOTE: If there is more than one absorbing state, a steady-state condition independent of initial state conditions does not exist. For example, in the above problem we cannot state what percentage of all inventory items eventually is discontinued or what percentage is put on clearance.

PROBLEM 4

The vice president of personnel at Jetair Aerospace has noticed that yearly shifts in personnel can be modeled by a Markov process. The transition matrix is:

		Next Year				
		Same Position	Promotion	Retire	Quit	Fired
Current Year	Same Position	.55	.10	.05	.20	.10
	Promotion	.70	.20	0	.10	0
	Retire	0	0	1	0	0
	Quit	0	0	0	1	0
	Fired	0	0	0	0	1

a) Write the transition matrix in the form of I, O, R, and Q submatrices.:

b) Compute the fundamental matrix, N, for this problem.

c) What is the probability of an employee who was just promoted eventually retiring? Quitting? Being fired?

SOLUTION 4

a) Because the identity matrix must be in the upper right corner, rewrite the transition matrix as follows:

		Next Year				
		Retire	Quit	Fired	Same	Promotion
Current Year	Retire	1	0	0	0	0
	Quit	0	1	0	0	0
	Fired	0	0	1	0	0
	Same	.05	.20	.10	.55	.10
	Promotion	0	.10	0	.70	.20

Note,

$$Q = \begin{bmatrix} .55 & .10 \\ .70 & .20 \end{bmatrix} \quad R = \begin{bmatrix} .05 & .20 & .10 \\ 0 & .10 & 0 \end{bmatrix}$$

b)
$$N = (I - Q)^{-1} = \begin{bmatrix} 1 & 0 \\ 0 & 1 \end{bmatrix} - \begin{bmatrix} .55 & .10 \\ .70 & .20 \end{bmatrix}^{-1} = \begin{bmatrix} .45 & -.10 \\ -.70 & .80 \end{bmatrix}^{-1}$$

The determinant, $d = a_{11}a_{22} - a_{21}a_{12} = (.45)(.80) - (-.70)(-.10) = .29$

Thus,

$$N = \begin{bmatrix} .80/.29 & .10/.29 \\ .70/.29 & .45/.29 \end{bmatrix} = \begin{bmatrix} 2.76 & .34 \\ 2.41 & 1.55 \end{bmatrix}$$

c) The probabilities of eventually moving to the absorbing states from the nonabsorbing states are given by:

$$NR = \begin{bmatrix} 2.76 & .34 \\ 2.41 & 1.55 \end{bmatrix} \times \begin{bmatrix} .05 & .20 & .10 \\ 0 & .10 & 0 \end{bmatrix}$$

$$= \begin{array}{c} \text{Same} \\ \text{Promotion} \end{array} \begin{array}{ccc} \text{Retire} & \text{Quit} & \text{Fired} \\ \begin{bmatrix} .14 & .59 & .28 \\ .12 & .64 & .24 \end{bmatrix} \end{array}$$

The probability of someone just being promoted eventually being in one of the absorbing states is given by the bottom row of this probability matrix. The answers are therefore:

Eventually Retiring = .12
Eventually Quitting = .64
Eventually Being Fired = .24

ANSWERED PROBLEMS

PROBLEM 5

Rent-To-Keep rents household furnishings by the month. At the end of a rental month a customer can: a) rent the item for another month, b) buy the item, or c) return the item. The matrix below describes the month-to-month transition probabilities for 32-inch stereo televisions the shop stocks.

		Next Month		
		Rent	Buy	Return
	Rent	.72	.10	.18
This Month	Buy	0	1	0
	Return	0	0	1

What is the probability that a customer who rented a TV this month will eventually buy it?

PROBLEM 6

The evening television news broadcast that individuals view on one evening is influenced by which broadcast they viewed previously. An executive at the C network has determined the following transition probability matrix describing this phenomenon.

		Next Network News Watched		
		A	C	N
Current Network News Watched	A	.80	.12	.08
	C	.08	.85	.07
	N	.08	.09	.83

a) Which network has the most loyal viewers?

b) What are the three networks' long-run market shares?

c) Suppose each of the three networks earns $1,250 in daily profit from advertising revenue for each 1,000,000 viewers it has. If on the average 40,000,000 people watch the evening television news, compute the long run average daily profit each network generates from its evening news broadcast.

PROBLEM 7

On any particular day an individual can take one of two routes to work. Route A has a 25% chance of being congested, whereas route B has a 40% chance of being congested.

The probability of the individual taking a particular route depends on his previous day's experience. If one day he takes route A and it is not congested, he will take route A again the next day with probability .8. If it is congested, he will take route B the next day with probability .7.

On the other hand, if on a day he takes route B and it is not congested, he will take route B again the next day with probability .9. Similarly if route B is congested, he will take route A the next day with probability .6.

a) Construct the transition matrix for this problem. (HINT: There are 4 states corresponding to the route taken and the congestion. The transition probabilities are products of the independent probabilities of congestion and next day choice.)

b) What is the long-run proportion of time that route A is taken?

PROBLEM 8

Mark is a specialist at repairing large metal-cutting machines that use laser technology. His repair territory consists of the cities of Austin, San Antonio, and Houston. His day-to-day repair assignment locations can be modeled as a Markov process. The transition matrix is as follows:

		Next Day's Location		
		Austin	San Antonio	Houston
This	Austin	.35	.35	.30
Day's	San Antonio	.25	.50	.25
Location	Houston	.10	.20	.70

a) Determine the probability of Mark working the next three days in Houston (where his girlfriend lives) if today he is working in Austin.

b) Find the long-run proportions of time Mark will work in each of the three cities.

PROBLEM 9

There are currently only two fast food restaurants in the Main Street area, Burger Prince and Feisty Fowl. The probability of a customer eating at one restaurant is dependent upon which restaurant he ate at previously. The probabilities can be summarized by the following transition matrix.

		Next Restaurant	
		Burger Prince	Feisty Fowl
Last	Burger Prince	.7	.3
Restaurant	Feisty Fowl	.4	.6

Colonel Mustard's Hot Dogs is planning to open a new restaurant next to Burger Prince. Burger Prince's management estimates the new restaurant will result in the transition matrix below. By how much will the long-run market share of Burger Prince and Feisty Fowl change with the opening of Colonel Mustard's.

		Next Restaurant		
		Burger Prince	Feisty Fowl	Colonel Mustard's
Last Restaurant	Burger Prince	.5	.3	.2
	Feisty Fowl	.3	.5	.2
	Col. Mustard's	.3	.3	.4

PROBLEM 10

Precision Craft, Inc. manufactures ornate pedestal sinks. On any day, the status of a given sink is either: a) somewhere in the normal manufacturing process, b) being reworked because of a detected flaw, c) finished successfully, or d) scrapped because a flaw could not be corrected. The transition matrix is:

		Tomorrow's Status			
		In-Process	Rework	Finished	Scrapped
Today's Status	In-Process	.30	.15	.50	.05
	Rework	.40	.10	.30	.20
	Finished	0	0	1	0
	Scrapped	0	0	0	1

a) What is the probability of a sink eventually being finished if it is currently in process?

b) What is the probability of a sink eventually being scrapped if it is currently in rework?

c) What is the probability that a sink currently in rework will have a "finished" status either tomorrow or the next day? (HINT: there are three ways this can happen.)

PROBLEM 11

Southside College has modeled its student loan program as a Markov process. Each year a student with a prior loan borrows again, defers repayment for a year, makes payments, pays the loan balance in full, or defaults on repayment. The transition matrix is as follows:

		Next Year				
		Borrowing	Deferring	Paying	Paid-Off	Default
This Year	Borrowing	.60	.30	0	.10	0
	Deferring	.15	0	.65	.10	.10
	Paying	0	0	.75	.15	.10
	Paid-Off	0	0	0	1	0
	Defaulted	0	0	0	0	1

a) If currently a student is making payments on his/her loan, what is the probability the loan will be paid in full eventually?

b) Is the probability of eventually defaulting greater for a student who is currently borrowing more or a student who is making payments?

c) What is the probability a student who is borrowing this year will repay the loan balance in full in two years or less?

PROBLEM 12

Three airlines compete on the route between New York and Los Angeles. Stanton Marketing has performed an analysis of first class business travelers to determine their airline choice. Stanton has modeled this choice as a Markov process and has determined the following transition probabilities.

		Next Airline		
		A	B	C
	A	.50	.30	.20
Last Airline	B	.30	.45	.25
	C	.10	.35	.55

a) Determine the long-run share for each of the three airlines.

b) Determine the probability of a passenger flying on three different airlines on his next three flights if his next flight is on: (1) A (2) B (3) C

c) The weekly profit of each airline is estimated to be $(y - .28)$ millions of dollars, where y is the airline's market share of the New York - Los Angeles traffic. The management of airline A is contemplating an advertising campaign that it believes will result in new business. Stanton Marketing has projected that such an advertising campaign will result in the following transition probability matrix.

		Next Airline		
		A	B	C
	A	.50	.25	.25
Last Airline	B	.35	.45	.20
	C	.15	.30	.55

What is the maximum amount that management should spend per week on this advertising campaign?

PROBLEM 13

A recent study done by an economist for the Small Business Administration investigated failures of small business. Failures were either classified as due to poor financing, poor management, or a poor product. The failure rates differed for new businesses (under one year old) versus established businesses (over one year old.)

As the result of the economist's study, the following probabilities were determined. For new businesses the probability of failure due to financing was .15, due to management .20, and due to product .05. The corresponding probabilities for established businesses were .10, .06, and .03 respectively.

a) Determine a five-state Markov Chain transition matrix with states for new, established, and each of the three failure states. Write it in the form of I, O, R, and Q submatrices.

b) Determine the probability that a new business will survive during the next three years.

c) What proportion of new businesses eventually fail due to:
(1) poor financing?   (2) poor management?   (3) poor product?

TRUE/FALSE

___ 14. The sum of the probabilities in a transition matrix equals the number of rows in the matrix.

___ 15. All Markov chains have steady-state probabilities.

___ 16. If the initial state probability distribution equals the Markov chain's steady-state probability distribution then all states will have this probability distribution.

___ 17. The steady-state probability distribution is a function of the initial state probability distribution.

___ 18. All Markov chain transition matrices have the same number of rows as columns.

___ 19. The fundamental matrix is used to calculate the probability of the process moving into each absorbing state.

___ 20. A state, i, is an absorbing state if, when $i = j$, $p_{ij} = 1$.

___ 21. A Markov chain cannot consist of all absorbing states.

___ 22. If a Markov chain has at least one absorbing state, steady-state probabilities cannot be calculated.

___ 23. In a Markov chain, p_{ij} must equal p_{ji}.

___ 24. Transition probabilities are joint probabilities.

___ 25. The sum of the "from" probabilities in any "to" column in a transition matrix must equal 1.

___ 26. The probability p_{ij}, when $i = j$, is a steady-state probability.

___ 27. The "memory-less" property of first-order Markov processes refers to not needing to know the state of the system prior to the current state in order to predict the future state.

___ 28. State j is an absorbing state if $p_{ij} = 1$.

CHAPTER 17

Multicriteria Decisions

KEY CONCEPTS

CONCEPT	ILLUSTRATED PROBLEMS	ANSWERED PROBLEMS
Goal Programming		
Formulations	1-3	9-11
Graphical Solution	1	10
Computer Solution	2	10
Scoring Model	4	12
Analytic Hierarchy Process		
Pairwise Comparison Matrices	5-8	13-17
Priority Vectors	5-8	13-17
Overall Priorities	7,8	14-17
Consistency	5	14,15
Hierarchy Construction	8	15,16

REVIEW

1. <u>Multicriteria decision making</u> involves two or more criteria that typically are in <u>conflict</u>. A <u>trade-off</u> among the criteria is usually necessary.

2. <u>Goal programming</u> may be used to solve linear programs with multiple objectives. Each objective may be viewed as a "goal".

3. An approach to goal programming is to satisfy goals in a <u>priority sequence</u>. Second-priority goals are pursued without reducing the first-priority goals; third-priority goals are pursued without reducing first- or second-priority goals, etc.

4. In goal programming, d_i^+ and d_i^- are the amounts a targeted goal i is <u>overachieved or underachieved</u>, respectively. Usually only one of d_i^+ or d_i^- is considered detrimental.

5. For each priority level, the objective function is to <u>minimize the (weighted) sum of the goal deviations</u>. Previous "optimal" achievements of goals are added to the constraint set so that they are not degraded while trying to achieve lesser priority goals. The goals themselves are added to the constraint set with d_i^+ and d_i^- acting as the surplus and slack variables.

6. <u>Infeasible linear programming problems</u> can be reformulated as goal programs so that some reasonable solution may be attained.

7. A <u>scoring model</u> is a quick and relatively easy way to identify the most desired decision alternative in a multicriteria decision problem.

8. The <u>steps required to develop a scoring model</u> are: (1) list the decision-making criteria; (2) assign a weight to, indicating the importance of, each criterion; (3) rate how well each decision alternative satisfies each criterion; (4) compute the score for each decision alternative; and (5) order the decision alternatives from highest score to lowest. The alternative with the highest score is the recommended alternative.

9. The <u>Analytic Hierarchy Process (AHP)</u>, is a procedure designed to quantify managerial judgments of the relative importance of each of several conflicting criteria used in the decision making process.

10. <u>Computer packages</u> can be used for evaluating AHPs. The AHP developed in this chapter is based on one such package, <u>EXPERT CHOICE</u>.

GOAL PROGRAMMING APPROACH

Assuming all of the objectives (goals) and functional constraints in the problem have been identified:

1. Decide the priority level of each goal.

2. If a priority level has more than one goal, for each goal i decide the weight, w_i, to be placed on the deviation(s), d_i^+ and/or d_i^-, from the goal.

3. Set up the initial linear program as follows:

   (Assume d_1^+ and d_2^- are the detrimental deviations from first priority goals 1 and 2, respectively.)

   $$\text{MIN } w_1 d_1^+ + w_2 d_2^-$$

   s.t. Functional Constraints, and
   Goal Constraints

4. Solve this linear program. (Assume that the optimal value of the objective function is k.) If there is a lower priority level, go to step 5; otherwise, a final optimal solution has been reached.

5. Consider the next-lower priority level goals and formulate a new objective function based on these goals. Add a constraint requiring the achievement of the next-higher priority level goals to be maintained. The new linear program is:

   (Assume d_3^+ and d_4^- are the detrimental deviations associated with the current priority level goals.)

   $$\text{MIN } w_3 d_3^+ + w_4 d_4^-$$

   s.t. Functional Constraints,
   Goal Constraints, and
   $w_1 d_1^+ + w_2 d_2^- = k$

Go to step 4. (Repeat steps 4 and 5 until all priority levels have been examined. Each time, solve a linear program with a lesser priority goal for the objective function and with the optimal achieved amounts of higher priority goals as added constraints.)

MULTICRITERIA DECISIONS 381

FLOW CHART FOR
GOAL PROGRAMMING SOLUTION PROCESS

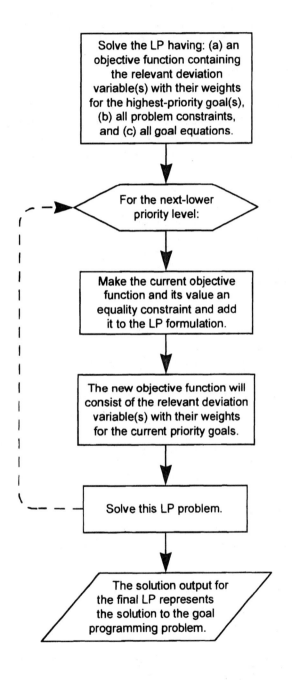

ANALYTIC HIERARCHY PROCESS

1. The first step is:
   (a) list an <u>overall goal</u> for the process;
   (b) list <u>criteria</u> that make up the relevant factors that contribute to achieving the goal;
   (c) list the <u>(n) possible decision alternatives</u> for each of the individual criterion.

   *** For each criterion, perform steps 2 through 5 ***

2. Develop a <u>pairwise comparison matrix</u> for a criterion by rating the relative importance between each pair of decision alternatives. The matrix lists the alternatives horizontally and vertically and has the numerical ratings comparing the horizontal (first) alternative with the vertical (second) alternative. Ratings are given as follows:

Compared to the second alternative, the first alternative is:	Numerical rating
extremely preferred	9
very strongly preferred	7
strongly preferred	5
moderately preferred	3
equally preferred	1

   Intermediate numeric ratings of 8, 6, 4, 2 can be assigned. A reciprocal rating (i.e. 1/9, 1/8, 1/7, etc.) is assigned when the second alternative is preferred to the first. The value of 1 is always assigned when comparing an alternative with itself.

3. Develop the <u>normalized matrix</u> by dividing each number in a column of the pairwise comparison matrix by its column sum.

4. Develop the <u>priority vector for the criterion</u> by averaging each row of the normalized matrix. These row averages form the <u>priority vector</u> of alternative preferences with respect to the particular criterion. The values in this vector sum to 1.

5. The consistency of the subjective input in the pairwise comparison matrix can be measured by calculating a <u>consistency ratio</u>. (See details below.) A consistency ratio of less than .1 is good. For consistency ratios which are greater than .1, the subjective input should be re-evaluated.

6. After steps 2 through 5 has been performed for all criteria, the results of step 4 are summarized in a <u>priority matrix</u> by listing the decision alternatives horizontally and the criteria vertically. The column entries are the priority vectors for each criterion.

7. Develop a <u>criteria pairwise development matrix</u> in the same manner as that used to construct alternative pairwise comparison matrices by using subjective ratings (step 2). Similarly, normalize the matrix (step 3) and develop a <u>criteria priority vector</u> (step 4).

8. Develop an <u>overall priority vector</u> by multiplying the criteria priority vector (from step 7) by the priority matrix (from step 6).

Determining the Consistency Ratio

1. For each row of the pairwise comparison matrix, determine a weighted sum by summing the multiples of the entries by the priority of its corresponding (column) alternative.

2. For each row, divide its weighted sum by the priority of its corresponding (row) alternative.

3. Determine the average, λ_{max}, of the results of step 2.

4. Compute the <u>consistency index</u>, CI, of the n alternatives by:

$$CI = (\lambda_{max} - n)/(n - 1).$$

5. Determine the <u>random index</u>, RI, from the the following chart:

Number of Decision Alternatives, n	Random Index, RI
3	0.58
4	0.90
5	1.12
6	1.24
7	1.32
8	1.41

6. Determine the <u>consistency ratio</u>, CR, as follows:

$$CR = CR/RI.$$

FLOW CHART OF ANALYTIC HIERARCHY PROCESS

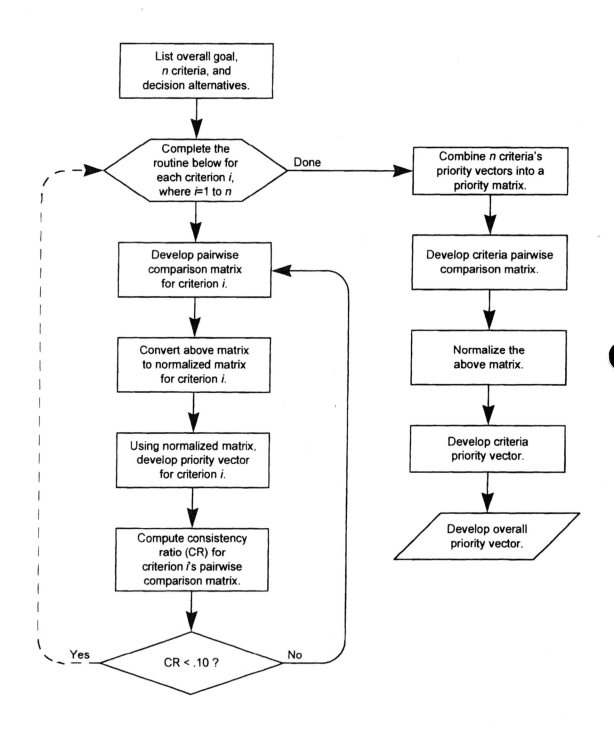

ILLUSTRATED PROBLEMS

> NOTE: Priorities and weights are not synonymous in goal programming. Think of priorities as absolute, firm, preemptive. No tradeoffs occur in the achievement of goals with different priorities. However, tradeoffs can occur in the achievement of goals with the same priority and different weights.

PROBLEM 1

Conceptual Products is a computer manufacturer that produces the CP2860 and the CP3860 computers. It takes one hour to manufacture a CP2860 and its profit is $200 and it takes one and one-half hours to manufacture a CP3860 and its profit is $500.

The two computer models differ in many ways, but they have two components in common, case and hard drive. The CP2860 model uses two hard drives and no DVD drive whereas the CP3860 models use one hard drive and one DVD drive. The hard drives, DVD drives, and cases are bought from vendors. There are 1000 hard drives, 500 DVD drives, and 600 cases available to Conceptual Products on a weekly basis

The company has four goals which are given below.

Priority 1: Meet a state contract of 200 CP2860 machines weekly. (Goal 1)
Priority 2: Make at least 500 total computers weekly. (Goal 2)
Priority 3: Make at least $250,000 weekly. (Goal 3)
Priority 4: Use no more than 400 man-hours per week. (Goal 4)

a) Formulate this problem as a goal program.

b) Solve graphically for the solution that best meets the goals of Conceptual Products as stated above.

c) Suppose Conceptual Products combined goals 3 and 4 into one priority level, Priority 3. What would be the recommendation under the following conditions:

  (1) Each goal was equally desirable.
  (2) Each $1000 underachieved from its profit goal was three times as important as an extra man-hour.
  (3) Each $1000 underachieved from its profit goal was five times as important as an extra man-hour.

> NOTE: A goal constraint is expressed as an equality (=) constraint having exactly one d^+ variable and one d^- variable (and one or more decision variables).

SOLUTION 1

a) Variables

x_1 = the number of CP2860 computers produced weekly
x_2 = the number of CP3860 computers produced weekly

Functional Constraints

Availability of hard drives: $\quad 2x_1 + x_2 \leq 1000$

Availability of DVD drives: $\quad x_2 \leq 500$

Availability of cases: $\quad x_1 + x_2 \leq 600$

Goals

Define: d_i^- = the amount the right hand side of goal i is deficient
d_i^+ = the amount the right hand side of goal i is exceeded

(1) 200 CP2860 computers weekly: $\quad x_1 + d_1^- - d_1^+ = 200$

(2) 500 total computers weekly: $\quad x_1 + x_2 + d_2^- - d_2^+ = 500$

(3) $250 (in thousands) profit: $\quad .2x_1 + .5x_2 + d_3^- - d_3^+ = 250$

(4) 400 total man-hours weekly: $\quad x_1 + 1.5x_2 + d_4^- - d_4^+ = 400$

Non-negativity: $x_1, x_2, d_i^-, d_i^+ \geq 0$ for all i

Objective Functions

Priority 1: Minimize the amount the state contract is not met: MIN d_1^-

Priority 2: Minimize the number under 500 computers produced weekly: MIN d_2^-

Priority 3: Minimize the amount under $250,000 earned weekly: MIN d_3^-

Priority 4: Minimize the man-hours over 400 used weekly: MIN d_4^+

MULTICRITERIA DECISIONS

Summary

Using the notation P_j to denote the priority level of the objective functions, the problem becomes:

$$\text{MIN } P_1(d_1^-) + P_2(d_2^-) + P_3(d_3^-) + P_4(d_4^+)$$

$$\begin{array}{llr}
\text{s. t.} & 2x_1 + x_2 & \leq 1000 \\
& x_2 & \leq 500 \\
& x_1 + x_2 & \leq 600 \\
& x_1 + d_1^- - d_1^+ & = 200 \\
& x_1 + x_2 + d_2^- - d_2^+ & = 500 \\
& .2x_1 + .5x_2 + d_3^- - d_3^+ & = 250 \\
& x_1 + 1.5x_2 + d_4^- - d_4^+ & = 400 \\
\end{array}$$

$$x_1, x_2, d_1^-, d_1^+, d_2^-, d_2^+, d_3^-, d_3^+, d_4^-, d_4^+ \geq 0$$

b) To solve graphically, first graph the functional constraints as below. Then graph the first goal: $x_1 = 200$ and note that there is a set of points that exceeds $x_1 = 200$, i.e. where $d_1^- = 0$.

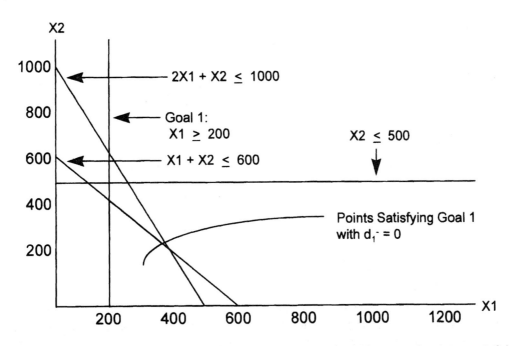

Now add goal 2: $x_1 + x_2 = 500$ as below. Note there is still a set of points satisfying the first goal that also satisfies this second goal, i.e. where $d_2^- = 0$.

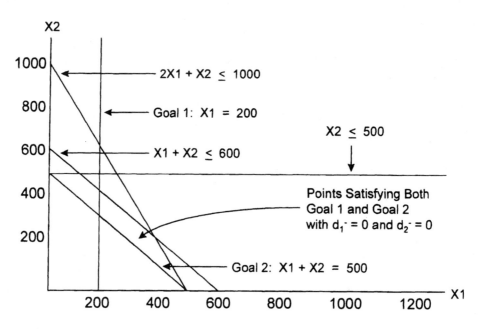

Now add goal 3: $.2x_1 + .5x_2 = 250$ as below. Note that no points satisfy the previous functional constraints and goals as well as this constraint. Thus to MIN d_3^-, this minimum value is achieved when we MAX $.2x_1 + .5x_2$. We note that this occurs at $x_1 = 200$, $x_2 = 400$, so that $.2x_1 + .5x_2 = 240$ or $d_3^- = 10$.

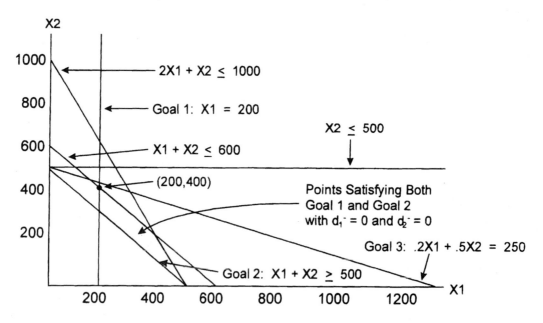

Since this is the only point the meets goal three with only a $10 (thousand) deficiency, priority 4's objective becomes irrelevant. This is because if we add the constraint: $.2x_1 + .5x_2 = 240$, there is only one feasible point.

This is our recommendation of $x_1 = 200$, $x_2 = 400$. Thus the recommendations is to make 200 CP2860 computers and 400 CP3860 computers weekly.

c) The first two priorities were met by a set of points and it was shown that there was no point that met the first two goals and goal 3. Similarly, by looking at the top graph on the previous page, it is seen that there are no points that meet the first two priorities and goal 4. Since both will not be achieved,

$$d_3^- = 250 - .2x_1 - .5x_2$$
$$d_4^+ = x_1 + 1.5x_2 - 400.$$

Case (1): MIN $d_3^- + d_4^+$ = MIN $.8x_1 + x_2 - 150$
Case (2): MIN $3d_3^- + d_4^+$ = MIN $.4x_1 + 350$
Case (3): MIN $5d_3^- + d_4^+$ = MIN $-x_2 + 850$

Ignoring the constant in each of the above, it can be seen in the next three graphs that:

Case (1) is minimized at $x_1 = 500$, $x_2 = 0$
Case (2) is minimized by all points on the line $x_1 = 200$ between $x_2 = 300$ and $x_2 = 400$
Case (3) is minimized where x_2 takes on its maximum value, i.e. at $x_1 = 200$, $x_2 = 400$

CASE (1)

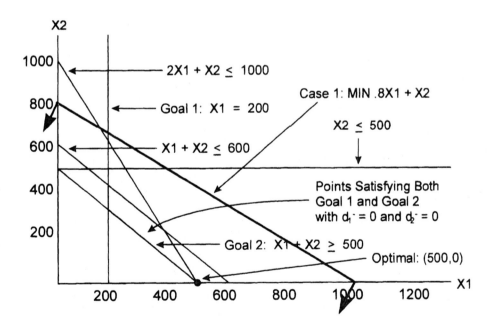

CASE (2)

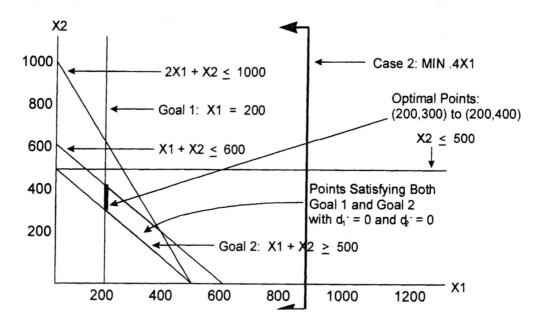

CASE (3)

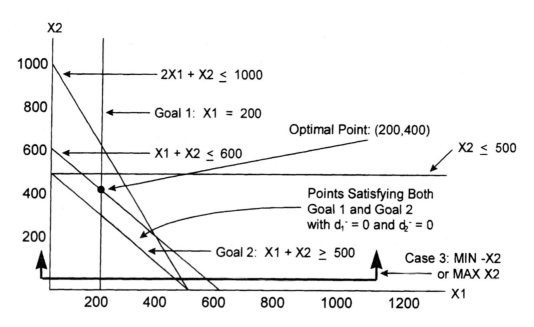

PROBLEM 2

Suppose in problem 1, Conceptual Products made a third computer, the CP4860 which requires two hard drives and a DVD drive. It takes two hours to manufacture and the profit is $900. Use a computer program such as *The Management Scientist* to solve the goal program of Case (3) in part (c) of problem 1.

> NOTE: A goal programming problem with p preemptive goal priorities can be solved as a series of p linear programming problems.

> NOTE: When we solve a goal programming problem as a series of LPs, the objective functions (always to be <u>minimized</u>) will include only relevant deviation variables and <u>not</u> decision variables.

SOLUTION 2

The goal programming formulation is now:

$$\text{MIN } P_1(d_1^-) + P_2(d_2^-) + P_3(5d_3^-) + P_3(d_4^+)$$

$$\begin{aligned}
\text{s.t.} \quad & 2x_1 + x_2 + 2x_3 && \leq 1000 && \text{(Hard drives)} \\
& x_2 + x_3 && \leq 500 && \text{(DVD drives)} \\
& x_1 + x_2 + x_3 && \leq 600 && \text{(Cases)}
\end{aligned}$$

(Priority 1) $\quad x_1 \qquad\qquad\qquad + d_1^- - d_1^+ = 200 \quad$ (Goal 1: contract)

(Priority 2) $\quad x_1 + x_2 + x_3 + d_2^- - d_2^+ = 500 \quad$ (Goal 2: total)

(Priority 3) $\quad .2x_1 + .5x_2 + .9x_3 + d_3^- - d_3^+ = 250 \quad$ (Goal 3: profit)

(Priority 3) $\quad x_1 + 1.5x_2 + 2x_3 + d_4^- - d_4^+ = 400 \quad$ (Goal 4: man-hours)

$$x_1, x_2, d_i^-, d_i^+ \geq 0 \text{ for all } i$$

Priority 1 Program

The following was input into *The Management Scientist*:

MIN D1MINUS

S.T. 2X1 + X2 + 2X3 < 1000

         X2 + X3 < 500

    X1 + X2 + X3 < 600

    X1 + D1MINUS − D1PLUS = 200

    X1 + X2 + X3 + D2MINUS − D2PLUS = 500

  .2X1 + .5X2 + .9X3 + D3MINUS − D3PLUS = 250

    X1 + 1.5X2 + 2X3 + D4MINUS − D4PLUS = 400

The following output was attained:

OBJECTIVE FUNCTION VALUE = 0.000

VARIABLE	VALUE	REDUCED COSTS
X1	200.000	0.000
X2	0.000	0.000
X3	233.333	0.000
D1MINUS	0.000	1.000
D1PLUS	0.000	0.000
D2MINUS	66.667	0.000
D2PLUS	0.000	0.000
D3MINUS	0.000	0.000
D3PLUS	0.000	0.000
D4MINUS	0.000	0.000
D4PLUS	266.667	0.000

Thus the priority 1 objective is met since the value of the objective function, D1MINUS = 0.

Priority 2 Program

Add to the constraints on the previous page a constraint requiring the priority 1 objective to be maintained at 0: D1MINUS = 0, <u>and</u> change the objective to the priority 2 objective: MIN D2MINUS.

The following output is generated:

OBJECTIVE FUNCTION VALUE = 0.000

VARIABLE	VALUE	REDUCED COSTS
X1	285.714	0.000
X2	0.000	0.000
X3	214.286	0.000
D1MINUS	0.000	0.000
D1PLUS	85.714	0.000
D2MINUS	0.000	1.000
D2PLUS	0.000	0.000
D3MINUS	0.000	0.000
D3PLUS	0.000	0.000
D4MINUS	0.000	0.000
D4PLUS	314.286	0.000

The priority 2 objective is met since the value of the objective function, D2MINUS, = 0.

Priority 3 Program

Add to the previous constraints, a constraint requiring that the priority 2 objective to be maintained at 0: D2MINUS = 0, and change the objective to the priority 3 objective: MIN 5 D3MINUS + D4PLUS.

The following output is generated:

OBJECTIVE FUNCTION VALUE = 314.286

VARIABLE	VALUE	REDUCED COSTS
X1	285.714	0.000
X2	0.000	0.071
X3	214.286	0.000
D1MINUS	0.000	0.000
D1PLUS	85.714	0.000
D2MINUS	0.000	0.000
D2PLUS	0.000	0.714
D3MINUS	0.000	3.571
D3PLUS	0.000	1.429
D4MINUS	0.000	1.000
D4PLUS	314.286	0.000

Thus the optimal recommendation is to produce 285.714286 CP2860 computers weekly and 214.285714 CP4860 computers weekly. All goals will be met except goal 4. 314.285714 extra man-hours or a total of 714.285714 man-hours will be used.

PROBLEM 3

The campaign headquarters of Jerry Black, a candidate for the Board of Supervisors, has 100 volunteers. With one week to go in the election, there are three major strategies remaining: media advertising, door-to-door canvassing, and telephone campaigning. It is estimated that each phone call will take approximately four minutes and each door-to-door personal contact will average seven minutes. These times include time between contacts for breaks, transportation, dialing, etc. Volunteers who work on advertising will not be able to handle any other duties. Each ad will utilize the talents of three workers for the entire week.

Volunteers are expected to work 12 hours per day during the final seven days of the campaign. At a minimum, Jerry Black feels he needs 30,000 phone contacts, 20,000 personal contacts, and three advertisements during the last week. However, he would like to see 50,000 phone contacts and 50,000 personal contacts made and five advertisements developed. It is felt that advertising is 50 times as important as personal contacts which in turn is twice as important as phone contacts.

Formulate and solve this goal programming problem with a single weighted priority to determine how the work should be distributed during the final week of the campaign.

SOLUTION 3

Define variables

x_1 = number of volunteers doing phone work during the week
x_2 = number of volunteers making personal contacts during week
x_3 = number of volunteers preparing advertising during the week

Define goals

1) 50,000 phone contacts:
   d_1^+ and d_1^- = the amount this quantity is overachieved and underachieved, respectively

2) 50,000 personal contacts:
   d_2^+ and d_2^- = the amount this quantity is overachieved and underachieved, respectively

3) 5 advertisements:
   d_3^+ and d_3^- = the amount this quantity is overachieved and underachieved, respectively

Define objective

It would not hurt Jerry Black if his goals were exceeded, however since the importance of his goals are in the ratio 1:2:100, the goal programming objective function would be:

MIN $d_1^- + 2d_2^- + 100d_3^-$

Define constraints

There are (7 days) x (12 hours per day) x (60 minutes per hour) = 5040 minutes per worker. Thus a phone worker could make 5040/4 = 1260 phone calls in the week. And a door-to-door canvasser could make 5040/7 = 720 personal contacts during the week.

Linear Programming Constraints

1) At least 30,000 phone contacts: $1260x_1 \geq 30{,}000$

2) At least 20,000 personal contacts: $720x_2 \geq 20{,}000$

3) At least 3 advertisements: $(1/3)x_3 \geq 3$

4) 100 volunteers: $x_1 + x_2 + x_3 = 100$

Goal Constraints

5) Make 50,000 phone contacts: $1260x_1 - d_1^+ + d_1^- = 50{,}000$

6) Make 50,000 personal contacts: $720x_2 - d_2^+ + d_2^- = 50{,}000$

7) Design 5 advertisements: $(1/3)x_3 - d_3^+ + d_3^- = 5$

Non-negativity of Variables

$x_1, x_2, x_3, d_1^+, d_1^-, d_2^+, d_2^-, d_3^+, d_3^- \geq 0$

PROBLEM 4

John Harris is interested in purchasing a new Harley-Davidson motorcycle. He has narrowed his choice to one of three models: Sportster Classic, Heritage Softtail, and Electra Glide. After much consideration, John has determined his decision-making criteria, assigned a weight to each criterion, and rated how well each model alternative satisfies each criterion.

Criterion	Weight	Sportster Classic	Heritage Softtail	Electra Glide
Wind protection	5	3	6	8
Fuel tank mileage	3	5	7	6
Passenger comfort	2	5	6	8
Seat height	3	8	5	6
Acceleration	4	8	5	3
Weight	3	8	6	3
Storage capacity	3	4	5	8

Using a scoring model, determine the recommended motorcycle model for John.

SOLUTION 4

Step 4: Compute the score for each decision alternative.
$$S_j = \sum_i w_i r_{ij}$$

$S_1 = 5(3)+3(5)+2(5)+3(8)+4(8)+3(8)+3(4) = 132$

$S_2 = 5(6)+3(7)+2(6)+3(5)+4(5)+3(6)+3(5) = 131$

$S_3 = 5(8)+3(6)+2(8)+3(6)+4(3)+3(3)+3(8) = \underline{137}$

Step 5: Order the decision alternatives from highest score to lowest score. The alternative with the highest score is the recommended alternative.

The <u>Electra Glide model</u> has the highest score and is the <u>recommended decision alternative</u>. Note that both the Sportster and the Electra Glide ranked first in 3 of 7 criteria, but due to the weights of the criteria the Electra Glide prevailed.

Spreadsheet showing data

	A	B	C	D	E
2			Sportster	Heritage	Electra
3	Criteria	Weight	Classic	Softtail	Glide
4	Wind protection	5	3	6	8
5	Fuel tank mileage	3	5	7	6
6	Passenger comfort	2	5	6	8
7	Seat height	3	8	5	6
8	Acceleration	4	8	5	3
9	Weight	3	8	6	3
10	Storage capacity	3	4	5	8

Spreadsheet showing formulas

	A	B	C	D	E
12					
13	SCORING CALCULATIONS				
14			Sportster	Heritage	Electra
15	Criteria		Classic	Softtail	Glide
16	Wind protection		=B4*C4	=B4*D4	=B4*E4
17	Fuel tank mileage		=B5*C5	=B5*D5	=B5*E5
18	Passenger comfort		=B6*C6	=B6*D6	=B6*E6
19	Seat height		=B7*C7	=B7*D7	=B7*E7
20	Acceleration		=B8*C8	=B8*D8	=B8*E8
21	Weight		=B9*C9	=B9*D9	=B9*E9
22	Storage capacity		=B10*C10	=B10*D10	=B10*E10
23	Score		=sum(C16:C22)		
24				=sum(D16:D22)	
25					=sum(E16:E22)

Spreadsheet showing solution

12	A	B	C	D	E
13	SCORING CALCULATIONS				
14			Sportster	Heritage	Electra
15	Criteria		Classic	Softtail	Glide
16	Wind protection		15	30	40
17	Fuel tank mileage		15	21	18
18	Passenger comfort		10	12	16
19	Seat height		24	15	18
20	Acceleration		32	20	12
21	Weight		24	18	9
22	Storage capacity		12	15	24
23	Score		132	131	137

PROBLEM 5

Designer Gill Glass must decide which of three manufacturers will develop his "signature" toothbrushes. Three factors seem important to Gill: (1) his costs; (2) reliability of the product; and, (3) delivery time of the orders.

The three manufacturers are Cornell Industries, Brush Pik, and Picobuy. Cornell Industries will sell toothbrushes to Gill Glass for $100 per gross, Brush Pik for $80 per gross, and Picobuy for $144 per gross. Gill has decided that in terms of price, Brush Pik is moderately preferred to Cornell and very strongly preferred to Picobuy. In turn Cornell is strongly to very strongly preferred to Picobuy.

a) Form the pairwise comparison matrix for cost.

b) Calculate the normalized matrix for cost.

c) Determine the priority vector for cost.

d) Are Gill Glass's responses to cost consistent? Explain.

SOLUTION 5

a) Because Brush Pik is moderately preferred to Cornell, Cornell's entry in the Brush Pik row is 3 and Brush Pik's entry in the Cornell row is 1/3. Because Brush Pik is very strongly preferred to Picobuy, Picobuy's entry in the Brush Pik row is 7 and Brush Pik's entry in the Picobuy row is 1/7. Because Cornell is strongly to very strongly preferred to Picobuy, Picobuy's entry in the Cornell row is 6 and Cornell's entry in the Picobuy row is 1/6.

All diagonal entries are 1. Hence:

Pairwise comparison matrix for cost

	Cornell	Brush Pik	Picobuy
Cornell	1	1/3	6
Brush Pik	3	1	7
Picobuy	1/6	1/7	1

b) To determine normalized matrix, divide each entry in the matrix by its corresponding column sum. For Cornell the column sum = 1 + 3 + 1/6 = 25/6. For Brush Pik the column sum is 1/3 + 1 + 1/7 = 31/21. For Picobuy the column sum is 6 + 7 + 1 = 14. This gives:

Normalized matrix for cost

	Cornell	Brush Pik	Picobuy
Cornell	6/25	7/31	6/14
Brush Pik	18/25	21/31	7/14
Picobuy	1/25	3/31	1/14

c) The priority vector is determined by averaging the row entries in the normalized matrix. Converting to decimals we get:

Priority vector for cost

Cornell: $(6/25 + 7/31 + 6/14)/3 = \begin{bmatrix} .298 \\ .632 \\ .069 \end{bmatrix}$
Brush Pik: $(18/25 + 21/31 + 7/14)/3 =$
Picobuy: $(1/25 + 3/31 + 1/14)/3 =$

d) To check consistency,

(1) Multiply each column of the pairwise comparison matrix by its priority:

$$.298 \begin{bmatrix} 1 \\ 3 \\ 1/6 \end{bmatrix} + .632 \begin{bmatrix} 1/3 \\ 1 \\ 1/7 \end{bmatrix} + .069 \begin{bmatrix} 6 \\ 7 \\ 1 \end{bmatrix} = \begin{bmatrix} .923 \\ 2.009 \\ .209 \end{bmatrix}$$

(2) Divide these number by their priorities to get:

$.923/.298 = 3.097$
$2.009/.632 = 3.179$
$.209/.069 = 3.029$

(3) Average the above results to get λ_{max}.

$\lambda_{max} = (3.097 + 3.179 + 3.029)/3 = 3.102$

(4) Compute the consistence index, CI, for two terms by:

$$CI = (\lambda_{max} - n)/(n - 1) = (3.102 - 3)/2 = .051.$$

(5) Compute the consistency ratio, CR, by CI/RI, where RI = .58 for 3 factors:

$$CR = CI/RI = .051/.58 = .088$$

Since the consistency ratio, CR, is less than .10, this is well within the acceptable range for consistency.

PROBLEM 6

Referring to problem (4), Gill Glass has determined that for reliability, Cornell is very strongly preferable to Brush Pik and equally to moderately preferable to Picobuy. Also, Picobuy is strongly preferable to Brush Pik.

Regarding delivery time, Cornell is equally preferred with Picobuy. Both Cornell and Picobuy are very strongly to extremely preferable to Brush Pik.

a) Construct pairwise comparison matrices for reliability and for delivery time.

b) Construct priority vectors for reliability and delivery time.

SOLUTION 6

a) **Pairwise comparison matrix for reliability**

	Cornell	Brush Pik	Picobuy
Cornell	1	7	2
Brush Pik	1/7	1	5
Picobuy	1/2	1/5	1

Pairwise comparison matrix for delivery time

	Cornell	Brush Pik	Picobuy
Cornell	1	8	1
Brush Pik	1/8	1	1/8
Picobuy	1	8	1

b) For reliability the column sums are 23/14, 41/5, and 8. Dividing each entry by its corresponding column sum, we get:

Normalized matrix for reliability

	Cornell	Brush Pik	Picobuy
Cornell	14/23	35/41	2/8
Brush Pik	2/23	5/41	5/8
Picobuy	7/23	1/41	1/8

Priority vector for reliability

$$\begin{array}{ll} \text{Cornell:} & (14/23 + 35/41 + 2/8)/3 \\ \text{Brush Pik:} & (2/23 + 5/41 + 5/8)/3 \\ \text{Picobuy:} & (7/23 + 1/41 + 1/8)/3 \end{array} = \begin{bmatrix} .571 \\ .278 \\ .151 \end{bmatrix}$$

For delivery time the column sums are 17/8, 17, and 17/8 respectively. Dividing each entry by its corresponding column sum gives:

Normalized matrix for delivery time

	Cornell	Brush Pik	Picobuy
Cornell	8/17	8/17	8/17
Brush Pik	1/17	1/17	1/17
Picobuy	8/17	8/17	8/17

Priority vector for delivery time

$$\begin{array}{ll} \text{Cornell:} & (8/17 + 8/17 + 8/17)/3 \\ \text{Brush Pik:} & (1/17 + 1/17 + 1/17)/3 \\ \text{Picobuy:} & (8/17 + 8/17 + 8/17)/3 \end{array} = \begin{bmatrix} .471 \\ .059 \\ .471 \end{bmatrix}$$

PROBLEM 7

The accounting department at Gill Glass (problems (4) and (5)) has determined that in terms of criteria, cost is extremely preferred to delivery time and very strongly preferred to reliability, and that reliability is very strongly preferred to delivery time.

a) Construct a pairwise comparison matrix for the criteria.

b) Construct a normalized pairwise matrix for the criteria.

c) Determine a priority vector for the criteria.

d) Determine an overall priority vector for the decision alternatives based on the criteria of the accounting dept.

SOLUTION 7

a) **Pairwise comparison matrix for criteria**

	Cost	Reliability	Delivery
Cost	1	7	9
Reliability	1/7	1	7
Delivery	1/9	1/7	1

b) The column sums are 79/63, 57/7, and 17 respectively. Dividing each entry by its corresponding column sum gives:

Normalized matrix for criteria

	Cost	Reliability	Delivery
Cost	63/79	49/57	9/17
Reliability	9/79	7/57	7/17
Delivery	7/79	1/57	1/17

c) Average the rows of the normalized matrix to get:

Priority vector for criteria

$$\begin{aligned}
\text{Cost:} & \quad (63/79 + 49/57 + 9/17)/3 \\
\text{Reliability:} & \quad (\ 9/79 + \ 7/57 + 7/17)/3 \\
\text{Delivery:} & \quad (\ 7/79 + \ 1/57 + 1/17)/3
\end{aligned} = \begin{bmatrix} .729 \\ .216 \\ .055 \end{bmatrix}$$

d) The overall priorities are determined by multiplying the priority vector of the criteria by the priorities for each decision alternative for each objective:

Priority Vector
for criteria -> [.729   .216   .055]

	Cost	Reliability	Delivery
Cornell	.298	.571	.471
Brush Pik	.632	.278	.059
Picobuy	.069	.151	.471

(Priority Matrix)

Overall priority vector

$$\begin{aligned}
\text{Cornell:} & \quad (.729)(.298) + (.216)(.571) + (.055)(.471) \\
\text{Brush Pik:} & \quad (.729)(.632) + (.216)(.278) + (.055)(.059) \\
\text{Picobuy:} & \quad (.729)(.069) + (.216)(.151) + (.055)(.471)
\end{aligned} = \begin{bmatrix} .366 \\ .524 \\ .109 \end{bmatrix}$$

Thus, Brush Pik appears to be the overall recommendation.

PROBLEM 8

A student has one quantitative elective left to select to complete his graduation requirements. The two quantitative electives that are available are an advanced management science class (MS) and an advanced statistics class (STAT). Two factors which are important to the student in his selection process are relevance (R) and difficulty (D). The student formulated the following pairwise consistency matrices:

Criteria	R	D
R	1	1/3
D	3	1

Relevance	MS	STAT
MS	1	1/3
STAT	3	1

Difficulty	MS	STAT
MS	1	5
STAT	1/5	1

a) Draw the hierarchy for this decision problem.

b) Compute the priorities for each of the pairwise comparison matrices.

c) Determine an overall priority for the course selection process.

SOLUTION 8

a) First, list the overall goal: Select the best course. Then, list the evaluation criteria: Relevance, Difficulty. Last, list the alternatives: Management Science, Statistics.

b) Priority Vectors:

Relevance

The column sums are 4 and 4/3 respectively. Dividing the entries by the column sums yields the following <u>normalized matrix</u>:

	MS	STAT
MS	1/4	1/4
STAT	3/4	3/4

Averaging the rows gives the following <u>priority vector</u> for relevance:

$$\begin{matrix} MS \\ STAT \end{matrix} \begin{bmatrix} 1/4 \\ 3/4 \end{bmatrix}$$

Difficulty

The column sums are 6/5 and 6 respectively. Dividing the entries by the column sums yields the following <u>normalized matrix</u>:

	MS	STAT
MS	5/6	5/6
STAT	1/6	1/6

Averaging the rows gives the following <u>priority vector</u> for difficulty:

$$\begin{array}{c} \text{MS} \\ \text{STAT} \end{array} \begin{bmatrix} 5/6 \\ 1/6 \end{bmatrix}$$

Criteria

The column sums are 4 and 4/3 respectively. Dividing the entries by the column sums yields the following <u>normalized matrix</u>:

	Rel.	Dif.
Rel.	1	1/3
Dif.	3	1

Averaging the rows gives the following <u>priority vector</u> for criteria:

$$\begin{array}{c} \text{Relevance} \\ \text{Difficulty} \end{array} \begin{bmatrix} 1/4 \\ 3/4 \end{bmatrix}$$

c) Determine the overall priorities for MS and STAT by multiplying their relevance and difficulty vectors by their priority:

Priority vector
for criteria --> [1/4   3/4]

	Rel.	Dif.
Rel.	1/4	5/6
Dif.	3/4	1/6

(Priority Matrix)

<u>Overall priority vector</u>

$$\begin{array}{l} \text{MS:} \ (.25)(.25) + (.75)(.833) = \\ \text{STAT:} \ (.25)(.75) + (.75)(.167) = \end{array} \begin{bmatrix} .6875 \\ .3125 \end{bmatrix}$$

Thus, MS appears to be the overall recommendation.

404 CHAPTER 17

ANSWERED PROBLEMS

PROBLEM 9

Alfax Industries is trying to promote a new product which it recently developed. It wishes to restrict advertising to television and radio ads. Television ads cost $50,000 each to produce and radio ads $15,000 each. Each television ad will require the use of three Alfax marketing employees and each radio ad will require one. There are 24 persons in the marketing department. Management requires a minimum of six total ads monthly.

Alfax has set the following goals for the production of ads:

(1) Do not exceed a monthly advertising budget of $250,000.
(2) Do not use more than 50% of its marketing personnel on this project.
(3) Produce at least 4 television ads monthly.
(4) Produce at least 4 radio ads monthly.

a) Using the goals as simple constraints, formulate the constraint set for this problem and show it is infeasible.

b) Solve the above goal program graphically with four levels of priority.

c) Suppose goals (3) and (4) were within the same priority level but with the goal of producing at least 4 television ads monthly deemed twice as important as making 4 radio ads. Resolve this problem with three levels of priority.

PROBLEM 10

Barry College has received a $200,000 donation for its scholarship fund to be used for $3000 athletic scholarships, $2500 minority scholarships and $2000 women's scholarships. The donor, an avid sports fan, has stipulated at least 20 athletic scholarships must be awarded.

The Board of Trustees of Barry College has three levels of priority for awarding the scholarships:

(1) It would like at least 80 total scholarships.
(2) It would like no more than 25% of the scholarships to be athletic scholarships.
(3) It would like at least 25 athletic scholarships, 40 minority scholarships, and 30 women's scholarships. (Meeting the target for minorities is viewed as three times as important as meeting the target for athletes or women.)

Formulate a goal program for Barry College and solve by a computer package such as the Management Scientist.

PROBLEM 11

Universal Electric (UE) has facilities all over the United States. Currently UE has 150,000 employees, 5,000 of which are in management positions. The government contends that UE is delinquent in its Affirmative Action policies and will take action if UE does not rectify the situation.

Although UE currently has 12,000 minority employees (8% of its total), only 50 are in management positions (or 1% of the management positions). UE has submitted a plan to the federal government which expresses that UE has a target of 20% minority employees by the end of the year. In negotiating with various minority groups, UE has promised that by the end of the year, at least 10% of its management positions will be held by minorities.

Attrition rates of all employees (both management and otherwise) are 8% for non-minorities, 4% for minorities. Because of a good year, UE will create 6,000 new positions, 200 of which will be in management. However, the mandate from the top is that it would be unwise from a community relations point of view, if more than 2/3 of the new employee positions and more than 2/3 of the promotions to management positions be minorities.

Assume all management positions will be filled in-house and hence all hiring will be for non-management positions. Further assume violations of the company mandate have the same weight as not meeting the goal of 10% minority management positions. However, not meeting federal standards is considered three times more serious.

Formulate and solve this problem as a goal program with a single weighted priority.

PROBLEM 12

Rudy's Barbecue is in need of an office management software package. After considerable research, Rudy has narrowed his choice to one of three packages: SoftTrack, VersaSuite, and N-able. He has determined his decision-making criteria, assigned a weight to each criterion, and rated how well each alternative satisfies each criterion.

		Decision Alternative		
Criterion	Weight	Soft-Track	Versa-Suite	N-Able
Ease of use	4	3	5	8
Report generation	3	8	7	6
Functional integration	5	5	8	6
On-line help	3	8	6	4
Entry error-checking	2	8	3	4
Price	4	4	7	5
Support cost	3	6	5	7

Using a scoring model, determine the recommended software package for Rudy's.

PROBLEM 13

Consider a beauty pageant in which there are three finalists. They are to be judged on (1) evening gown; (2) swim suit; and, (3) answering a corny question posed by an equally corny MC. The three finalists are Miss Northern State, Miss Central State, and Miss Southern State.

It is known that Judge Jones believes that swim suit is strongly more important than evening gown appearance and extremely more important than answering the question. Further, he feels that the evening gown appearance is very strongly more important than answering the question.

In the evening gown competition, Miss Central State performed very strongly compared to Miss Southern State and strongly compared to Miss Northern State. Miss Northern State compared equally to moderately more strongly than Miss Southern State.

In the swim suit competition, Miss Southern State was extremely preferred to both Miss Northern State or Miss Central State and Miss Northern State was moderately preferred to Miss Central State.

In the question competition, Miss Northern State was extremely preferred to Miss Southern State and strongly preferred to Miss Central State. Miss Central State was moderately preferred to Miss Southern State.

Judge Jones has 100 points to distribute among the three finalists. Using his observations and preferences, how should he divide his 100 points between the three finalists?

PROBLEM 14

The Drezners have a choice of three neighborhood supermarkets: Gamma Delta, Bill's, and Hewes. Five factors are important to the Drezners: (1) Location; (2) Overall Prices; (3) Cleanliness; (4) Ease of Parking; and, (5) Selection/Quality. The pairwise comparison matrices are shown below:

Location:

	GD	B	H
GD	1	1/2	4
B	2	1	3
H	1/4	1/3	1

Overall Prices:

	GD	B	H
GD	1	1/2	4
B	2	1	3
H	1/4	1/3	1

Cleanliness:

	GD	B	H
GD	1	1/2	4
B	2	1	3
H	1/4	1/3	1

Ease of Parking:

	GD	B	H
GD	1	1/2	4
B	2	1	3
H	1/4	1/3	1

Selection/Quality:

	GD	B	H
GD	1	1/2	4
B	2	1	3
H	1/4	1/3	1

a) Determine the priority vector for each of the criteria.

b) If all five criteria were equally important, what would be the overall supermarket priority vector?

Now assume the Drezners feel that:
1) Location and ease of parking have equal priority.
2) Overall price is moderately more important than cleanliness or selection, and is very strongly more important than location or ease of parking.
3) Cleanliness is strongly more important than ease of parking, very strongly more important than location, and moderately more important than selection.
4) Selection/quality are moderately more important than location and very strongly more important than parking.

c) Given these preferences, construct a criteria pairwise comparison matrix.

d) Calculate the consistency ratio for the criteria and comment.

e) Determine the overall priority vector for the Drezners.

PROBLEM 15

Abraham L. Ford is a lifelong Republican who has a dilemma in the upcoming election. In general, Mr. Ford very strongly prefers Republicans over Democrats and strongly prefers Republicans over Independents. He moderately favors Independents over Democrats.

However, in the upcoming election, on the issues he strongly to very strongly favors Democrat Fritz Carter over Republican Ron Nixon and strongly favors Fritz Carter over Independent George Anderson. He moderately favors Anderson over Nixon. He has decided that issues are moderately to strongly more important to him than party.

a) Determine the hierarchy for this problem.

b) Determine the pairwise comparison matrix for each alternative.

c) Determine the priority vector for the alternatives for each criterion.

d) Determine the consistency ratio for the issues criterion.

e) Determine a priority vector for the criteria.

f) Determine Mr. Ford's overall priority vector for the candidates.

PROBLEM 16

Terry's Trucking is trying to determine which database package to purchase for its microcomputer. It has narrowed its choices to BASE 8 and DATA RECORD. BASE 8 is very strongly to extremely preferred in price, but DATA RECORD is strongly to very strongly preferred for ease of use.

a) Determine the hierarchy for Terry's Trucking's decision.

b) Construct the pairwise comparison matrix for each criterion.

c) Determine the priority vector for the alternatives for each criterion.

d) Determine the overall priority vector if:
   (1) price is extremely preferred to ease of use; or if
   (2) ease of use is extremely preferred to price; or if
   (3) price and ease of use are equally preferred.

e) Approximately what relation between price and ease of use would give BASE 8 and DATA RECORD equal priority in an overall priority vector?

PROBLEM 17

Consider the following military application regarding the so called "star wars" or "strategic defense initiative". The overall goal is to maintain peace.
The criteria are: (1) strong defense; (2) international image; (3) U.S. - Soviet relations. The decision alternatives are: (1) continue with "star wars" development; (2) scale down "star wars" development; (3) abandon "star wars" development.

a) Make up your own preferences and feelings regarding the criteria and alternatives. Determine an overall priority vector of the alternatives.

b) Comment on a possible limitation of this model with regard to the scale used for the pairwise comparison matrices.

TRUE/FALSE

___ 18. A goal programming problem will always have more than one priority level.

___ 19. The number of linear programs that must be sequentially solved to develop the solution to a goal program is determined by the number of goals in the objective function.

___ 20. For each goal in a goal program, two variables are added: one for underachieving the goal and one for overachieving the goal.

___ 21. It is not possible to have two penalties for the same goal, one for underachieving the goal and another for overachieving it.

___ 22. For a particular goal program, the Phase 1 objective was Minimize d_1^+. The objective was not satisfied but underachieved by an amount of 50. Any solution to the Phase 2 problem of Minimize d_2^- must have $d_1^+ = 50$.

___ 23. In the process of satisfying second-priority goals, improvement in the satisfaction of first-priority goals sometimes occurs.

___ 24. Trade-offs among goals at the same priority level will not occur unless the goals have the same relative weight.

___ 25. In the analytic hierarchy process, if the entry in row A for decision B is 2, the entry in row B for decision A is 1/2.

___ 26. In the analytic hierarchy process, a consistency ratio of .99 is extremely good.

___ 27. If the consistency index is .90, the consistency ratio is .90.

___ 28. The analytic hierarchy process utilizes pairwise comparisons to establish priority measures for both the criteria and the decision alternatives.

___ 29. The criteria priority vector for gum is: Price: .3, Taste: .7. For Price, BRAND A's priority is .8 and BRAND B's is .2. For Taste, BRAND A's priority is .4 and BRAND B's is .6. Overall BRAND A has a higher priority than BRAND B.

___ 30. On the 1-to-9 pairwise comparison scale, a numerical rating of 9 means extremely preferred and 5 means equally preferred.

___ 31. In the analytic hierarchy process, the hierarchy from top to bottom is overall goal, decision alternatives, and criteria.

___ 32. A consistency ratio measures the consistency of a decision alternative's relative rank from one criterion to another.

Answers to Problems and True/False

CHAPTER 1

6) a) Management science can provide a quantitative methodology for deciding the job-machine pairings so that total job processing time is minimized.
   b) How long it takes to process each job on each machine, and any job-machine pairings that are unacceptable.
   c) <u>Decision variables</u>: one for each job-machine pairing, taking on a value of 1 if the pairing is used and 0 otherwise.
   <u>Objective function</u>: minimize total job processing time.
   <u>Constraints</u>: each job is assigned to exactly one machine, and each machine be assigned no more than one job.
   d) Stochastic: job processing times vary due to varying machine set-up times, variable operator performance, and more.
   e) Assume that processing times are deterministic (known/fixed).

7) a) The company because of its size and the newness of its product should probably select the simplified model because of the reduced costs and the time period for attaining results. This first pass at the problem most likely will give a "ballpark" figure. The model can then be refined as more experience with production and demand is attained.
   b) Because of the large volume and the national distribution, perhaps the more complete study should be done in this case. Additionally, since the firm is a conglomerate, a more complete justification of its policy to the stockholders may be required.

8) a) MAX $10s_1 + 10s_2 + 25s_3$
   S.T. $25s_1 \leq 5{,}000$
   $50s_2 \leq 5{,}000$
   $100s_3 \leq 5{,}000$
   $25s_1 \geq 1{,}000$
   $50s_2 \geq 1{,}000$
   $100s_3 \geq 1{,}000$
   $25s_1 + 50s_2 + 100s_3 \leq 10{,}000$
   $s_1, s_2, s_3 \geq 0$

   b) MAX $1.4x_1 + 1.2x_2 + 1.25x_3$
   S.T. $x_1 \leq 5{,}000$
   $x_2 \leq 5{,}000$
   $x_3 \leq 5{,}000$
   $x_1 \geq 1{,}000$
   $x_2 \geq 1{,}000$
   $x_3 \geq 1{,}000$
   $x_1 + x_2 + x_3 \leq 10{,}000$
   $x_1, x_2, x_3 \geq 0$

   c) Both give the same result.

9) a) $N = L/25$; $C = 8C_gN$; $C = .32C_gL$
   b) $TC = C_n + .32C_gL$; if $L = 50$, $TC = \$260$; if $L = 65$, $N = 2.6$ guards, so it must be decided whether to use 2 or 3 guards during the day.

10) a) X_1 = number of seats CPI manufactures in quarter 1
    X_2 = number of seats subcontracted in quarter 1
    X_3 = number of seats held in inventory from quarter 1 to quarter 2
    X_4 = number of seats CPI manufactures in quarter 2
    X_5 = number of seats subcontracted in quarter 2
    b) Minimize $10.25X_1 + 12.50X_2 + 1.50X_3 + 10.25X_4 + 13.75X_5$

ANSWERS

CHAPTER 1 continued

11) a) $X_1 \leq 3800$ (1)
    $X_4 \leq 3800$ (2)
    $X_1 + X_2 - X_3 = 3700$ (3)
    $X_3 \leq 300$ (4)
    $X_3 + X_4 + X_5 = 4200$ (5)

  b) The real problem is stochastic

  c) Three decision variables must be introduced:
      X_6 = number of seats held in inventory from quarter 2 to quarter 3
      X_7 = number of seats CPI manufactures in quarter 3
      X_8 = number of seats subcontracted in quarter 3
    Addition to objective function: $+1.50X_6 + 10.25X_7 + 13.75X_8$
    Adjustment to constraint (5) above: $X_3 + X_4 + X_5 - X_6 = 4200$
    Three additional constraints:
      $X_6 \leq 300$
      $X_7 \leq 3800$
      $X_6 + X_7 + X_8$ = 3rd quarter demand

12) a) $50s + 30c \leq 800$
      $s \geq 5$
      $c \geq 5$

  b) (1) Max $s + c$; (2) Max $.03s + .05c$; (3) Max $6s + 5c$

13) a) Variable costs: Assembly labor = \$8, Packaging labor = \$1,
      Packaging material = \$11, Components = \$630
    Fixed costs: Lease, utilities, etc. = \$5,000
    Cost function: $c(x) = 650x + 5,000$

  b) $r(x) = 700x$
  c) 100
  d) \$650
  e) \$20,000

14) False   15) True   16) True   17) True   18) False
19) True   20) False   21) False   22) False   23) False
24) False   25) False   26) True   27) True   28) False

CHAPTER 2

7) a) Let E_j = (# poor, # average, # excellent).
    $E_1 = (4,0,0)$, $E_2 = (3,1,0)$, $E_3 = (3,0,1)$, $E_4 = (2,2,0)$, $E_5 = (2,1,1)$, $E_6 = (2,0,2)$,
    $E_7 = (1,3,0)$, $E_8 = (1,2,1)$, $E_9 = (1,1,2)$, $E_{10} = (1,0,3)$, $E_{11} = (0,4,0)$, $E_{12} = (0,3,1)$,
    $E_{13} = (0,2,2)$, $E_{14} = (0,1,3)$, $E_{15} = (0,0,4)$.

  b) $\{E_7, E_8\}$
  c) $\{E_9\}$
  d) yes
  e) no
  f) yes

CHAPTER 2 continued

8) a) .357
   b) .583

9) a) .15
   b) .167
   c) .56

10) a) .162
    b) .37
    c) .408
    d) no

11) a) lower .095; same .469; higher .437
    b) .604
    c) .384

12) a) 6
    b) {(G,S,B), (G,B,S)}
    c) {(G,S,B)}
    d) {(G,B,S)}

13) a) .25
    b) .49
    c) .12

14) a) .4225
    b) .325
    c) .06
    d) .1925

15) .304

16) a) (1) .0121; (2) .7921 (3) .1958
    b) .15
    c) .571
    d) No
    e) P(Married Prefers) = P(Husband Prefers)P(husband) + P(Wife Prefers)P(Wife) = .145; more.

17) a) .234
    b) .829
    c) .133
    d) yes
    e) .987

18) True   19) False   20) False   21) True   22) False
23) False   24) False   25) True   26) True   27) False
28) False   29) False   30) True   31) False   32) True

CHAPTER 3

9) a) (1) P(3) = 1/12, P(4) = 1/12, P(6) = 1/12, P(10) = 1/6, P(16) = 1/4, P(20) = 1/3, all other P(x) = 0.
      (2) P(0) = 1/12, P(1) =1/6, P(2) = 1/6, P(3) = 1/6, P(4) = 1/3, all other P(y) = 0
      (3) P(0) = 1/12, P(.05) = 1/12, P(.1875) = 1/6, P(.20) = 1/3, P(.25) = 1/6, P(.3) = 1/12,
          P(.333) = 1/12, all other P(z) = 0
   b) all distributions are for discrete random variables
   c) (1) the third; (2) the first; (3) the second.
   d) 1/2   e) 5/12   f) 13.417   g) 2.583   h) 17.15%   i) No

10) a) Relief m = 15, s^2 = 150; Comfort m = 15.25, s^2 = 6.19
    b) Relief
    c) Comfort

CHAPTER 3 continued

11) a) .3277
    b) .2048
    c) 300 minutes

12) a) 40 mph
    b) .9332

13) a) 1/2
    b) 1/2
    c) 1
    d) 0
    e) (1) 0 (2) 1/2
    f) (1) 1/2 (2) 1
    g) 8:10

14) M- .6247
    T- .3218
    W- .1587
    R- .3721
    F- .6247
    S- .2620

15) a) $30
    b) $45.50
    c) Yes, new daily profit = $71

16) a) $\mu = 4$, $\sigma^2 = 2.4$
    b) .3822
    c) 6
    d) .2335

17) a) .0120
    b) .0183

18) a) f(x) = 1.25 for 31.8 < x < 32.6,
         = 0 otherwise
    b) 0, .375, 0
    c) no

19) a) 0, .5, .5, .1587
    b) $2

20) a) Store 1
    b) Store 1 - .1587, Store 2 - .2266
    c) No - Store has larger standard deviation

21) a) .0183, .0733, .1465, .7619
    b) .0107
    c) .8647

22) False   23) True   24) True   25) False   26) False
27) True   28) False   29) False   30) False   31) True
32) True   33) True   34) False   35) False   36) True

CHAPTER 4

6) a) Stock C -- it dominates Stock B
   b) (1) Stock A; (2) Stock C; (3) Stock D
   c) Stock D

7) a)

| | | Rail Freight | |
		Bid $470,000	Doesn't Bid
Transrail	Bid $500,000	0	$100,000
	Bid $460,00	$60,000	$ 60,000

   b) Bid $460,000
   c) Bid $500,000
   d) $15,000

CHAPTER 4 continued

8) a)

	Sales (in 1,000,000's)		
	100	50	1
Introduce	$1,000,000	$200,000	-$2,000,000
Do Not Introduce	-$400,000	-$400,000	-$400,000

b)

	Sales (in 1,000,000's)		
	100	50	1
Introduce	$0	$0	$1,600,000
Do Not Introduce	$1,400,000	$600,000	$0

c) (1) do not introduce; (2) introduce; (3) do not introduce   d) Yes   e) No

9) a) Introduce root beer; $p \leq .483$
   b) EVPI = $112,000
      NOTE: The answers to (c)-(e) are very sensitive to roundoff error. Figures in parentheses are for two decimal places only.
   c) Stanton: EVSI = $13,200 ($11,862)
      Efficiency = .118 (.106)
      New World: EVSI = $6,400 ($6,424)
      Efficiency = .057 (.057)
   d) hire Stanton (Stanton)
   e) hire New World (Stanton)

10) a)

Plan	Number of Subscribers					
	10,000	20,000	30,000	40,000	50,000	60,000
I	-550	-400	-250	-100	50	200
II	-520	-340	-160	20	200	380
III	-500	-300	-100	100	300	500
IV	-460	-220	20	260	500	740

(table in $1000's)

b) (1) Plan IV; (2) Plan IV; (3) Plan IV   c) Plan II -- Expected Value = $11,000

11) a)

Number of Computers Manufactured	Number of Clients Purchasing a Computer				
		0	1	2	3	4
	1	-40	50	30	10	-10
	2	-70	20	110	90	70
	3	-80	10	100	190	170
	4	-70	20	110	200	290

(Payoffs are in $1,000's)

b) Build 4 computers
c) Nothing

12) a) Yes, Dollar should sell store   b) EVPI = $8,000   c) No; survey cost exceeds EVPI

CHAPTER 4 continued

13) a)

	Number of Copies (in 1,000's per month)					
Plan	12.6	14.4	16.2	18.0	19.8	21.6
I	341.6	370.4	399.2	428.0	456.8	485.6
II	351.2	372.8	394.4	416.0	437.6	459.2
III	317.2	356.8	396.4	436.0	475.6	515.2
IV	363.0	372.0	381.0	390.0	399.0	408.0

b) (1) Plan III; (2) Plan IV
c) Plan IV

14)

		Demand For Ovens			
		0	1	2	3
Ovens Ordered	0	0	-25	-50	-75
	1	-70	80	55	30
	2	-140	10	160	135
	3	-210	-60	90	240

a) Order one oven -- EV = $25.00
b) EVPI = $63
c) Favorable: order 2; Unfavorable: order 0; No opinion: order 1
d) EVSI = $9.10

15) False   16) True   17) False   18) False   19) False
20) False   21) True   22) False   23) True   24) False
25) True   26) True   27) True   28) False   29) False

CHAPTER 5

8) a) $1,075
   b) $5,000

9) a) Risk averse
   b) Produce root beer as long as $p \geq 60/105 = .571$
   c) Choose Stanton

10) a) Risk Taker -- Second Vice President
       Risk Avoider -- First Vice President
    b) First Vice President -- System I
       Second Vice President -- System III
    c) Risk Neutral Vice President -- System I

CHAPTER 5 continued

11) a) A risk avoider

b)
Amount	Utility
-$40,000	32
$20,000	56
$100,000	72

c)
Amount	Utility
-$40,000	16
$20,000	88
$100,000	136

d) Decision is d_2; EV criterion decision would be d_1
e) Paul should accept the offer since his utility of $20,000 is greater than the expected utility under the optimal decision

12) Buy 3 leases

13) b) I – risk taker; II – risk neutral; III – risk avoider
c) Risk avoider would pay 400; Risk taker would pay 200
d) I – d_1; II – d_1; III – d_1

14) Depends on your personal values

15) Depends on your personal values; Note in (e) -- the utility for $d and $x is the same and hence their values should be identical.

16) Franklin should select Hillsboro; Lincoln should select Fremont. Value of game = 2,000 customers

17) Optimal pure strategy does not exist. Strategy a_1 is dominated by a_3. Then, b_1 is dominated by b_2. Optimal mixed-strategy probabilities: .8 for a_2, .2 for a_3, .5 for b_2, .5 for b_3. Value of game = 1.

18) True	19) False	20) True	21) True	22) False
23) True	24) False	25) False	26) True	27) True
28) True	29) True	30) False	31) True	32) False

CHAPTER 6

6) a) 4 period weighted moving average; MSE = 408
b) exponential smoothing -- $298.48
weighted moving average -- $294.27

7) a) $\sigma = .6$ is better MSE (9.12); $\sigma = .2$ has MSE = 10.82
b) For $\sigma = .6$, $F_9 = 22.86$; for $\sigma = .2$, $F_9 = 21.72$

8) Year 6: Quarter 1 -- 56; Quarter 2 -- 48; Quarter 3 -- 79; Quarter 4 -- 40

9) a) The 3 week moving average gives the better forecast (MSE = 16,337) (compared to a MSE = 17,911 for the 4 week moving average)
b) 3 week moving average forecast for week 11 = 1,200
4 week moving average forecast for week 11 = 1,158

CHAPTER 6 continued

10) a) Yes; y = 11.93 + 1.0046x
    b) Week 17 -- 29.0; Week 18 -- 30.0; Week 19 -- 31.0; Week 20 -- 32.0

11) b) 141.44, 144.18, 119.44, 157.72, 111.93

12) a) $F_t = 34.80 - 1.329t$
    b) 26 months
    c) After 27 months sales will be approximately -1 cars; this is clearly impossible; the assumption of a continued linear decline must be in error.

13) 42.15, 70.14, 52.17; 43.46, 72.30, 53.76

14) a) y = 80.8757 + 5.3605(x)
    b) y = 80.8757 + 5.3605(33.5) = 260.45245 hours for a 3350 sq. ft. house.

15) True	16) True	17) False	18) True	19) False
20) True	21) True	22) False	23) True	24) True
25) False	26) True	27) False	28) False	29) False

CHAPTER 7

9) $X_1 = 0, X_2 = 4; Z = 48$

10) a) $X_1 = 4, X_2 = 3/2; Z = 19$ b) All points on line $2X_1 + X_2 = 6$ between (3,0) and (24/13,30/13).

11) a-b) extreme points: (0,4), (1,2), (3,1) -- feasible region is unbounded.
    c) (1) optimal solution $X_1 = 3, X_2 = 1, Z = 1$
       (2) alternate optimal solutions on $X_1 - 2X_2 = 1$ above (3,1)
       (3) unbounded linear program

12) a) $X_1 = 3, X_2 = 4; Z = 29$ b) No change. If only a single point is feasible, objective function's slope is inconsequential.

13) a) $X_1 = 10, X_2 = 15, Z = 230$
    b) Any point satisfying all the other constraints will also satisfy constraint 1
    c) $X_1 = 20, X_2 = 15, Z = 310$

14) a) MAX $60X_1 + 43X_2$
    S.T. $X_1 + 3X_2 - S_1 = 9$
          $6X_1 - 2X_2 = 12$
          $X_1 + 2X_2 + S_3 = 10$
          $X_1, X_2, S_1, S_3 \geq 0$
    b) line segment of $6X_1 - 2X_2 = 12$ between (22/7,24/7) and (27/10,21/10).
    c) Extreme points: (22/7,24/7) and (27/10,21/10). First one is optimal giving Z = 336.

15) a) (35,0) Z = $630,000   b) (5,12) Z = $630,000   c) (10,10) Z = $630,000

CHAPTER 7 continued

16) a) 100 dozen childs, 66 2/3 doz. adults, profit = $12,333.33; both slack variables = 0.
    b) this point is an interior point in the original formulation

17) True	18) True	19) True	20) False	21) False
22) False	23) False	24) False	25) False	26) True
27) True	28) True	29) False	30) True	31) False

CHAPTER 8

7) a) 1500 shares of airlines stock, 800 shares of insurance stock; the expected return is $5400
   b) Between $0 and $2.40

8) a) 6 product 1, 4 product 2, Profit = $540
   b) Between $50 and $75; at $70 the profit is $580
   c) No -- total % change is 83 1/3% < 100%
   d) Dual prices are the shadow prices for the resources; since there was unused copper (because $S_2 = 2$), extra copper is worth $0
   e) $30
   f) $10; this is the amount extra man-hours are worth
   g) The shadow price is the "premium" for aluminum -- would be willing to pay up to $10 + $30 = $40 for extra aluminum

9) a) Standard -- 33 1/3, Slim-Line -- 266 2/3, Z = $1100
   b) $5.60
   c) Standard -- 86 2/3, Slim-Line -- 213 1/3, Z = $1180

10) a) No   b) No   c) Yes -- Sum of % changes > 100%   d) $10 -- Value of extra man-hours
    e) $1 -- this is the "premium" value for the boxes; hence extra boxes are worth $1.25
    f) They will stay the same -- Sum of % changes < 100%

11) a) 200 shares of James, 20 shares of QM, 40 shares of Delic. Total gain = $3200
    b) Each dollar increase in the allowed minimum investment in QM will result in a $.05 decrease in the total $ gain.
    c) Constraint #1 dual price (.25) X 1000 = $250
    d) No. It would be James Ind. (only stock with + dual price)
    e) Constraint #2 dual price (.15) X 1000 = $150
    f) Allowed max. investment could be raised to $8000
    g) No, cumulative change does not exceed 100%

12) a) $X_1 = 2$, $X_2 = 4$, Z = 32   b) $5 \leq C_1 \leq 10$; $3 \leq C_2 \leq 6$ c) $X_1 = 0$, $X_2 = 6$
    d) 1; improvement in the objective function for an extra unit of iron
    e) Zinc's shadow price = 0 as long as this line does not determine the optimal point (its RHS ≤ 2).

13) a) 120 containers of jade figurines, 60 containers of linen placemats; Profit = $13,200
    b) Between $30 and $120   c) (1) $6.67; (2) $16.67; (3) $0

ANSWERS 421

CHAPTER 8 continued

14) a) 105 minivans, 75 trailers; $673,500 profit
    b) $2812.50 $\leq$ minivan profit $\leq$ no upper limit, no lower limit $\leq$ trailer profit $\leq$ $5120.00
    c) Dual price for yardman hours is $1600. Thus, an increase of $1600 in monthly profit can be gained for each hour increase in the monthly availability of yardmen!

15) a) 10,000 mi. on Harley; 35,000 mi. on Hauler   b) $15.00   c) 7.5 cents

16) True   17) True   18) True   19) True   20) False
21) False   22) True   23) True   24) True   25) True
26) True   27) True   28) True   29) False   30) True

CHAPTER 9

11) a) MAX $2X_1 + X_2 + .75X_3 + .5X_4 + .75X_5 + 1.5X_6$
    S.T. $20X_1 + 16X_2 + 13X_3 + 9X_4 + 12X_5 + 17X_6 \leq 550$
    $X_1 + X_2 + X_3 + X_4 + X_5 + X_6 \geq 40$
    $X_1 + X_2 + X_3 - 2X_4 - 2X_5 - 2X_6 \geq 0$
    $X_1 + X_2 + X_5 + X_6 \geq 20$
    $X_j \geq 0 \; j = 1,...,6$

Solution: $X_1 = 5.833$, $X_2 = 0.833$, $X_3 = 20$, $X_4 = 0$, $X_5 = 13.333$, $X_6 = 0$; Total = 37.5 tons

b) The variables must be integers.

12) P_i = the number of producers in month i (where i = 1,2,3)
    T_i = the number of trainers in month i (where i = 1,2)
    A_i = the number of apprentices in month i (where i = 2,3)
    R_i = the number of recruits in month i (where i = 1,2)

MIN $3000P_1 + 3300T_1 + 2200R_1 + 3000P_2 + 3300T_2 + 2600A_2 + 2200R_2 + 3000P_3 + 2600A_3$
s.t. $.6P_1 + .3T_1 + .05R_1 \geq 20$
    $.6P_1 + .3T_1 + .05R_1 + .6P_2 + .3T_2 + .4A_2 + .05R_2 \geq 44$
    $.6P_1 + .3T_1 + .05R_1 + .6P_2 + .3T_2 + .4A_2 + .05R_2 + .6P_3 + .4A_3 \geq 74$
    $P_1 - P_2 + T_1 - T_2 = 0$
    $P_2 - P_3 + T_2 + A_2 = 0$
    $A_2 - R_1 = 0$
    $A_3 - R_2 = 0$
    $2T_1 - R_1 \geq 0$
    $2T_2 - R_2 \geq 0$
    $P_1 + T_1 = 100$
    $P_3 + A_3 \geq 140$
    $P_j, T_j, A_j, R_j \geq 0$ for all j

Solution: $P_1 = 100$, $T_1 = 0$, $R_1 = 0$, $P_2 = 80$, $T_2 = 20$, $A_2 = 0$, $R_2 = 40$, $P_3 = 100$, $A_3 = 40$
Total cost = $1,098,000.

CHAPTER 9 continued

13) X_j = the number of instrument j produced; where j = 1(deluxe trumpet); 2(prof. trumpet); 3(deluxe cornet); 4(prof. cornet)

$$\begin{aligned}
\text{MAX } & 80X_1 + 160X_2 + 60X_3 + 120X_4 \\
\text{s.t. } & 2X_1 + 1.5X_2 + 1.5X_3 + X_4 \le 2000 \\
& X_1 + 1.5X_2 + X_3 + 1.5X_4 \le 1800 \\
& X_1 \ge 500 \\
& X_3 \ge 300 \\
& X_2 \le 150 \\
& X_4 \le 100 \\
& X_1 + X_2 - 2X_3 - 2X_4 = 0 \\
& X_j \ge 0 \quad j = 1,\ldots,4
\end{aligned}$$

Solution: $X_1 = 620$, $X_2 = 150$, $X_3 = 300$, $X_4 = 85$; Total profit = $101,800.

14) a) $$\begin{aligned}
\text{MAX } & 5X_1 + 12X_2 + 25X_3 + 4X_4 + 10X_5 + 18X_6 + 9X_7 + 16X_8 \\
\text{S.T. } & 1.8X_1 + 2.2X_2 + 3X_3 + .74X_4 + 1.6X_5 + 2.2X_6 + X_7 + 1.5X_8 \le 300 \\
& 1.8X_1 + 2.2X_2 + 3X_3 \ge 75 \\
& 1.8X_1 + 2.2X_2 + 3X_3 \le 120 \\
& .74X_4 + 1.6X_5 + 2.2X_6 \ge 75 \\
& .74X_4 + 1.6X_5 + 2.2X_6 \le 120 \\
& X_7 + 1.5X_8 \ge 30 \\
& X_7 + 1.5X_8 \le 75 \\
& .75X_1 - .25X_2 - .25X_3 + .75X_4 - .25X_5 - .25X_6 - .25X_7 - .25X_8 \ge 0 \\
& X_j \ge 0 \quad j = 1,\ldots,8
\end{aligned}$$

b) $X_1 = 0$, $X_2 = 0$, $X_3 = 40$, $X_4 = 41.28$, $X_5 = 0$, $X_6 = 33.84$, $X_7 = 0$, $X_8 = 50$, $Z = 2{,}574.28$
c) The variables must be integers.

15) X_{11} = pounds of chocolate used in Chompers
X_{21} = pounds of chocolate used in Smerks
X_{31} = pounds of chocolate used in Delicious Chocolate
X_{12} = pounds of caramel used in Chompers
X_{22} = pounds of caramel used in Smerks
X_{23} = pounds of peanuts used in Smerks
Y_1 = number of one ounce Chompers bars produced daily
Y_2 = number of one ounce Smerks bars produced daily
Y_3 = number of one ounce Delicious Choc. bars produced daily
Y_4 = number of one pound Delicious Choc. bags produced daily

ANSWERS   423

CHAPTER 9 continued

15) continued

MAX $.128Y_1 + .148Y_2 + .138Y_3 + 2.261Y_4 - 1.60X_{11} - 1.60X_{21} - 1.60X_{31} - .95X_{12} - .95X_{22} - 1.40X_{23}$
s.t.
$(1/16)Y_3 + Y_4 = X_{31}$
$(1/16)Y_1 = X_{11} + X_{12}$
$(1/16)Y_2 = X_{21} + X_{22} + X_{23}$
$X_{12} \geq .18(X_{11} + X_{12})$
$X_{12} \leq .28(X_{11} + X_{12})$
$X_{22} = X_{23}$
$X_{21} \geq .20(X_{21} + X_{22} + X_{23})$
$X_{21} \leq .40(X_{21} + X_{22} + X_{23})$
$Y_1 + Y_2 + Y_3 \leq 20,000$
$Y_4 \leq 1,000$
$Y_1 \geq 3,000$
$Y_2 \geq 3,000$
$Y_3 \geq 3,000$
$Y_1 - Y_2 \leq .10(Y_1 + Y_2)$
$Y_2 - Y_1 \leq .10(Y_1 + Y_2)$
$X_{11} + X_{21} + X_{31} \geq 1,000$
$X_{12} + X_{22} = 350$
$X_{23} \leq 500$
$X_{ij} \geq 0 \quad i = 1,2,3; j = 1,2,3$
$Y_j \geq 0 \quad j = 1,2,3,4$

Solution: $Y_1 = 7650, Y_2 = 9350, Y_3 = 3000, Y_4 = 1000$
$X_{11} = 361.875, X_{12} = 116.25$
$X_{21} = 116.875, X_{22} = 233.75, X_{23} = 233.75$
$X_{31} = 1187.5$; Total daily profit = $1,712.25

16) X_1 = number of bulldozers purchased for the year
X_2 = number of bulldozers leased for the year
annual cost = (purchasing) + (leasing) − (salvage) − (interest)
   = $(40,000X_1) + (8000X_2) − (20,000X_1) − 0.08(1,000,000 − 40,000X_1 − 8,000X_2)$
   = $23,200X_1 + 8,640X_2 − 80,000$. Thus,

MIN $23,200X_1 + 8,640X_2$
s.t. $4,000X_1 + 8,000X_2 \leq 1,000,000$
      $8X_1 + 5X_2 \geq 240$
      $X_1, X_2 \geq 0$

Solution: Buy 0 bulldozers, lease 48; Annual cost = $334,720

17) X_j = $ invested in investment j; where j = 1(Uni Eq.), 2(Col. Must.), 3(1st Gen REIT), 4(Met. Elec.),
5(Uni Debt), 6(Lem. Trans.), 7(Fair. Apt.), 8(T-Bill),
9(Money Market), 10(All Saver's)

MAX $.15X_1 + .17X_2 + .175X_3 + .118X_4 + .122X_5 + .12X_6 + .22X_7 + .096X_8 + .105X_9 + .126X_{10}$
S.T. $X_1 + X_2 + X_3 + X_4 + X_5 + X_6 + X_7 + X_8 + X_9 + X_{10} = 400,000$
$100X_1 + 100X_2 + 100X_3 + 95X_4 + 92X_5 + 79X_6 + 80X_8 + 100X_9 \geq$
$65(X_1 + X_2 + X_3 + X_4 + X_5 + X_6 + X_7 + X_8 + X_9 + X_{10})$
$60X_1 + 70X_2 + 75X_3 + 20X_4 + 30X_5 + 22X_6 + 50X_7 + 10X_9 \leq$
$55(X_1 + X_2 + X_3 + X_4 + X_5 + X_6 + X_7 + X_8 + X_9 + X_{10})$

CHAPTER 9 continued

17) continued

$$X_1 + X_5 \leq 60{,}000$$
$$X_1 + X_2 + X_3 \leq 160{,}000$$
$$X_4 + X_5 + X_6 \leq 160{,}000$$
$$X_3 + X_7 \leq 160{,}000$$
$$X_1 \leq 80{,}000$$
$$X_2 \leq 80{,}000$$
$$X_3 \leq 80{,}000$$
$$X_4 \leq 80{,}000$$
$$X_5 \leq 80{,}000$$
$$X_6 \leq 80{,}000$$
$$X_7 \leq 80{,}000$$
$$X_8 \leq 80{,}000$$
$$X_9 \geq 1{,}000$$
$$X_{10} \leq 15{,}000$$
$$X_4 + X_5 + X_6 \geq 90{,}000$$
$$X_8 \geq 10{,}000$$
$$X_j \geq 0 \quad j = 1,\ldots,10$$

Solution: $X_1 = 0$; $X_2 = 80{,}000$; $X_3 = 80{,}000$; $X_4 = 0$; $X_5 = 60{,}000$; $X_6 = 74{,}000$; $X_7 = 80{,}000$; $X_8 = 10{,}000$; $X_9 = 1{,}000$; $X_{10} = 15{,}000$; Total return = \$64,355.

18) X_1 = amount invested in new soda advertising
X_2 = amount invested in traditional soda advertising

MAX $X_1 + 4X_2$ <==== .02(50X_1) + .04(100X_2)
S.T. $X_1 + X_2 \leq 10{,}000{,}000$
$X_1 \geq 5{,}000{,}000$
$X_2 \geq 2{,}000{,}000$
$50X_1 + 100X_2 \geq 750{,}000{,}000$
$X_1, X_2 \geq 0$

Answer: spend \$5,000,000 on new soda ad, spend \$5,000,000 on traditional ad, profit = \$25 mil

19) X_1-X_3 = number of students from NE to McHale,McCallum,McBride
X_4-X_6 = number of students from SE to McHale,McCallum,McBride
X_7-X_9 = number of students from SW to McHale,McCallum,McBride
X_{10}-X_{12} = number of students from NW to McHale,McCallum,McBride
X_{13}-X_{15} = number of students from Central to McHale,McCallum,McBride

MIN $1.5X_1 + 2.5X_2 + .5X_3 + 4X_4 + 1.5X_5 + 3X_6 + 2.5X_7 + 3X_8$
$+ 3.5X_9 + .5X_{10} + 4X_{11} + 1.5X_{12} + 1X_{13} + 2X_{14} + 1X_{15}$
S.T. $X_1 + X_2 + X_3 = 700$
$X_4 + X_5 + X_6 = 1100$
$X_7 + X_8 + X_9 = 900$
$X_{10} + X_{11} + X_{12} = 600$
$X_{13} + X_{14} + X_{15} = 800$
$X_1 + X_4 + X_{10} + X_{13} \leq 1500$
$X_2 + X_5 + X_{11} + X_{14} \leq 1800$
$X_3 + X_6 + X_{12} + X_{15} \leq 1100$
$X_j \geq 0 \quad j = 1,2,\ldots,15$

CHAPTER 9 continued

19) continued
Solution: 700 students from NE to McBride, 1100 from SE to McCallum
500 from SW to McHale, 400 from SW to McCallum, 600 from NW to McHale,
400 from Central to McHale, and 400 from Central to McBride.

20) E = .968; Alabama store appears moderately inefficient

21) X_1-X_4 = number of full-time clerks starting at 8,9,10,11am
X_5-X_{11} = number of part-time clerks starting at 8,9,10,11am and 12,1,2pm

$$\text{MIN } 63X_1+63X_2+63X_3+63X_4+26X_5+26X_6+26X_7+26X_8+26X_9+26X_{10}+26X_{11}$$

S.T.
$$X_1 \geq 1$$
$$X_4 \geq 1$$
$$X_1 + X_2 + X_3 + X_4 \geq 4$$
$$X_1 + X_5 \geq 5$$
$$X_1 + X_2 + X_5 + X_6 \geq 4$$
$$X_1 + X_2 + X_3 + X_5 + X_6 + X_7 \geq 6$$
$$X_2 + X_3 + X_4 + X_5 + X_6 + X_7 + X_8 \geq 8$$
$$X_1 + X_3 + X_4 + X_6 + X_7 + X_8 + X_9 \geq 10$$
$$X_1 + X_2 + X_4 + X_7 + X_8 + X_9 + X_{10} \geq 9$$
$$X_1 + X_2 + X_3 + X_8 + X_9 + X_{10} + X_{11} \geq 7$$
$$X_2 + X_3 + X_4 + X_9 + X_{10} + X_{11} \geq 4$$
$$X_3 + X_4 + X_{10} + X_{11} \geq 7$$
$$X_4 + X_{11} \geq 5$$
$$X_j \geq 0 \quad j = 1,2,...,11$$

Solution: $X_1 = 1$, $X_2 = 0$, $X_3 = 0$, $X_4 = 3$, $X_5 = 4$,
$X_6 = 3$, $X_7 = 0$, $X_8 = 3$, $X_9 = 0$, $X_{10} = 2$, $X_{11} = 2$, Total labor cost = $616.

22) False 23) False 24) True 25) False 26) True
27) True   28) True   29) True 30) False 31) False
32) False 33) False 34) False 35) False 36) False

CHAPTER 10

8. a)

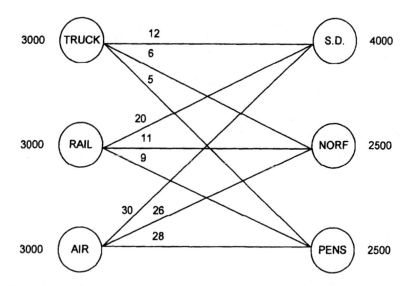

b) Let: P_{TS}, P_{TN}, P_{TP} = lbs. of material shipped by truck to San Diego, Norfolk, Pensacola
P_{RS}, P_{RN}, P_{RP} = lbs. of material shipped by railroad to San Diego, Norfolk, Pensacola
P_{AS}, P_{AN}, P_{AP} = lbs. of material shipped by airline to San Diego, Norfolk, Pensacola

Min $12P_{TS} + 6P_{TN} + 5P_{TP} + 20P_{RS} + 11P_{RN} + 9P_{RP} + 30P_{AS} + 26P_{AN} + 28P_{AP}$
s.t. $P_{TS} + P_{TN} + P_{TP} \leq 3000$
$P_{RS} + P_{RN} + P_{RP} \leq 3000$
$P_{AS} + P_{AN} + P_{AP} \leq 3000$
$P_{TS} + P_{RS} + P_{AS} = 4000$
$P_{TN} + P_{RN} + P_{AN} = 2500$
$P_{TP} + P_{RP} + P_{AP} = 2500$
$P_{TS}, P_{TN}, P_{TP}, P_{RS}, P_{RN}, P_{RP}, P_{AS}, P_{AN}, P_{AP} \geq 0$

c) Truck – S.D. 1000; Truck – Norf. 2000; Rail – Norf. 500; Rail – Pens. 2500; Air – S.D. 3000

9. a) Let: A_{N1}, A_{N2}, A_{N3}, A_{N4} = number of newspaper ads by MR1, MR2, MR3, MR4
A_{T1}, A_{T2}, A_{T3}, A_{T4} = number of television ads by MR1, MR2, MR3, MR4
A_{R1}, A_{R2}, A_{R3}, A_{R4} = number of radio ads by MR1, MR2, MR3, MR4

Min $16A_{N1} + 10A_{N2} + 12A_{N3} + 12A_{N4} + 26A_{T1} + 20A_{T2} + 30A_{T3} + 21A_{T4} + 22A_{R1} + 15A_{R2}$
$+ 23A_{R3} + 14A_{R4}$

s.t. $A_{N1} + A_{T1} + A_{R1} \leq 15$
$A_{N2} + A_{T2} + A_{R2} \leq 25$
$A_{N3} + A_{T3} + A_{R3} \leq 10$
$A_{N4} + A_{T4} + A_{R4} \leq 20$
$A_{N1} + A_{N2} + A_{N3} + A_{N4} = 30$
$A_{T1} + A_{T2} + A_{T3} + A_{T4} = 15$
$A_{R1} + A_{R2} + A_{R3} + A_{R4} = 25$ and all A's ≥ 0

CHAPTER 10 continued

9) b) Newspaper - MR1 15; Newspaper - MR2 5; Newspaper - MR3 10; TV - MR2 15; Radio - MR2 5; Radio - MR4 20; Total Cost = $1,065,000
   c) Newspaper - MR2 20; Newspaper - MR3 10; TV - MR1 15; Radio - MR2 5; Radio - MR4 20; Total cost = $1,065,000
   d) Many answers. One with TV-MR1 = 5 is: Newspaper - MR1 10; Newspaper - MR2 10; Newspaper - MR3 10; TV - MR1 5; TV - MR2 10; Radio - MR2 5; Radio - MR4 20; Total cost = $1,065,000

10) Ace: 25 Sanitation and 5 Police; Band: 25 Parks and 15 Administration; QM: 20 Police and 10 Administration; Cost $1,035,000

11) LA-Denver 25; LA-NY 10; CHI-NY 30; NY-ATL 35; Profit $2,165,000

12) Abbey-Cab3, Babbs-Cab4, Carla-Cab2, Diane-Cab5, Ellsa-Cab1; Performance rating total = 35

13) a) (diagram below)
    b) Computer Town - P5; Computer World - P4; Universal - P1; Local Computer - P2 and P3; Total Cost $71,000
    c) +M placed in matrix in row 2, column 5; no change in solution
    d) +M's placed in assignment matrix row 2 column 4, and in row 2 column 5; Computer Town - P5; Computer World - P2; Universal - P1; Local Computer - P3 and P4; Total cost = $77,000

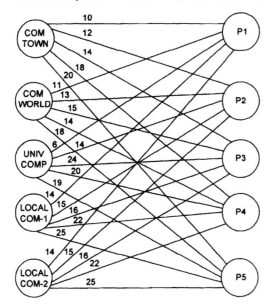

14) Division 1 - Bats; Division 2 - Golf Clubs; Division 3 - Racquetball Rackets; Division 4 - Tennis Rackets; Total 410

428 ANSWERS

CHAPTER 10 continued

15) a)

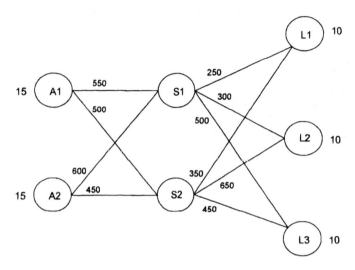

15) b) Denote A1 as node 1, A2 as node 2, S1 as node 3, S2 as node 4,
L1 as node 5, L2 as node 6, and L3 as node 7

$$\text{MIN } 550X_{13} + 500X_{14} + 600X_{23} + 450X_{24} + 250X_{35} + 300X_{36} + 500X_{37}$$
$$+ 350X_{45} + 650X_{46} + 450X_{47}$$

S.T.
$$X_{13} + X_{14} \leq 15$$
$$X_{23} + X_{24} \leq 15$$
$$X_{13} + X_{23} - X_{35} - X_{36} - X_{37} = 0$$
$$X_{14} + X_{24} - X_{45} - X_{46} - X_{47} = 0$$
$$X_{35} + X_{45} = 10$$
$$X_{36} + X_{46} = 10$$
$$X_{37} + X_{47} = 10$$
$$X_{ij} \geq 0 \text{ for all } i,j$$

Solution: $X_{13} = 15$, $X_{14} = 0$, $X_{23} = 0$, $X_{24} = 15$,
$X_{35} = 5$, $X_{36} = 10$, $X_{37} = 0$, $X_{45} = 5$, $X_{46} = 0$, $X_{47} = 10$, Total cost = \$25,500.

16) True   17) True   18) False   19) True   20) True
21) False 22) False 23) False   24) True   25) True
26) True   27) False 28) True   29) False   30) True

CHAPTER 11

6) a) $X_1 = 2.8$, $X_2 = 3.4$, $Z = 48.8$
   b) $X_1 = 3$, $X_2 = 3$ -- infeasible
   c) $X_1 = 2$, $X_2 = 3$ -- feasible but not optimal
   d) Optimal $X_1 = 2$, $X_2 = 9$, $Z = 48 < 48.8$
   e) Additional integer constraints restrict the feasible region further;
      optimal solution values: ILP $\leq$ Mixed ILP $\leq$ LP.

CHAPTER 11 continued

7) a) $X_1 = 7/3$, $X_2 = 1/3$, $Z = 31/3$
   b) $X_1 = 2$, $X_2 = 0$, $Z = 6$
   c) L.P. -- optimal solution changes slightly; ILP -- infeasible

8) a) $X_1 = 1.5$, $X_2 = 2.5$, $Z = 17.5$; upper bound
   b) $X_1 = 1$, $X_2 = 3$, $Z = 17$
   c) $X_1 = 1$, $X_2 = 3$, $Z = 14$

9) MAX $Y_1 + 1.8Y_2 + 2Y_3 + 1.5Y_4 + 3.6Y_5 + 2.2Y_6$
   S.T. $20Y_1 + 55Y_2 + 47Y_3 + 38Y_4 + 90Y_5 + 63Y_6 \leq 175$
   $15Y_1 + 45Y_2 + 50Y_3 + 40Y_4 + 70Y_5 + 70Y_6 \leq 150$
   $Y_4 - Y_6 = 0$
   $Y_1 - Y_2 \geq 0$
   $Y_3 + Y_5 \leq 1$
   $Y_1 + Y_2 + Y_3 + Y_4 + Y_5 + Y_6 \leq 3$
   $Y_i = $ 0 or 1

   Solution: $Y_1 = 1$, $Y_2 = 1$, $Y_3 = 0$, $Y_4 = 0$, $Y_5 = 1$, $Y_6 = 0$, Total profit = $6.4 million.

10) MAX $5.2X_1 + 3.6X_2 + 3.2X_3 + 2.8X_4$
    S.T. $.35X_1 + .50X_2 + .35X_3 + .50X_4 \leq 0.85$ (First Year)
    $.55X_1 + .50X_2 + .40X_3 \leq 1.00$ (Second Year)
    $.75X_1 + .45X_3 \leq 1.20$ (Third Year)
    $X_1 + X_2 + X_3 + X_4 = 2$
    $-X_2 + X_3 \geq 0$
    $X_i = $ 0 or 1

    Solution: $X_1 = 1$, $X_2 = 0$, $X_3 = 1$, $X_4 = 0$, Total projected benefits = $8.4 million.

11) a) $X_1 = 3$, $X_2 = 4.25$, $Z = 49$ b) $X_1 = 4$, $X_2 = 4$, $Z = 52$

12) a) 4/5 container of Grain A, 2 containers of Grain B, $3960
    b) 2 containers of Grain A, 1 container of Grain B, $3900
    c) (1) (0,2) feasible, not optimal; (2-3) (1,2) infeasible

13) a) $X_1 = 3$, $X_2 = 3.2$, $Z = 34.4$
    b) $X_1 = 2$, $X_2 = 4$, $Z = 36$

14) False	15) True	16) False	17) False	18) True
19) False	20) True	21) True	22) True	23) False
24) False	25) True	26) True	27) False	28) True

CHAPTER 12

6) a)

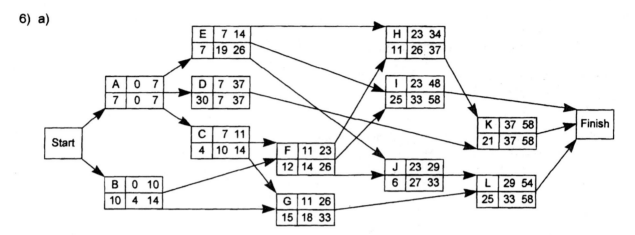

Expected Completion Time = 58; Critical Path = A - D - K

b) Yes
c) Yes
d) Slack on G = 7 hours; Slack on L = 4 hours
e) 4 hours; the slack times are not independent as G and L are on the same path.

7) a)

Activity	ES	EF	LS	LF	Slack
A	0:00	4:00	0:00	4:00	0
B	0:00	3:00	6:00	9:00	6
C	4:00	9:00	17:00	22:00	13
D	4:00	7:00	8:00	11:00	4
E	4:00	9:00	4:00	9:00	0
F	9:00	11:00	9:00	11:00	0
G	9:00	10:00	15:00	16:00	6
H	11:00	13:00	20:00	22:00	9
I	11:00	14:00	20:00	23:00	9
J	11:00	16:00	11:00	16:00	0
K	13:00	14:00	22:00	23:00	9
L	16:00	23:00	16:00	23:00	0

b) A - E - F - J - L; 23 hours
c) There are many; here is one. Note that for Man 2, activity D must precede activity C (D's LS is 8:00 and C's EF is 9:00).

Intern 1	Intern 2
A 0:00 - 4:00	B 0:00 - 3:00
E 4:00 - 9:00	D 4:00 - 7:00
F 9:00 - 11:00	C 7:00 - 12:00
J 11:00 - 16:00	G 12:00 - 13:00
L 16:00 - 23:00	H 13:00 - 15:00
	I 15:00 - 18:00
	K 18:00 - 19:00

CHAPTER 12 continued

8) a)

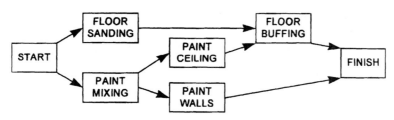

b) 8 hrs.
c) .8264

9) a)

Activity	ES	EF	LS	LF	Slack
A	0	6	2	8	2
B	0	8	0	8	0
C	6	7	8	9	2
D	6	9	9	12	3
E	8	11	9	12	1
F	8	12	9	13	1
G	8	18	8	18	0
H	8	13	10	15	2
I	11	14	12	15	1
J	12	13	14	15	2
K	12	17	13	18	1
L	14	17	15	18	1

b) Critical Path: B - G; Expected completion time = 18 weeks   c) Train its own employees

10) a) 16 weeks = Aug. 21   b) About Sept. 22 (20.66 weeks from May 1)   c) Do not accept the offer

11) a) Expected Project Completion Time = 32; Standard deviation = 3.16
    b) Do not spend the money
    c) Activities off the critical path may vary enough so that a new critical path could be formed.

12) a) For this problem label the activities as follows:
    A = Feasibility Study   D = Advert. Staff Selected   G = Prototype Manufactured
    B = Building Purchased   E = Materials Purchased   H = Production Run of 100
    C = Project Leader Hired   F = Manufacturing Staff Hired   I = Advertising Campaign

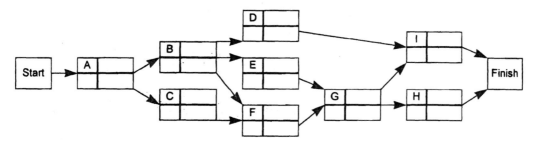

CHAPTER 12 continued

12) b) Define X_i = earliest finish time for activity i Y_i = the amount of time activity i is crashed

MIN $20Y_A + 50Y_C + 50Y_D + 70Y_E + 60Y_F + 350Y_H + 75Y_I$ (in $000)

S.T.
$$X_A \geq 0 + (6 - Y_A)$$
$$Y_A \leq 1$$
$$Y_C \leq 1$$
$$Y_D \leq 3$$
$$Y_E \leq 1$$
$$Y_F \leq 3$$
$$Y_H \leq 1$$
$$Y_I \leq 4$$
$$X_B \geq X_A + 4$$
$$X_C \geq X_A + (3 - Y_C)$$
$$X_D \geq X_B + (6 - Y_D)$$
$$X_E \geq X_B + (3 - Y_E)$$
$$X_F \geq X_B + (10 - Y_F)$$
$$X_F \geq X_C + (10 - Y_F)$$
$$X_G \geq X_E + 2$$
$$X_G \geq X_F + 2$$
$$X_H \geq X_G + (6 - Y_H)$$
$$X_I \geq X_D + (8 - Y_I)$$
$$X_I \geq X_G + (8 - Y_I)$$
$$X_H \leq 26$$
$$X_I \leq 26$$
$$X_i, Y_j \geq 0 \text{ for all i}$$

Solution: $X_A = 5, X_B = 9, X_C = 9, X_D = 18, X_E = 16, X_F = 16, X_G = 18, X_H = 24, X_I = 26$,
$Y_A = 1, Y_C = 0, Y_D = 0, Y_E = 0, Y_F = 3, Y_H = 0, Y_I = 0$, Total crash cost = $200,000.

c) The time reduction for each activity is proportional to the crashing money spent on that activity.
d) Per-week reduction cost is shadow price for $X_H \leq 26$ or $X_I \leq 26$.

13) The PERT network for this problem is:

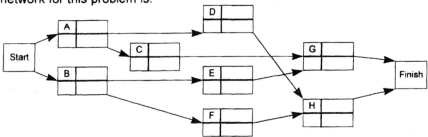

a) Define X_i = earliest finish time for activity i Y_i = the amount of time activity i is crashed

MIN $2000Y_A + 2000Y_C + 500Y_D + 500Y_E + 2000Y_F + 5000Y_G + 6000Y_H$

S.T.
$$X_A \geq 0 + (6 - Y_A)$$
$$Y_A \leq 2$$
$$Y_C \leq 3$$
$$Y_D \leq 2$$
$$Y_E \leq 1$$
$$Y_F \leq 2$$
$$Y_G \leq 1$$
$$Y_H \leq 2$$
$$X_B \geq 0 + 3$$
$$X_C \geq X_A + (9 - Y_C)$$
$$X_D \geq X_A + (4 - Y_D)$$
$$X_E \geq X_B + (2 - Y_E)$$
$$X_F \geq X_B + (3 - Y_F)$$
$$X_G \geq X_C + (5 - Y_G)$$
$$X_G \geq X_E + (5 - Y_G)$$
$$X_H \geq X_D + (7 - Y_H)$$
$$X_H \geq X_F + (7 - Y_H)$$
$$X_G \leq 16$$
$$X_H \leq 16$$
$$X_i, Y_j \geq 0 \text{ for all i}$$

Solution: $X_A = 4, X_B = 6, X_C = 11, X_D = 9, X_E = 11, X_F = 9, X_G = 16, X_H = 16$,
$Y_A = 2, Y_C = 2, Y_D = 0, Y_E = 0, Y_F = 0, Y_G = 0, Y_H = 0$, Total crash cost = $8,000.

b) Same formulation as (a) except: objective function is now MIN X_I with constraints $X_I \geq X_G$ and $X_I \geq X_H$ and $1250Y_A + 2000Y_C + 500Y_D + 500Y_E + 2000Y_F + 5000Y_G + 6000Y_H \leq 50,000$

CHAPTER 12 continued

14) [Activity, ES, EF, LS, LF, Slack]
[A, 0, 4, 0, 4, 0]; [B, 4, 7, 4, 7, 0]; [C, 4, 8, 6, 10, 2]; [D, 7, 9, 8, 10, 1]; [E, 7, 12, 7, 12, 0]
[F, 9, 11, 10, 12, 1]; [G, 12, 17, 12, 17, 0]; [H, 9, 15, 11, 17, 2]

15) True   16) True   17) False   18) False   19) True
20) True   21) False   22) False   23) False   24) True
25) True   26) True   27) False   28) True   29) True

CHAPTER 13

9) a) 75   b) $Q^* = 273.861$, order 274   c) $164.32   d) Order every 4 mos. based on forecast

10) a) 1) Buy: $Q^* = 800$; Tot. Annual Var. Cost = $1200;   Tot. Annual Cost = $76,200
      2) Make: $Q^* = 5345$; Tot. Annual Var. Cost = $5,986.65; Tot. Annual Cost = $75,986.65.
   b) Harrison should manufacture displays; total cost is cheaper.

11) a) $Q^* = 109,544.5$ or 109,545   b) 13.33   c) 115 days = 365 - (9.13)(13.33 + 14)

12) Order 2000 every week

13) a) (1) $Q^* = 16$ (2) 20 or when inventory reaches 4 and one order is pending   (3) $47,840
    b) No; Yearly net profit under this policy is only $45,774

14) $Q^* = 233$, Total annual cost = $64,695.54

15) a) Order 648 every 3.08 weeks when supply reaches 42   b) 12 spark plugs   c) $6.60

16) 196

17) $Z = 1.43$ ==> Pr(Stockout) = .0764

18) Yes; Annual manufacture = $38,600; annual purchase = $40,800

19) a) Machine I Profit = $40,000 (operating at full capacity with no inventory); Machine II = $41,520.
    b) For machine II, $Q^* = 92,582$, resulting in 3.24 cycles/year each lasting 112.6 days. Production time is 33.8 days, so some of the meat sold will be as old as 78 days. Using Machine II, he would need more production runs per year thus incurring more setups and reducing his profit below the $40,000 from using machine I.

20) a) Order 32 when inventory reaches 7   b) Total annual variable cost = $1,301

21) a) .0951   b) 299

22) a) 71,600 loaves   b) 61,280 loaves

23) a) 46 pumpkins   b) .525

CHAPTER 13 continued

24) Order 1000 tubes of lipstick each time

25) Choose Option III

26) a) order 100 units when the supply on hand reaches 172   b) $24,024   c) 272 units

27) Order 555 bags; safety stock = 105 bags

28) True	29) False	30) False	31) False	32) True
33) False	34) True	35) True	36) False	37) False
38) True	39) False	40) False	41) True	42) True

CHAPTER 14

10) a) $P(X = 0) = .1353$
    b) $P(T < 45 \text{ min.}) = .6321$
    c) $W_q = 2\ 1/4$ hours
    d) $P_3 = 27/256 = .105$

11) a) Yes; Total Cost = $70/day for computer vs. $90/day without computer
    b) Better off with computer; Total Cost for second employee is $86.18/day

12) 30 percent

13) a) $\lambda = 40$ per hour
    b) $L_q = 4/3$
    c) $L = 2$
    d) $1 - P_0 = \lambda/\mu = 2/3$
    e) $\mu = 97.2$ per hour

14) Purchase digital router; total cost = $204/day ($255/day without router)

15) a) $L_q = 1.04$ at each desk; Consolidated $L_q = .80$
    b) $P_0 = (3/8)(3/8) = .14$; Consolidated $P_0 = .23$
    c) $W = 1/3$ hour = 20 minutes; Consolidated W = 12.31 minutes
    d) P(both busy) = $(1-P_0)(1-P_0) = (5/8)(5/8) = .39$; Consolidated $P_W = .48$
    e) Yes if it will not result in a substantial loss in business. (Average wait time (part c) decreases by 38% when two servers are consolidated.)

16) No system: $49.20; System A: $50.00; System B: $47.20. Hence, install B.

17) a) George: W = 20 minutes; John: W = 3 1/3 minutes
    b) Total Cost: George -- $46.00; John -- $18.67. Hire John.

18) a) W = 6.22 minutes so total time in the store is 61.22 min.
    b) Yes; W = 6.01 minutes

CHAPTER 14 continued

19) a) .667
    b) .076 hrs = 4.56 mins.
    c) 1.43 vehicles
    d) .9772

20) a) no, 49.18%
    b) 18.03%

21) a) k = 4; P_W = .021
    b) P_0 = .331

22) a) L_q = .15
    b) W = .028 hr. = 1.7 minutes
    c) P_W = .417

23) a) L_q = .7094
    b) W_q = 4.96 days
    c) $1 - P_0$ = .7151

24) True 25) False 26) False 27) False 28) True
29) True 30) False 31) False 32) False 33) False
34) True 35) True   36) False 37) False 38) True

CHAPTER 15

FOR ALL PROBLEMS IN THIS CHAPTER, IT IS ASSUMED THAT THE RANDOM NUMBERS ARE 00 - 99. THE COLUMN HEADINGS FOR THE EACH SIMULATION ARE GIVEN TOGETHER WITH THE FIRST ROW IN THE SIMULATION. THEN THE ANSWER BASED ON THE SIMULATION IS GIVEN.

7)

	Routes 5 - 57				Routes 55 - 91		
RN	Time On 5	RN	Time On 57	RN	Time On 55	RN	Time On 91
63	7	59	6	71	8	51	4

Freeways 5-57 have 62 mins.; Freeways 55-91 have 61 mins.; Select 55-91.

8)

Period	RN	Number of Arrivals	Service Free?	Number Waiting	RN	Service Time	Period When Service Completed
1	63	1	Yes	0	59	1	2

The average number of customers waiting in line for service is the average of the entries in the Number Waiting column. This average is 1.

CHAPTER 15 continued

9) a) Three critical paths with 17 weeks: B-C-F-G-H; B-D-G-H; A-F-G-H.
   b) Five complete PERT analyses must be done. The first is as follows:

Job	RN	Change	Completion Time
A	14	-1	6
B	41	0	4
C	35	0	3
D	38	0	6
E	91	+1	6
F	78	+1	4
G	90	+1	4
H	18	-1	3

Critical path is B-C-F-G-H completion time = 18. Repeating 4 more times:

Trial	Critical Path	Completion Time
2	B-D-G-H	17
3	B-C-F-G-H	17
4	A-F-G-H or B-C-F-G-H	17
5	A-E-H	17

The average completion time of the five trials is 17.2 weeks.

10) a) Order 8 when inventory level reaches 2 units

b)

Day	Beg. Invent.	RN	Demand	End. Invent.	Order?	RN	Deliv. Time	Inv. Cost
1	6	59	2	4	No	--	--	$8

The average inventory cost per day is $13.00

11) a)

Week	Beg. Inv.	RN	Demand	End. Inv.	Avg. Inv.	Hold. Cost	Reord. Cost	Lead RN	Lead Time	Stkout. Cost
1	5	51	3	2	3.5	$7	$40	80	2	--

The total cost for 8 weeks = $168.00

b)

Week	Beg. Inv.	RN	Demand	End. Inv.	Avg. Inv.	Hold. Cost	Service Charge	Stockout Cost
1	5	51	3	2	3.5	$7	$12	--

The total cost for 8 weeks = $194.00; Since this is more than $168.00, the company should keep its current policy.

CHAPTER 15 continued

12) a)

Per.	RN	# of Cust.	Cust.	RN	Here/ToGo?	RN	Serv. Time	Compl. in Per.	# Cust. Pres.	Profit
1	63	0	--	--	--	--	--	--	0	--
2	88	1	1	59	ToGo	71	4	6	1	$.90

b) $3.04   c) 64%   d) .84

13)

Day	RN	# of Cust.	Cust. Number	RN	Type Purch.	In Stock?	RN	Switch?	Remain. Invent. A L T
1	58	2	1	93	T	Yes	--	--	3 2 3
			2	63	L	Yes	--	--	3 1 3

Archie will sell out in 8 days in this simulation.

14) a)

Day	RN	Number Rented	Car Rented	RN	Day Returned	Cost of Unused Cars	Shortage Cost
1	63	3	1	59	3		
			2	09	2		
			3	57	3	$5	--

The 10-day profit of this simulation is $885.

b)

Day	RN	Number Rented	Car Rented	RN	Day Returned	Cost of Unused Cars	Shortage Cost
1	63	3	1	59	3		
			2	09	2		
			3	57	3	$10	--

The 10-day profit of this simulation is $970. Take the extra car.

15)

Per.	RN	# of Arr.	Serv. Area Free?	RN	Type Purch.	Serv. Time	# of Lost Cust.	# of Cust. Wait.	Serv. Compl. This Per.?	Profit
1	63	1	Yes	59	Cone	30	0	0	Yes	$.12

a) Sum last column = $3.34
b) Divide total number of lost customers by total number of arrivals = 7/35 = 20%.
c) Divide the total # of customers waiting by 30 periods = 65/30 = 2.17.

438 ANSWERS

CHAPTER 15 continued

16) a)

	A Last	B Last	C Last
A	00 - 49	00 - 29	00 - 09
Next B	50 - 79	30 - 74	10 - 44
C	80 - 99	75 - 99	45 - 99

b)

Flight	RN	Next Airline
1	71	C
2	95	C

20% of flights on Airline A
48% of flights on Airline B
32% of flights on Airline C

17) False   18) True   19) False   20) False   21) True
22) True   23) True   24) False   25) False   26) True
27) False   28) FALSE 29) True   30) False   31) False

CHAPTER 16

5) P(buy) = .357   P(return) = .643

6) a) Network C   b) (p_A, p_C, p_N) = (.29, .41, .30)
   c) Network A: $14,300; Network C: $20,550; Network N: $15,200

7) a)

This Day		Next Day			
		1	2	3	4
Route A Not Congested	1	.600	.200	.120	.080
Route A Congested	2	.225	.075	.420	.280
Route B Not Congested	3	.075	.025	.540	.360
Route B Congested	4	.450	.150	.240	.160

   b) .36 + .12 = .48

8) a) .16   b) Austin: .199; San Antonio: .328; Houston: .473

9) B. Prince and F. Fowl's shares will drop by .196 and .054

10) a) .868   b) .281   c) .3 + .03 + .2 = .53

11) a) .6
    b) Making payments has higher probability (.4 versus .304)
    c) .06 + .03 + .10 = .19

12) a) (p_A, p_B, p_C) = (.29, .37, .34)
    b) (1) .15   (2) .09   (3) .14
    c) $42,333

CHAPTER 16 continued

13) a)

	Fail-F	Fail-M	Fail-P	New	Est.
Failure - Failure	1	0	0	0	0
Failure - Management	0	1	0	0	0
Failure - Product	0	0	1	0	0
New	.15	.20	.05	0	.60
Established	.10	.06	.03	0	.81

b) .394

c) (1) .47   (2) .39   (3) .14

14) True   15) False   16) True   17) False   18) True
19) True   20) True   21) False   22) True   23) False
24) False   25) False   26) False   27) True   28) False

CHAPTER 17

9) a) Linear Programming constraints:
$$3X_1 + X_2 \le 24 \text{ (total marketing employees)}$$
$$X_1 + X_2 \ge 6 \text{ (minimum required ads)}$$
$$50X_1 + 15X_2 \le 250 \text{ (Goal 1: Budget in \$1000's)}$$
$$3X_1 + X_2 \le 12 \text{ (Goal 2: 50\% of marketing employees)}$$
$$X_1 \ge 4 \text{ (Goal 3: TV ads)}$$
$$X_2 \ge 4 \text{ (Goal 4: Radio ads)}$$
$$X_1, X_2 \ge 0$$
There are no feasible points.

b) MIN $P_1(d_1^+) + P_2(d_2^+) + P_3(d_3^-) + P_4(d_4^-)$
S.T.
$$3X_1 + X_2 \le 24$$
$$X_1 + X_2 \ge 6$$
$$50X_1 + 15X_2 + d_1^- - d_1^+ = 250$$
$$3X_1 + X_2 + d_2^- - d_2^+ = 12$$
$$X_1 + d_3^- - d_3^+ = 4$$
$$X_2 + d_4^- - d_4^+ = 4$$
$$X_1, X_2, d_j^-, d_j^+ \ge 0 \text{ for all } j$$

Optimal G.P. solution: 3 TV ads, 3 radio ads.
Priorities: $P_1(d_1^+) = 0$, $P_2(d_2^+) = 0$, $P_3(d_3^-) = 1$, $P_4(d_4^-) = 1$

c) The new objective function is:
MIN $P_1(d_1^+) + P_2(d_2^+) + P_3(2d_3^-) + P_3(d_4^-)$
Optimal G.P. solution: Produce 2.67 TV ads, 4 radio ads. Goal 3 is missed by 1.33 TV ads while the other goals are achieved.

CHAPTER 17 continued

10) MIN $P_1(d_1^-) + P_2(d_2^+) + P_3(d_3^-) + P_3(3d_4^-) + P_3(d_5^-)$
    S.T. $3X_1 + 2.5X_2 + 2X_3 \le 200$ (donation)
    $X_1 \ge 20$ (athletic sch.)
    $X_1 + X_2 + X_3 + d_1^- - d_1^+ = 80$ (P1, Goal 1: Total sch.)
    $.75X_1 - .25X_2 - .25X_3 + d_2^- - d_2^+ = 0$ (P2, Goal 2: 25% ath. sch.)
    $X_1 + d_3^- - d_3^+ = 25$ (P3, Goal 3: desired ath. sch.)
    $X_2 + d_4^- - d_4^+ = 40$ (P3, Goal 4: desired min. sch.)
    $X_3 + d_5^- - d_5^+ = 20$ (P3, Goal 5: desired women sch.)
    $X_j, d_j^-, d_j^+ \ge 0$ for all j

    Optimal: Award 20 athletic, 40 minority, and 20 women's scholarships.
    Goals 1, 2, and 4 are met. Goal 3 is underachieved by 5 athletic scholarships and
    Goal 5 is underachieved by 10 women's scholarships.

11) X_1 = number of new non-minority hires
    X_2 = number of new minority hires
    X_3 = number of non-minority management promotions
    X_4 = number of minority management promotions

    MIN $3d_1^- + d_2^- + d_3^+ + d_4^+$
    S.T. $X_1 + X_2 = 17,520$
    $X_3 + X_4 = 598$
    $X_1 - d_1^+ + d_1^- = 19,680$
    $X_3 - d_2^+ + d_2^- = 472$
    $X_1 - d_3^+ + d_3^- = 11,680$
    $X_3 - d_4^+ + d_4^- = 399$
    $X_j, d_j^+, d_j^- \ge 0 \quad j = 1,2,3,4$

    Solution: $X_1 = 17520, X_2 = 0, X_3 = 472, X_4 = 126$,
    d1minus = 2160, d3plus = 5840, d4plus = 73, all other d's = 0,
    Obj. function = 12393.

12) SoftTrack = 135, <u>VersaSuite = 148</u>, N-Able = 141

13) Miss Northern State 18; Miss Central State 24; Miss Southern State 58.

14) a) Location: BG - .360; B - .512; H - .128   b) GD - .255; B - .482; H - .264
    Price: BG - .589; B - .252; H - .159
    Cleanliness: BG - .123; B - .320; H - .557
    Parking: BG - .102; B - .612; H - .286
    Selection: BG - .100; B - .713; H - .187

    c)

	LOC	PRI	CLE	PAR	SEL
LOC	1	1/7	1/7	1	1/3
PRI	7	1	3	7	3
CLE	7	1/3	1	5	3
PAR	1	1/7	1/5	1	1/7
SEL	3	1/3	1/3	7	1

    d) CR = .073 -- good consistency   e) GD - .340; B - .380; H - .280

CHAPTER 17 continued

15) a) <u>Goal</u>: Choose the right candidate   <u>Criteria</u>: Party, Issue   <u>Alternatives</u>: Carter, Nixon, Anderson

b)
          Party   Issues

	C	N	A
C	1	1/7	1/3
N	7	1	5
A	3	1/5	1

	C	N	A
C	1	6	5
N	1/6	1	1/3
A	1/5	3	1

c) Party: C - .083; N - .723; A - .193   Issues: C - .707; N - .092; A - .201

d) .082 – good consistency   e) Party - .2; Issues - .8   f) C - .582; N - .218; A - .199

16) a) Goal: Choose the best database program
     Criteria: Price; Ease of Use
     Alternatives: BASE 8; DATA RECORD

b)   Price   Ease Of Use

	B8	DR
B8	1	8
DR	1/8	1

	B8	DR
B8	1	1/6
DR	6	1

c) Price: B8 - 8/9; DR - 1/9   Ease of Use: B8 - 1/7; DR - 6/7

d) (1) B8 - .81; DR - .19   (2) B8 - .22; DR - .78   (3) B8 - .516; DR - .484

e) Ease of use is slightly preferred to price.

18) False   19) False   20) True   21) False   22) True
23) False   24) False   25) True   26) False   27) False
28) True   29) False   30) False   31) False   32) False

Appendices

APPENDIX A: BINOMIAL PROBABILITIES

Entries in the table give the probability of x successes in n trials of a binomial experiment, where p is the probability of a success on one trial. For example, with n = 6 trials and p = 0.40, the probability of x = 2 successes is 0.3110.

						p					
n	x	0.05	0.10	0.15	0.20	0.25	0.30	0.35	0.40	0.45	0.50
1	0	0.9500	0.9000	0.8500	0.8000	0.7500	0.7000	0.6500	0.6000	0.5500	0.5000
	1	0.0500	0.1000	0.1500	0.2000	0.2500	0.3000	0.3500	0.4000	0.4500	0.5000
2	0	0.9025	0.8100	0.7225	0.6400	0.5625	0.4900	0.4225	0.3600	0.3025	0.2500
	1	0.0950	0.1800	0.2550	0.3200	0.3750	0.4200	0.4550	0.4800	0.4950	0.5000
	2	0.0025	0.0100	0.0225	0.0400	0.0625	0.0900	0.1225	0.1600	0.2025	0.2500
3	0	0.8574	0.7290	0.6141	0.5120	0.4219	0.3430	0.2746	0.2160	0.1664	0.1250
	1	0.1354	0.2430	0.3251	0.3840	0.4219	0.4410	0.4436	0.4320	0.4084	0.3750
	2	0.0071	0.0270	0.0574	0.0960	0.1406	0.1890	0.2389	0.2880	0.3341	0.3750
	3	0.0001	0.0010	0.0034	0.0080	0.0156	0.0270	0.0429	0.0640	0.0911	0.1250
4	0	0.8145	0.6561	0.5220	0.4096	0.3164	0.2401	0.1785	0.1296	0.0915	0.0625
	1	0.1715	0.2916	0.3685	0.4096	0.4219	0.4116	0.3845	0.3456	0.2995	0.2500
	2	0.0135	0.0486	0.0975	0.1536	0.2109	0.2646	0.3105	0.3456	0.3675	0.3750
	3	0.0005	0.0036	0.0115	0.0256	0.0469	0.0756	0.1115	0.1536	0.2005	0.2500
	4	0.0000	0.0001	0.0005	0.0016	0.0039	0.0081	0.0150	0.0256	0.0410	0.0695
5	0	0.7738	0.5905	0.4437	0.3277	0.2373	0.1681	0.1160	0.0778	0.0503	0.0312
	1	0.2036	0.3280	0.3915	0.4096	0.3955	0.3602	0.3124	0.2592	0.2059	0.1562
	2	0.0214	0.0729	0.1382	0.2048	0.2637	0.3087	0.3364	0.3456	0.3369	0.3125
	3	0.0011	0.0081	0.0244	0.0512	0.0879	0.1323	0.1811	0.2304	0.2757	0.3125
	4	0.0000	0.0004	0.0022	0.0064	0.0146	0.0284	0.0488	0.0768	0.1128	0.1562
	5	0.0000	0.0000	0.0001	0.0003	0.0010	0.0024	0.0053	0.0102	0.0185	0.0312
6	0	0.7351	0.5314	0.3771	0.2621	0.1780	0.1176	0.0754	0.0467	0.0277	0.0156
	1	0.2321	0.3543	0.3993	0.3932	0.3560	0.3025	0.2437	0.1866	0.1359	0.0938
	2	0.0305	0.0984	0.1762	0.2458	0.2966	0.3241	0.3280	0.3110	0.2780	0.2344
	3	0.0021	0.0146	0.0415	0.0819	0.1318	0.1852	0.2355	0.2765	0.3032	0.3125
	4	0.0001	0.0012	0.0055	0.0154	0.0330	0.0595	0.0951	0.1382	0.1861	0.2344
	5	0.0000	0.0001	0.0004	0.0015	0.0044	0.0102	0.0205	0.0369	0.0609	0.0938
	6	0.0000	0.0000	0.0000	0.0001	0.0002	0.0007	0.0018	0.0041	0.0083	0.0156
7	0	0.6983	0.4783	0.3206	0.2097	0.1335	0.0824	0.0490	0.0280	0.0152	0.0078
	1	0.2573	0.3720	0.3960	0.3670	0.3115	0.2471	0.1848	0.1306	0.0872	0.0547
	2	0.0406	0.1240	0.2097	0.2753	0.3115	0.3177	0.2985	0.2613	0.2140	0.1641
	3	0.0036	0.0230	0.0617	0.1147	0.1730	0.2269	0.2679	0.2903	0.2918	0.2734
	4	0.0002	0.0026	0.0109	0.0287	0.0577	0.0972	0.1442	0.1935	0.2388	0.2734
	5	0.0000	0.0002	0.0012	0.0043	0.0115	0.0250	0.0466	0.0774	0.1172	0.1641
	6	0.0000	0.0000	0.0001	0.0004	0.0013	0.0036	0.0084	0.0172	0.0320	0.0547
	7	0.0000	0.0000	0.0000	0.0000	0.0001	0.0002	0.0006	0.0016	0.0037	0.0078

Binomial Probabilities (Continued)

		p									
n	x	0.05	0.10	0.15	0.20	0.25	0.30	0.35	0.40	0.45	0.50
8	0	0.6634	0.4305	0.2725	0.1678	0.1001	0.0576	0.0319	0.0168	0.0084	0.0039
	1	0.2793	0.3826	0.3847	0.3355	0.2670	0.1977	0.1373	0.0896	0.0548	0.0312
	2	0.0515	0.1488	0.2376	0.2936	0.3115	0.2965	0.2587	0.2090	0.1569	0.1094
	3	0.0054	0.0331	0.0839	0.1468	0.2076	0.2541	0.2786	0.2787	0.2568	0.2188
	4	0.0004	0.0046	0.0185	0.0459	0.0865	0.1361	0.1875	0.2322	0.2627	0.2734
	5	0.0000	0.0004	0.0026	0.0092	0.0231	0.0467	0.0808	0.1239	0.1719	0.2188
	6	0.0000	0.0000	0.0002	0.0011	0.0038	0.0100	0.0217	0.0413	0.0703	0.1094
	7	0.0000	0.0000	0.0000	0.0001	0.0004	0.0012	0.0033	0.0079	0.0164	0.0312
	8	0.0000	0.0000	0.0000	0.0000	0.0000	0.0001	0.0002	0.0007	0.0017	0.0039
9	0	0.6302	0.3874	0.2316	0.1342	0.0751	0.0404	0.0207	0.0101	0.0046	0.0020
	1	0.2985	0.3874	0.3679	0.3020	0.2253	0.1556	0.1004	0.0605	0.0339	0.0176
	2	0.0629	0.1722	0.2597	0.3020	0.3003	0.2668	0.2162	0.1612	0.1110	0.0703
	3	0.0077	0.0446	0.1069	0.1762	0.2336	0.2668	0.2716	0.2508	0.2119	0.1641
	4	0.0006	0.0074	0.0283	0.0661	0.1168	0.1715	0.2194	0.2508	0.2600	0.2461
	5	0.0000	0.0008	0.0050	0.0165	0.0389	0.0735	0.1181	0.1672	0.2128	0.2461
	6	0.0000	0.0001	0.0006	0.0028	0.0087	0.0210	0.0424	0.0743	0.1160	0.1641
	7	0.0000	0.0000	0.0000	0.0003	0.0012	0.0039	0.0098	0.0212	0.0407	0.0703
	8	0.0000	0.0000	0.0000	0.0000	0.0001	0.0004	0.0013	0.0035	0.0083	0.0176
	9	0.0000	0.0000	0.0000	0.0000	0.0000	0.0000	0.0001	0.0003	0.0008	0.0020
10	0	0.5987	0.3487	0.1969	0.1074	0.0563	0.0282	0.0135	0.0060	0.0025	0.0010
	1	0.3151	0.3874	0.3474	0.2684	0.1877	0.1211	0.0725	0.0403	0.0207	0.0098
	2	0.0746	0.1937	0.2759	0.3020	0.2816	0.2335	0.1757	0.1209	0.0763	0.0439
	3	0.0105	0.0574	0.1298	0.2013	0.2503	0.2668	0.2522	0.2150	0.1665	0.1172
	4	0.0010	0.0112	0.0401	0.0881	0.1460	0.2001	0.2377	0.2508	0.2384	0.2051
	5	0.0001	0.0015	0.0085	0.0264	0.0584	0.1029	0.1536	0.2007	0.2340	0.2461
	6	0.0000	0.0001	0.0012	0.0055	0.0162	0.0368	0.0689	0.1115	0.1596	0.2051
	7	0.0000	0.0000	0.0001	0.0008	0.0031	0.0090	0.0212	0.0425	0.0746	0.1172
	8	0.0000	0.0000	0.0000	0.0001	0.0004	0.0014	0.0043	0.0106	0.0229	0.0439
	9	0.0000	0.0000	0.0000	0.0000	0.0000	0.0001	0.0005	0.0016	0.0042	0.0098
	10	0.0000	0.0000	0.0000	0.0000	0.0000	0.0000	0.0000	0.0001	0.0003	0.0010
12	0	0.5404	0.2824	0.1422	0.0687	0.0317	0.0138	0.0057	0.0022	0.0008	0.0002
	1	0.3413	0.3766	0.3012	0.2062	0.1267	0.0712	0.0368	0.0174	0.0075	0.0029
	2	0.0988	0.2301	0.2924	0.2835	0.2323	0.1678	0.1088	0.0639	0.0339	0.0161
	3	0.0173	0.0853	0.1720	0.2362	0.2581	0.2397	0.1954	0.1419	0.0923	0.0537
	4	0.0021	0.0213	0.0683	0.1329	0.1936	0.2311	0.2367	0.2128	0.1700	0.1208
	5	0.0002	0.0038	0.0193	0.0532	0.1032	0.1585	0.2039	0.2270	0.2225	0.1934
	6	0.0000	0.0005	0.0040	0.0155	0.0401	0.0792	0.1281	0.1766	0.2124	0.2256
	7	0.0000	0.0000	0.0006	0.0033	0.0115	0.0291	0.0591	0.1009	0.1489	0.1934
	8	0.0000	0.0000	0.0001	0.0005	0.0024	0.0078	0.0199	0.0420	0.0762	0.1208
	9	0.0000	0.0000	0.0000	0.0001	0.0004	0.0015	0.0048	0.0125	0.0277	0.0537
	10	0.0000	0.0000	0.0000	0.0000	0.0000	0.0002	0.0008	0.0025	0.0068	0.0161
	11	0.0000	0.0000	0.0000	0.0000	0.0000	0.0000	0.0001	0.0003	0.0010	0.0029
	12	0.0000	0.0000	0.0000	0.0000	0.0000	0.0000	0.0000	0.0000	0.0001	0.0002

Binomial Probabilities (Continued)

n	x	p=0.05	0.10	0.15	0.20	0.25	0.30	0.35	0.40	0.45	0.50
15	0	0.4633	0.2059	0.0874	0.0352	0.0134	0.0047	0.0016	0.0005	0.0001	0.0000
	1	0.3658	0.3432	0.2312	0.1319	0.0668	0.0305	0.0126	0.0047	0.0016	0.0005
	2	0.1348	0.2669	0.2856	0.2309	0.1559	0.0916	0.0476	0.0219	0.0090	0.0032
	3	0.0307	0.1285	0.2184	0.2501	0.2252	0.1700	0.1110	0.0634	0.0318	0.0139
	4	0.0049	0.0428	0.1156	0.1876	0.2252	0.2186	0.1792	0.1268	0.0780	0.0417
	5	0.0006	0.0105	0.0449	0.1032	0.1651	0.2061	0.2123	0.1859	0.1404	0.0916
	6	0.0000	0.0019	0.0132	0.0430	0.0917	0.1472	0.1906	0.2066	0.1914	0.1527
	7	0.0000	0.0003	0.0030	0.0138	0.0393	0.0811	0.1319	0.1771	0.2013	0.1964
	8	0.0000	0.0000	0.0005	0.0035	0.0131	0.0348	0.0710	0.1181	0.1647	0.1964
	9	0.0000	0.0000	0.0001	0.0007	0.0034	0.0116	0.0298	0.0612	0.1048	0.1527
	10	0.0000	0.0000	0.0000	0.0001	0.0007	0.0030	0.0096	0.0245	0.0515	0.0916
	11	0.0000	0.0000	0.0000	0.0000	0.0001	0.0006	0.0024	0.0074	0.0191	0.0417
	12	0.0000	0.0000	0.0000	0.0000	0.0000	0.0001	0.0004	0.0016	0.0052	0.0139
	13	0.0000	0.0000	0.0000	0.0000	0.0000	0.0000	0.0001	0.0003	0.0010	0.0032
	14	0.0000	0.0000	0.0000	0.0000	0.0000	0.0000	0.0000	0.0000	0.0001	0.0005
	15	0.0000	0.0000	0.0000	0.0000	0.0000	0.0000	0.0000	0.0000	0.0000	0.0000
18	0	0.3972	0.1501	0.0536	0.0180	0.0056	0.0016	0.0004	0.0001	0.0000	0.0000
	1	0.3763	0.3002	0.1704	0.0811	0.0338	0.0126	0.0042	0.0012	0.0003	0.0001
	2	0.1683	0.2835	0.2556	0.1723	0.0958	0.0458	0.0190	0.0069	0.0022	0.0006
	3	0.0473	0.1680	0.2406	0.2297	0.1704	0.1046	0.0547	0.0246	0.0095	0.0031
	4	0.0093	0.0700	0.1592	0.2153	0.2130	0.1681	0.1104	0.0614	0.0291	0.0117
	5	0.0014	0.0218	0.0787	0.1507	0.1988	0.2017	0.1664	0.1146	0.0666	0.0327
	6	0.0002	0.0052	0.0301	0.0816	0.1436	0.1873	0.1941	0.1655	0.1181	0.0708
	7	0.0000	0.0010	0.0091	0.0350	0.0820	0.1376	0.1792	0.1892	0.1657	0.1214
	8	0.0000	0.0002	0.0022	0.0120	0.0376	0.0811	0.1327	0.1734	0.1864	0.1669
	9	0.0000	0.0000	0.0004	0.0033	0.0139	0.0386	0.0794	0.1284	0.1694	0.1855
	10	0.0000	0.0000	0.0001	0.0008	0.0042	0.0149	0.0385	0.0771	0.1248	0.1669
	11	0.0000	0.0000	0.0000	0.0001	0.0010	0.0046	0.0151	0.0374	0.0742	0.1214
	12	0.0000	0.0000	0.0000	0.0000	0.0002	0.0012	0.0047	0.0145	0.0354	0.0708
	13	0.0000	0.0000	0.0000	0.0000	0.0000	0.0002	0.0012	0.0045	0.0134	0.0327
	14	0.0000	0.0000	0.0000	0.0000	0.0000	0.0000	0.0002	0.0011	0.0039	0.0117
	15	0.0000	0.0000	0.0000	0.0000	0.0000	0.0000	0.0000	0.0002	0.0009	0.0031
	16	0.0000	0.0000	0.0000	0.0000	0.0000	0.0000	0.0000	0.0000	0.0001	0.0006
	17	0.0000	0.0000	0.0000	0.0000	0.0000	0.0000	0.0000	0.0000	0.0000	0.0001
	18	0.0000	0.0000	0.0000	0.0000	0.0000	0.0000	0.0000	0.0000	0.0000	0.0000

Binomial Probabilities (Continued)

		p									
n	x	0.05	0.10	0.15	0.20	0.25	0.30	0.35	0.40	0.45	0.50
20	0	0.3585	0.1216	0.0388	0.0115	0.0032	0.0008	0.0002	0.0000	0.0000	0.0000
	1	0.3774	0.2702	0.1368	0.0576	0.0211	0.0068	0.0020	0.0005	0.0001	0.0000
	2	0.1887	0.2852	0.2293	0.1369	0.0669	0.0278	0.0100	0.0031	0.0008	0.0002
	3	0.0596	0.1901	0.2428	0.2054	0.1339	0.0716	0.0323	0.0123	0.0040	0.0011
	4	0.0133	0.0898	0.1821	0.2182	0.1897	0.1304	0.0738	0.0350	0.0139	0.0046
	5	0.0022	0.0319	0.1028	0.1746	0.2023	0.1789	0.1272	0.0746	0.0365	0.0148
	6	0.0003	0.0089	0.0454	0.1091	0.1686	0.1916	0.1712	0.1244	0.0746	0.0370
	7	0.0000	0.0020	0.0160	0.0545	0.1124	0.1643	0.1844	0.1659	0.1221	0.0739
	8	0.0000	0.0004	0.0046	0.0222	0.0609	0.1144	0.1614	0.1797	0.1623	0.1201
	9	0.0000	0.0001	0.0011	0.0074	0.0271	0.0654	0.1158	0.1597	0.1771	0.1602
	10	0.0000	0.0000	0.0002	0.0020	0.0099	0.0308	0.0686	0.1171	0.1593	0.1762
	11	0.0000	0.0000	0.0000	0.0005	0.0030	0.0120	0.0336	0.0710	0.1185	0.1602
	12	0.0000	0.0000	0.0000	0.0001	0.0008	0.0039	0.0136	0.0355	0.0727	0.1201
	13	0.0000	0.0000	0.0000	0.0000	0.0002	0.0010	0.0045	0.0146	0.0366	0.0739
	14	0.0000	0.0000	0.0000	0.0000	0.0000	0.0002	0.0012	0.0049	0.0150	0.0370
	15	0.0000	0.0000	0.0000	0.0000	0.0000	0.0000	0.0003	0.0013	0.0049	0.0148
	16	0.0000	0.0000	0.0000	0.0000	0.0000	0.0000	0.0000	0.0003	0.0013	0.0046
	17	0.0000	0.0000	0.0000	0.0000	0.0000	0.0000	0.0000	0.0000	0.0002	0.0011
	18	0.0000	0.0000	0.0000	0.0000	0.0000	0.0000	0.0000	0.0000	0.0000	0.0002
	19	0.0000	0.0000	0.0000	0.0000	0.0000	0.0000	0.0000	0.0000	0.0000	0.0000
	20	0.0000	0.0000	0.0000	0.0000	0.0000	0.0000	0.0000	0.0000	0.0000	0.0000

Binomial Probabilities (Continued)

		p								
n	x	0.55	0.60	0.65	0.70	0.75	0.80	0.85	0.90	0.95
2	0	0.2025	0.1600	0.1225	0.0900	0.0625	0.0400	0.0225	0.0100	0.0025
	1	0.4950	0.4800	0.4550	0.4200	0.3750	0.3200	0.2550	0.1800	0.0950
	2	0.3025	0.3600	0.4225	0.4900	0.5625	0.6400	0.7225	0.8100	0.9025
3	0	0.0911	0.0640	0.0429	0.0270	0.0156	0.0080	0.0034	0.0010	0.0001
	1	0.3341	0.2880	0.2389	0.1890	0.1406	0.0960	0.0574	0.0270	0.0071
	2	0.4084	0.4320	0.4436	0.4410	0.4219	0.3840	0.3251	0.2430	0.1354
	3	0.1664	0.2160	0.2746	0.3430	0.4219	0.5120	0.6141	0.7290	0.8574
4	0	0.0410	0.0256	0.0150	0.0081	0.0039	0.0016	0.0005	0.0001	0.0000
	1	0.2005	0.1536	0.1115	0.0756	0.0469	0.0256	0.0115	0.0036	0.0005
	2	0.3675	0.3456	0.3105	0.2646	0.2109	0.1536	0.0975	0.0486	0.0135
	3	0.2995	0.3456	0.3845	0.4116	0.4219	0.4096	0.3685	0.2916	0.1715
	4	0.0915	0.1296	0.1785	0.2401	0.3164	0.4096	0.5220	0.6561	0.8145
5	0	0.0185	0.0102	0.0053	0.0024	0.0010	0.0003	0.0001	0.0000	0.0000
	1	0.1128	0.0768	0.0488	0.0284	0.0146	0.0064	0.0022	0.0005	0.0000
	2	0.2757	0.2304	0.1811	0.1323	0.0879	0.0512	0.0244	0.0081	0.0011
	3	0.3369	0.3456	0.3364	0.3087	0.2637	0.2048	0.1382	0.0729	0.0214
	4	0.2059	0.2592	0.3124	0.3601	0.3955	0.4096	0.3915	0.3281	0.2036
	5	0.0503	0.0778	0.1160	0.1681	0.2373	0.3277	0.4437	0.5905	0.7738
6	0	0.0083	0.0041	0.0018	0.0007	0.0002	0.0001	0.0000	0.0000	0.0000
	1	0.0609	0.0369	0.0205	0.0102	0.0044	0.0015	0.0004	0.0001	0.0000
	2	0.1861	0.1382	0.0951	0.0595	0.0330	0.0154	0.0055	0.0012	0.0001
	3	0.3032	0.2765	0.2355	0.1852	0.1318	0.0819	0.0415	0.0146	0.0021
	4	0.2780	0.3110	0.3280	0.3241	0.2966	0.2458	0.1762	0.0984	0.0305
	5	0.1359	0.1866	0.2437	0.3025	0.3560	0.3932	0.3993	0.3543	0.2321
	6	0.0277	0.0467	0.0754	0.1176	0.1780	0.2621	0.3771	0.5314	0.7351
7	0	0.0037	0.0016	0.0006	0.0002	0.0001	0.0000	0.0000	0.0000	0.0000
	1	0.0320	0.0172	0.0084	0.0036	0.0013	0.0004	0.0001	0.0000	0.0000
	2	0.1172	0.0774	0.0466	0.0250	0.0115	0.0043	0.0012	0.0002	0.0000
	3	0.2388	0.1935	0.1442	0.0972	0.0577	0.0287	0.0109	0.0026	0.0002
	4	0.2918	0.2903	0.2679	0.2269	0.1730	0.1147	0.0617	0.0230	0.0036
	5	0.2140	0.2613	0.2985	0.3177	0.3115	0.2753	0.2097	0.1240	0.0406
	6	0.0872	0.1306	0.1848	0.2471	0.3115	0.3670	0.3960	0.3720	0.2573
	7	0.0152	0.0280	0.0490	0.0824	0.1335	0.2097	0.3206	0.4783	0.6983
8	0	0.0017	0.0007	0.0002	0.0001	0.0000	0.0000	0.0000	0.0000	0.0000
	1	0.0164	0.0079	0.0033	0.0012	0.0004	0.0001	0.0000	0.0000	0.0000
	2	0.0703	0.0413	0.0217	0.0100	0.0038	0.0011	0.0002	0.0000	0.0000
	3	0.1719	0.1239	0.0808	0.0467	0.0231	0.0092	0.0026	0.0004	0.0000
	4	0.2627	0.2322	0.1875	0.1361	0.0865	0.0459	0.0185	0.0046	0.0004
	5	0.2568	0.2787	0.2786	0.2541	0.2076	0.1468	0.0839	0.0331	0.0054
	6	0.1569	0.2090	0.2587	0.2965	0.3115	0.2936	0.2376	0.1488	0.0515
	7	0.0548	0.0896	0.1373	0.1977	0.2670	0.3355	0.3847	0.3826	0.2793
	8	0.0084	0.0168	0.0319	0.0576	0.1001	0.1678	0.2725	0.4305	0.6634

Binomial Probabilities (Continued)

		p								
n	x	0.55	0.60	0.65	0.70	0.75	0.80	0.85	0.90	0.95
9	0	0.0008	0.0003	0.0001	0.0000	0.0000	0.0000	0.0000	0.0000	0.0000
	1	0.0083	0.0035	0.0013	0.0004	0.0001	0.0000	0.0000	0.0000	0.0000
	2	0.0407	0.0212	0.0098	0.0039	0.0012	0.0003	0.0000	0.0000	0.0000
	3	0.1160	0.0743	0.0424	0.0210	0.0087	0.0028	0.0006	0.0001	0.0000
	4	0.2128	0.1672	0.1181	0.0735	0.0389	0.0165	0.0050	0.0008	0.0000
	5	0.2600	0.2508	0.2194	0.1715	0.1168	0.0661	0.0283	0.0074	0.0006
	6	0.2119	0.2508	0.2716	0.2668	0.2336	0.1762	0.1069	0.0446	0.0077
	7	0.1110	0.1612	0.2162	0.2668	0.3003	0.3020	0.2597	0.1722	0.0629
	8	0.0339	0.0605	0.1004	0.1556	0.2253	0.3020	0.3679	0.3874	0.2985
	9	0.0046	0.0101	0.0207	0.0404	0.0751	0.1342	0.2316	0.3874	0.6302
10	0	0.0003	0.0001	0.0000	0.0000	0.0000	0.0000	0.0000	0.0000	0.0000
	1	0.0042	0.0016	0.0005	0.0001	0.0000	0.0000	0.0000	0.0000	0.0000
	2	0.0229	0.0106	0.0043	0.0014	0.0004	0.0001	0.0000	0.0000	0.0000
	3	0.0746	0.0425	0.0212	0.0090	0.0031	0.0008	0.0001	0.0000	0.0000
	4	0.1596	0.1115	0.0689	0.0368	0.0162	0.0055	0.0012	0.0001	0.0000
	5	0.2340	0.2007	0.1536	0.1029	0.0584	0.0264	0.0085	0.0015	0.0001
	6	0.2384	0.2508	0.2377	0.2001	0.1460	0.0881	0.0401	0.0112	0.0010
	7	0.1665	0.2150	0.2522	0.2668	0.2503	0.2013	0.1298	0.0574	0.0105
	8	0.0763	0.1209	0.1757	0.2335	0.2816	0.3020	0.2759	0.1937	0.0746
	9	0.0207	0.0403	0.0725	0.1211	0.1877	0.2684	0.3474	0.3874	0.3151
	10	0.0025	0.0060	0.0135	0.0282	0.0563	0.1074	0.1969	0.3487	0.5987
12	0	0.0001	0.0000	0.0000	0.0000	0.0000	0.0000	0.0000	0.0000	0.0000
	1	0.0010	0.0003	0.0001	0.0000	0.0000	0.0000	0.0000	0.0000	0.0000
	2	0.0068	0.0025	0.0008	0.0002	0.0000	0.0000	0.0000	0.0000	0.0000
	3	0.0277	0.0125	0.0048	0.0015	0.0004	0.0001	0.0000	0.0000	0.0000
	4	0.0762	0.0420	0.0199	0.0078	0.0024	0.0005	0.0001	0.0000	0.0000
	5	0.1489	0.1009	0.0591	0.0291	0.0115	0.0033	0.0006	0.0000	0.0000
	6	0.2124	0.1766	0.1281	0.0792	0.0401	0.0155	0.0040	0.0005	0.0000
	7	0.2225	0.2270	0.2039	0.1585	0.1032	0.0532	0.0193	0.0038	0.0002
	8	0.1700	0.2128	0.2367	0.2311	0.1936	0.1329	0.0683	0.0213	0.0021
	9	0.0923	0.1419	0.1954	0.2397	0.2581	0.2362	0.1720	0.0852	0.0173
	10	0.0339	0.0639	0.1088	0.1678	0.2323	0.2835	0.2924	0.2301	0.0988
	11	0.0075	0.0174	0.0368	0.0712	0.1267	0.2062	0.3012	0.3766	0.3413
	12	0.0008	0.0022	0.0057	0.0138	0.0317	0.0687	0.1422	0.2824	0.5404

Binomial Probabilities (Continued)

		\multicolumn{9}{c}{p}								
n	x	0.55	0.60	0.65	0.70	0.75	0.80	0.85	0.90	0.95
15	0	0.0000	0.0000	0.0000	0.0000	0.0000	0.0000	0.0000	0.0000	0.0000
	1	0.0001	0.0000	0.0000	0.0000	0.0000	0.0000	0.0000	0.0000	0.0000
	2	0.0010	0.0003	0.0001	0.0000	0.0000	0.0000	0.0000	0.0000	0.0000
	3	0.0052	0.0016	0.0004	0.0001	0.0000	0.0000	0.0000	0.0000	0.0000
	4	0.0191	0.0074	0.0024	0.0006	0.0001	0.0000	0.0000	0.0000	0.0000
	5	0.0515	0.0245	0.0096	0.0030	0.0007	0.0001	0.0000	0.0000	0.0000
	6	0.1048	0.0612	0.0298	0.0116	0.0034	0.0007	0.0001	0.0000	0.0000
	7	0.1647	0.1181	0.0710	0.0348	0.0131	0.0035	0.0005	0.0000	0.0000
	8	0.2013	0.1771	0.1319	0.0811	0.0393	0.0138	0.0030	0.0003	0.0000
	9	0.1914	0.2066	0.1906	0.1472	0.0917	0.0430	0.0132	0.0019	0.0000
	10	0.1404	0.1859	0.2123	0.2061	0.1651	0.1032	0.0449	0.0105	0.0006
	11	0.0780	0.1268	0.1792	0.2186	0.2252	0.1876	0.1156	0.0428	0.0049
	12	0.0318	0.0634	0.1110	0.1700	0.2252	0.2501	0.2184	0.1285	0.0307
	13	0.0090	0.0219	0.0476	0.0916	0.1559	0.2309	0.2856	0.2669	0.1348
	14	0.0016	0.0047	0.0126	0.0305	0.0668	0.1319	0.2312	0.3432	0.3658
	15	0.0001	0.0005	0.0016	0.0047	0.0134	0.0352	0.0874	0.2059	0.4633
18	0	0.0000	0.0000	0.0000	0.0000	0.0000	0.0000	0.0000	0.0000	0.0000
	1	0.0000	0.0000	0.0000	0.0000	0.0000	0.0000	0.0000	0.0000	0.0000
	2	0.0001	0.0000	0.0000	0.0000	0.0000	0.0000	0.0000	0.0000	0.0000
	3	0.0009	0.0002	0.0000	0.0000	0.0000	0.0000	0.0000	0.0000	0.0000
	4	0.0039	0.0011	0.0002	0.0000	0.0000	0.0000	0.0000	0.0000	0.0000
	5	0.0134	0.0045	0.0012	0.0002	0.0000	0.0000	0.0000	0.0000	0.0000
	6	0.0354	0.0145	0.0047	0.0012	0.0002	0.0000	0.0000	0.0000	0.0000
	7	0.0742	0.0374	0.0151	0.0046	0.0010	0.0001	0.0000	0.0000	0.0000
	8	0.1248	0.0771	0.0385	0.0149	0.0042	0.0008	0.0001	0.0000	0.0000
	9	0.1694	0.1284	0.0794	0.0386	0.0139	0.0033	0.0004	0.0000	0.0000
	10	0.1864	0.1734	0.1327	0.0811	0.0376	0.0120	0.0022	0.0002	0.0000
	11	0.1657	0.1892	0.1792	0.1376	0.0820	0.0350	0.0091	0.0010	0.0000
	12	0.1181	0.1655	0.1941	0.1873	0.1436	0.0816	0.0301	0.0052	0.0002
	13	0.0666	0.1146	0.1664	0.2017	0.1988	0.1507	0.0787	0.0218	0.0014
	14	0.0291	0.0614	0.1104	0.1681	0.2130	0.2153	0.1592	0.0700	0.0093
	15	0.0095	0.0246	0.0547	0.1046	0.1704	0.2297	0.2406	0.1680	0.0473
	16	0.0022	0.0069	0.0190	0.0458	0.0958	0.1723	0.2556	0.2835	0.1683
	17	0.0003	0.0012	0.0042	0.0126	0.0338	0.0811	0.1704	0.3002	0.3763
	18	0.0000	0.0001	0.0004	0.0016	0.0056	0.0180	0.0536	0.1501	0.3972

Binomial Probabilities (Continued)

		p								
n	x	0.55	0.60	0.65	0.70	0.75	0.80	0.85	0.90	0.95
20	0	0.0000	0.0000	0.0000	0.0000	0.0000	0.0000	0.0000	0.0000	0.0000
	1	0.0000	0.0000	0.0000	0.0000	0.0000	0.0000	0.0000	0.0000	0.0000
	2	0.0000	0.0000	0.0000	0.0000	0.0000	0.0000	0.0000	0.0000	0.0000
	3	0.0002	0.0000	0.0000	0.0000	0.0000	0.0000	0.0000	0.0000	0.0000
	4	0.0013	0.0003	0.0000	0.0000	0.0000	0.0000	0.0000	0.0000	0.0000
	5	0.0049	0.0013	0.0003	0.0000	0.0000	0.0000	0.0000	0.0000	0.0000
	6	0.0150	0.0049	0.0012	0.0002	0.0000	0.0000	0.0000	0.0000	0.0000
	7	0.0366	0.0146	0.0045	0.0010	0.0002	0.0000	0.0000	0.0000	0.0000
	8	0.0727	0.0355	0.0136	0.0039	0.0008	0.0001	0.0000	0.0000	0.0000
	9	0.1185	0.0710	0.0336	0.0120	0.0030	0.0005	0.0000	0.0000	0.0000
	10	0.1593	0.1171	0.0686	0.0308	0.0099	0.0020	0.0002	0.0000	0.0000
	11	0.1771	0.1597	0.1158	0.0654	0.0271	0.0074	0.0011	0.0001	0.0000
	12	0.1623	0.1797	0.1614	0.1144	0.0609	0.0222	0.0046	0.0004	0.0000
	13	0.1221	0.1659	0.1844	0.1643	0.1124	0.0545	0.0160	0.0020	0.0000
	14	0.0746	0.1244	0.1712	0.1916	0.1686	0.1091	0.0454	0.0089	0.0003
	15	0.0365	0.0746	0.1272	0.1789	0.2023	0.1746	0.1028	0.0319	0.0022
	16	0.0139	0.0350	0.0738	0.1304	0.1897	0.2182	0.1821	0.0898	0.0133
	17	0.0040	0.0123	0.0323	0.0716	0.1339	0.2054	0.2428	0.1901	0.0596
	18	0.0008	0.0031	0.0100	0.0278	0.0669	0.1369	0.2293	0.2852	0.1887
	19	0.0001	0.0005	0.0020	0.0068	0.0211	0.0576	0.1368	0.2702	0.3774
	20	0.0000	0.0000	0.0002	0.0008	0.0032	0.0115	0.0388	0.1216	0.3585

APPENDIX B: POISSON PROBABILITIES

Entries in the table give the probability of x occurrences for a Poisson process with a mean λ. For example, when $\lambda = 2.5$, the probability of $x = 4$ occurrences is 0.1336.

					λ					
x	0.1	0.2	0.3	0.4	0.5	0.6	0.7	0.8	0.9	1.0
0	0.9048	0.8187	0.7408	0.6703	0.6065	0.5488	0.4966	0.4493	0.4066	0.3679
1	0.0905	0.1637	0.2222	0.2681	0.3033	0.3293	0.3476	0.3595	0.3659	0.3679
2	0.0045	0.0164	0.0333	0.0536	0.0758	0.0988	0.1217	0.1438	0.1647	0.1839
3	0.0002	0.0011	0.0033	0.0072	0.0126	0.0198	0.0284	0.0383	0.0494	0.0613
4	0.0000	0.0001	0.0002	0.0007	0.0016	0.0030	0.0050	0.0077	0.0111	0.0153
5	0.0000	0.0000	0.0000	0.0001	0.0002	0.0004	0.0007	0.0012	0.0020	0.0031
6	0.0000	0.0000	0.0000	0.0000	0.0000	0.0000	0.0001	0.0002	0.0003	0.0005
7	0.0000	0.0000	0.0000	0.0000	0.0000	0.0000	0.0000	0.0000	0.0000	0.0001

					λ					
x	1.1	1.2	1.3	1.4	1.5	1.6	1.7	1.8	1.9	2.0
0	0.3329	0.3012	0.2725	0.2466	0.2231	0.2019	0.1827	0.1653	0.1496	0.1353
1	0.3662	0.3614	0.3543	0.3452	0.3347	0.3230	0.3106	0.2975	0.2842	0.2707
2	0.2014	0.2169	0.2303	0.2417	0.2510	0.2584	0.2640	0.2678	0.2700	0.2707
3	0.0738	0.0867	0.0998	0.1128	0.1255	0.1378	0.1496	0.1607	0.1710	0.1804
4	0.0203	0.0260	0.0324	0.0395	0.0471	0.0551	0.0636	0.0723	0.0812	0.0902
5	0.0045	0.0062	0.0084	0.0111	0.0141	0.0176	0.0216	0.0260	0.0309	0.0361
6	0.0008	0.0012	0.0018	0.0026	0.0035	0.0047	0.0061	0.0078	0.0098	0.0120
7	0.0001	0.0002	0.0003	0.0005	0.0008	0.0011	0.0015	0.0020	0.0027	0.0034
8	0.0000	0.0000	0.0001	0.0001	0.0001	0.0002	0.0003	0.0005	0.0006	0.0009
9	0.0000	0.0000	0.0000	0.0000	0.0000	0.0000	0.0001	0.0001	0.0001	0.0002

					λ					
x	2.1	2.2	2.3	2.4	2.5	2.6	2.7	2.8	2.9	3.0
0	0.1225	0.1108	0.1003	0.0907	0.0821	0.0743	0.0672	0.0608	0.0550	0.0498
1	0.2572	0.2438	0.2306	0.2177	0.2052	0.1931	0.1815	0.1703	0.1596	0.1494
2	0.2700	0.2681	0.2652	0.2613	0.2565	0.2510	0.2450	0.2384	0.2314	0.2240
3	0.1890	0.1966	0.2033	0.2090	0.2138	0.2176	0.2205	0.2225	0.2237	0.2240
4	0.0992	0.1082	0.1169	0.1254	0.1336	0.1414	0.1488	0.1557	0.1622	0.1680
5	0.0417	0.0476	0.0538	0.0602	0.0668	0.0735	0.0804	0.0872	0.0940	0.1008
6	0.0146	0.0174	0.0206	0.0241	0.0278	0.0319	0.0362	0.0407	0.0455	0.0540
7	0.0044	0.0055	0.0068	0.0083	0.0099	0.0118	0.0139	0.0163	0.0188	0.0216
8	0.0011	0.0015	0.0019	0.0025	0.0031	0.0038	0.0047	0.0057	0.0068	0.0081
9	0.0003	0.0004	0.0005	0.0007	0.0009	0.0011	0.0014	0.0018	0.0022	0.0027
10	0.0001	0.0001	0.0001	0.0002	0.0002	0.0003	0.0004	0.0005	0.0006	0.0008
11	0.0000	0.0000	0.0000	0.0000	0.0000	0.0001	0.0001	0.0001	0.0002	0.0002
12	0.0000	0.0000	0.0000	0.0000	0.0000	0.0000	0.0000	0.0000	0.0000	0.0001

Poisson Probabilities (Continued)

					λ					
x	3.1	3.2	3.3	3.4	3.5	3.6	3.7	3.8	3.9	4.0
0	0.0450	0.0408	0.0369	0.0344	0.0302	0.0273	0.0247	0.0224	0.0202	0.0183
1	0.1397	0.1304	0.1217	0.1135	0.1057	0.0984	0.0915	0.0850	0.0789	0.0733
2	0.2165	0.2087	0.2008	0.1929	0.1850	0.1771	0.1692	0.1615	0.1539	0.1465
3	0.2237	0.2226	0.2209	0.2186	0.2158	0.2125	0.2087	0.2046	0.2001	0.1954
4	0.1734	0.1781	0.1823	0.1858	0.1888	0.1912	0.1931	0.1944	0.1951	0.1954
5	0.1075	0.1140	0.1203	0.1264	0.1322	0.1377	0.1429	0.1477	0.1522	0.1563
6	0.0555	0.0608	0.0662	0.0716	0.0771	0.0826	0.0881	0.0936	0.0989	0.1042
7	0.0246	0.0278	0.0312	0.0348	0.0385	0.0425	0.0466	0.0508	0.0551	0.0595
8	0.0095	0.0111	0.0129	0.0148	0.0169	0.0191	0.0215	0.0241	0.0269	0.0298
9	0.0033	0.0040	0.0047	0.0056	0.0066	0.0076	0.0089	0.0102	0.0116	0.0132
10	0.0010	0.0013	0.0016	0.0019	0.0023	0.0028	0.0033	0.0039	0.0045	0.0053
11	0.0003	0.0004	0.0005	0.0006	0.0007	0.0009	0.0011	0.0013	0.0016	0.0019
12	0.0001	0.0001	0.0001	0.0002	0.0002	0.0003	0.0003	0.0004	0.0005	0.0006
13	0.0000	0.0000	0.0000	0.0000	0.0001	0.0001	0.0001	0.0001	0.0002	0.0002
14	0.0000	0.0000	0.0000	0.0000	0.0000	0.0000	0.0000	0.0000	0.0000	0.0001

					λ					
x	4.1	4.2	4.3	4.4	4.5	4.6	4.7	4.8	4.9	5.0
0	0.0166	0.0150	0.0136	0.0123	0.0111	0.0101	0.0091	0.0082	0.0074	0.0067
1	0.0679	0.0630	0.0583	0.0540	0.0500	0.0462	0.0427	0.0395	0.0365	0.0337
2	0.1393	0.1323	0.1254	0.1188	0.1125	0.1063	0.1005	0.0948	0.0894	0.0842
3	0.1904	0.1852	0.1798	0.1743	0.1687	0.1631	0.1574	0.1517	0.1460	0.1404
4	0.1951	0.1944	0.1933	0.1917	0.1898	0.1875	0.1849	0.1820	0.1789	0.1755
5	0.1600	0.1633	0.1662	0.1687	0.1708	0.1725	0.1738	0.1747	0.1753	0.1755
6	0.1093	0.1143	0.1191	0.1237	0.1281	0.1323	0.1362	0.1398	0.1432	0.1462
7	0.0640	0.0686	0.0732	0.0778	0.0824	0.0869	0.0914	0.0959	0.1002	0.1044
8	0.0328	0.0360	0.0393	0.0428	0.0463	0.0500	0.0537	0.0575	0.0614	0.0653
9	0.0150	0.0168	0.0188	0.0209	0.0232	0.0255	0.0280	0.0307	0.0334	0.0363
10	0.0061	0.0071	0.0081	0.0092	0.0104	0.0118	0.0132	0.0147	0.0164	0.0181
11	0.0023	0.0027	0.0032	0.0037	0.0043	0.0049	0.0056	0.0064	0.0073	0.0082
12	0.0008	0.0009	0.0011	0.0014	0.0016	0.0019	0.0022	0.0026	0.0030	0.0034
13	0.0002	0.0003	0.0004	0.0005	0.0006	0.0007	0.0008	0.0009	0.0011	0.0013
14	0.0001	0.0001	0.0001	0.0001	0.0002	0.0002	0.0003	0.0003	0.0004	0.0005
15	0.0000	0.0000	0.0000	0.0000	0.0001	0.0001	0.0001	0.0001	0.0001	0.0002

Poisson Probabilities (Continued)

	λ									
x	5.1	5.2	5.3	5.4	5.5	5.6	5.7	5.8	5.9	6.0
0	0.0061	0.0055	0.0050	0.0045	0.0041	0.0037	0.0033	0.0030	0.0027	0.0025
1	0.0311	0.0287	0.0265	0.0244	0.0225	0.0207	0.0191	0.0176	0.0162	0.0149
2	0.0793	0.0746	0.0701	0.0659	0.0618	0.0580	0.0544	0.0509	0.0477	0.0446
3	0.1348	0.1293	0.1239	0.1185	0.1133	0.1082	0.1033	0.0985	0.0938	0.0892
4	0.1719	0.1681	0.1641	0.1600	0.1558	0.1515	0.1472	0.1428	0.1383	0.1339
5	0.1753	0.1748	0.1740	0.1728	0.1714	0.1697	0.1678	0.1656	0.1632	0.1606
6	0.1490	0.1515	0.1537	0.1555	0.1571	0.1587	0.1594	0.1601	0.1605	0.1606
7	0.1086	0.1125	0.1163	0.1200	0.1234	0.1267	0.1298	0.1326	0.1353	0.1377
8	0.0692	0.0731	0.0771	0.0810	0.0849	0.0887	0.0925	0.0962	0.0998	0.1033
9	0.0392	0.0423	0.0454	0.0486	0.0519	0.0552	0.0586	0.0620	0.0654	0.0688
10	0.0200	0.0220	0.0241	0.0262	0.0285	0.0309	0.0334	0.0359	0.0386	0.0413
11	0.0093	0.0104	0.0116	0.0129	0.0143	0.0157	0.0173	0.0190	0.0207	0.0225
12	0.0039	0.0045	0.0051	0.0058	0.0065	0.0073	0.0082	0.0092	0.0102	0.0113
13	0.0015	0.0018	0.0021	0.0024	0.0028	0.0032	0.0036	0.0041	0.0046	0.0052
14	0.0006	0.0007	0.0008	0.0009	0.0011	0.0013	0.0015	0.0017	0.0019	0.0022
15	0.0002	0.0002	0.0003	0.0003	0.0004	0.0005	0.0006	0.0007	0.0008	0.0009
16	0.0001	0.0001	0.0001	0.0001	0.0001	0.0002	0.0002	0.0002	0.0003	0.0003
17	0.0000	0.0000	0.0000	0.0000	0.0000	0.0001	0.0001	0.0001	0.0001	0.0001

	λ									
x	6.1	6.2	6.3	6.4	6.5	6.6	6.7	6.8	6.9	7.0
0	0.0022	0.0020	0.0018	0.0017	0.0015	0.0014	0.0012	0.0011	0.0010	0.0009
1	0.0137	0.0126	0.0116	0.0106	0.0098	0.0090	0.0082	0.0076	0.0070	0.0064
2	0.0417	0.0390	0.0364	0.0340	0.0318	0.0296	0.0276	0.0258	0.0240	0.0223
3	0.0848	0.0806	0.0765	0.0726	0.0688	0.0652	0.0617	0.0584	0.0552	0.0521
4	0.1294	0.1249	0.1205	0.1162	0.1118	0.1076	0.1034	0.0992	0.0952	0.0912
5	0.1579	0.1549	0.1519	0.1487	0.1454	0.1420	0.1385	0.1349	0.1314	0.1277
6	0.1605	0.1601	0.1595	0.1586	0.1575	0.1562	0.1546	0.1529	0.1511	0.1490
7	0.1399	0.1418	0.1435	0.1450	0.1462	0.1472	0.1480	0.1486	0.1489	0.1490
8	0.1066	0.1099	0.1130	0.1160	0.1188	0.1215	0.1240	0.1263	0.1284	0.1304
9	0.0723	0.0757	0.0791	0.0825	0.0858	0.0891	0.0923	0.0954	0.0985	0.1014
10	0.0441	0.0469	0.0498	0.0528	0.0558	0.0588	0.0618	0.0649	0.0679	0.0710
11	0.0245	0.0265	0.0285	0.0307	0.0330	0.0353	0.0377	0.0401	0.0426	0.0452
12	0.0124	0.0137	0.0150	0.0164	0.0179	0.0194	0.0210	0.0227	0.0245	0.0264
13	0.0058	0.0065	0.0073	0.0081	0.0089	0.0098	0.0108	0.0119	0.0130	0.0142
14	0.0025	0.0029	0.0033	0.0037	0.0041	0.0046	0.0052	0.0058	0.0064	0.0071
15	0.0010	0.0012	0.0014	0.0016	0.0018	0.0020	0.0023	0.0025	0.0029	0.0033
16	0.0004	0.0005	0.0005	0.0006	0.0007	0.0008	0.0010	0.0011	0.0013	0.0014
17	0.0001	0.0002	0.0002	0.0002	0.0003	0.0003	0.0004	0.0004	0.0005	0.0006
18	0.0000	0.0001	0.0001	0.0001	0.0001	0.0001	0.0001	0.0002	0.0002	0.0002
19	0.0000	0.0000	0.0000	0.0000	0.0000	0.0000	0.0000	0.0001	0.0001	0.0001

Poisson Probabilities (Continued)

x	λ 7.1	7.2	7.3	7.4	7.5	7.6	7.7	7.8	7.9	8.0
0	0.0008	0.0007	0.0007	0.0006	0.0006	0.0005	0.0005	0.0004	0.0004	0.0003
1	0.0059	0.0054	0.0049	0.0045	0.0041	0.0038	0.0035	0.0032	0.0029	0.0027
2	0.0208	0.0194	0.0180	0.0167	0.0156	0.0145	0.0134	0.0125	0.0116	0.0107
3	0.0492	0.0464	0.0438	0.0413	0.0389	0.0366	0.0345	0.0324	0.0305	0.0286
4	0.0874	0.0836	0.0799	0.0764	0.0729	0.0696	0.0663	0.0632	0.0602	0.0573
5	0.1241	0.1204	0.1167	0.1130	0.1094	0.1057	0.1021	0.0986	0.0951	0.0916
6	0.1468	0.1445	0.1420	0.1394	0.1367	0.1339	0.1311	0.1282	0.1252	0.1221
7	0.1489	0.1486	0.1481	0.1474	0.1465	0.1454	0.1442	0.1428	0.1413	0.1396
8	0.1321	0.1337	0.1351	0.1363	0.1373	0.1382	0.1388	0.1392	0.1395	0.1396
9	0.1042	0.1070	0.1096	0.1121	0.1144	0.1167	0.1187	0.1207	0.1224	0.1241
10	0.0740	0.0770	0.0800	0.0829	0.0858	0.0887	0.0914	0.0941	0.0967	0.0993
11	0.0478	0.0504	0.0531	0.0558	0.0585	0.0613	0.0640	0.0667	0.0695	0.0722
12	0.0283	0.0303	0.0323	0.0344	0.0366	0.0388	0.0411	0.0434	0.0457	0.0481
13	0.0154	0.0168	0.0181	0.0196	0.0211	0.0227	0.0243	0.0260	0.0278	0.0296
14	0.0078	0.0086	0.0095	0.0104	0.0113	0.0123	0.0134	0.0145	0.0157	0.0169
15	0.0037	0.0041	0.0046	0.0051	0.0057	0.0062	0.0069	0.0075	0.0083	0.0090
16	0.0016	0.0019	0.0021	0.0024	0.0026	0.0030	0.0033	0.0037	0.0041	0.0045
17	0.0007	0.0008	0.0009	0.0010	0.0012	0.0013	0.0015	0.0017	0.0019	0.0021
18	0.0003	0.0003	0.0004	0.0004	0.0005	0.0006	0.0006	0.0007	0.0008	0.0009
19	0.0001	0.0001	0.0001	0.0002	0.0002	0.0002	0.0003	0.0003	0.0003	0.0004
20	0.0000	0.0000	0.0001	0.0001	0.0001	0.0001	0.0001	0.0001	0.0001	0.0002
21	0.0000	0.0000	0.0000	0.0000	0.0000	0.0000	0.0000	0.0000	0.0001	0.0001

x	λ 8.1	8.2	8.3	8.4	8.5	8.6	8.7	8.8	8.9	9.0
0	0.0003	0.0003	0.0002	0.0002	0.0002	0.0002	0.0002	0.0002	0.0001	0.0001
1	0.0025	0.0023	0.0021	0.0019	0.0017	0.0016	0.0014	0.0013	0.0012	0.0011
2	0.0100	0.0092	0.0086	0.0079	0.0074	0.0068	0.0063	0.0058	0.0054	0.0050
3	0.0269	0.0252	0.0237	0.0222	0.0208	0.0195	0.0183	0.0171	0.0160	0.0150
4	0.0544	0.0517	0.0491	0.0466	0.0443	0.0420	0.0398	0.0377	0.0357	0.0337
5	0.0882	0.0849	0.0816	0.0784	0.0752	0.0722	0.0692	0.0663	0.0635	0.0607
6	0.1191	0.1160	0.1128	0.1097	0.1066	0.1034	0.1003	0.0972	0.0941	0.0911
7	0.1378	0.1358	0.1338	0.1317	0.1294	0.1271	0.1247	0.1222	0.1197	0.1171
8	0.1395	0.1392	0.1388	0.1382	0.1375	0.1366	0.1356	0.1344	0.1332	0.1318
9	0.1256	0.1269	0.1280	0.1290	0.1299	0.1306	0.1311	0.1315	0.1317	0.1318
10	0.1017	0.1040	0.1063	0.1084	0.1104	0.1123	0.1140	0.1157	0.1172	0.1186
11	0.0749	0.0776	0.0802	0.0828	0.0853	0.0878	0.0902	0.0925	0.0948	0.0970
12	0.0505	0.0530	0.0555	0.0579	0.0604	0.0629	0.0654	0.0679	0.0703	0.0728
13	0.0315	0.0334	0.0354	0.0374	0.0395	0.0416	0.0438	0.0459	0.0481	0.0504
14	0.0182	0.0196	0.0210	0.0225	0.0240	0.0256	0.0272	0.0289	0.0306	0.0324
15	0.0098	0.0107	0.0116	0.0126	0.0136	0.0147	0.0158	0.0169	0.0182	0.1094
16	0.0050	0.0055	0.0060	0.0066	0.0072	0.0079	0.0086	0.0093	0.0101	0.0109
17	0.0024	0.0026	0.0029	0.0033	0.0036	0.0040	0.0044	0.0048	0.0053	0.0058
18	0.0011	0.0012	0.0014	0.0015	0.0017	0.0019	0.0021	0.0024	0.0026	0.0029
19	0.0005	0.0005	0.0006	0.0007	0.0008	0.0009	0.0010	0.0011	0.0012	0.0014
20	0.0002	0.0002	0.0002	0.0003	0.0003	0.0004	0.0004	0.0005	0.0005	0.0006
21	0.0001	0.0001	0.0001	0.0001	0.0001	0.0002	0.0002	0.0002	0.0002	0.0003
22	0.0000	0.0000	0.0000	0.0000	0.0001	0.0001	0.0001	0.0001	0.0001	0.0001

Poisson Probabilities (Continued)

x	λ									
	9.1	9.2	9.3	9.4	9.5	9.6	9.7	9.8	9.9	10
0	0.0001	0.0001	0.0001	0.0001	0.0001	0.0001	0.0001	0.0001	0.0001	0.0000
1	0.0010	0.0009	0.0009	0.0008	0.0007	0.0007	0.0006	0.0005	0.0005	0.0005
2	0.0046	0.0043	0.0040	0.0037	0.0034	0.0031	0.0029	0.0027	0.0025	0.0023
3	0.0140	0.0131	0.0123	0.0115	0.0107	0.0100	0.0093	0.0087	0.0081	0.0076
4	0.0319	0.0302	0.0285	0.0269	0.0254	0.0240	0.0226	0.0213	0.0201	0.0189
5	0.0581	0.0555	0.0530	0.0506	0.0483	0.0460	0.0439	0.0418	0.0398	0.0378
6	0.0881	0.0851	0.0822	0.0793	0.0764	0.0736	0.0709	0.0682	0.0656	0.0631
7	0.1145	0.1118	0.1091	0.1064	0.1037	0.1010	0.0982	0.0955	0.0928	0.0901
8	0.1302	0.1286	0.1269	0.1251	0.1232	0.1212	0.1191	0.1170	0.1148	0.1126
9	0.1317	0.1315	0.1311	0.1306	0.1300	0.1293	0.1284	0.1274	0.1263	0.1251
10	0.1198	0.1210	0.1219	0.1228	0.1235	0.1241	0.1245	0.1249	0.1250	0.1251
11	0.0991	0.1012	0.1031	0.1049	0.1067	0.1083	0.1098	0.1112	0.1125	0.1137
12	0.0752	0.0776	0.0799	0.0822	0.0844	0.0866	0.0888	0.0908	0.0928	0.0948
13	0.0526	0.0549	0.0572	0.0594	0.0617	0.0640	0.0662	0.0685	0.0707	0.0729
14	0.0342	0.0361	0.0380	0.0399	0.0419	0.0439	0.0459	0.0479	0.0500	0.0521
15	0.0208	0.0221	0.0235	0.0250	0.0265	0.0281	0.0297	0.0313	0.0330	0.0347
16	0.0118	0.0127	0.0137	0.0147	0.0157	0.0168	0.0180	0.0192	0.0204	0.0217
17	0.0063	0.0069	0.0075	0.0081	0.0088	0.0095	0.0103	0.0111	0.0119	0.0128
18	0.0032	0.0035	0.0039	0.0042	0.0046	0.0051	0.0055	0.0060	0.0065	0.0071
19	0.0015	0.0017	0.0019	0.0021	0.0023	0.0026	0.0028	0.0031	0.0034	0.0027
20	0.0007	0.0008	0.0009	0.0010	0.0011	0.0012	0.0014	0.0015	0.0017	0.0019
21	0.0003	0.0003	0.0004	0.0004	0.0005	0.0006	0.0006	0.0007	0.0008	0.0009
22	0.0001	0.0001	0.0002	0.0002	0.0002	0.0002	0.0003	0.0003	0.0004	0.0004
23	0.0000	0.0001	0.0001	0.0001	0.0001	0.0001	0.0001	0.0001	0.0002	0.0002
24	0.0000	0.0000	0.0000	0.0000	0.0000	0.0000	0.0000	0.0001	0.0001	0.0001

Poisson Probabilities (Continued)

	λ									
x	11	12	13	14	15	16	17	18	19	20
0	0.0000	0.0000	0.0000	0.0000	0.0000	0.0000	0.0000	0.0000	0.0000	0.0000
1	0.0002	0.0001	0.0000	0.0000	0.0000	0.0000	0.0000	0.0000	0.0000	0.0000
2	0.0010	0.0004	0.0002	0.0001	0.0000	0.0000	0.0000	0.0000	0.0000	0.0000
3	0.0037	0.0018	0.0008	0.0004	0.0002	0.0001	0.0000	0.0000	0.0000	0.0000
4	0.0102	0.0053	0.0027	0.0013	0.0006	0.0003	0.0001	0.0001	0.0000	0.0000
5	0.0224	0.0127	0.0070	0.0037	0.0019	0.0010	0.0005	0.0002	0.0001	0.0001
6	0.0411	0.0255	0.0152	0.0087	0.0048	0.0026	0.0014	0.0007	0.0004	0.0002
7	0.0646	0.0437	0.0281	0.0174	0.0104	0.0060	0.0034	0.0018	0.0010	0.0005
8	0.0888	0.0655	0.0457	0.0304	0.0194	0.0120	0.0072	0.0042	0.0024	0.0013
9	0.1085	0.0874	0.0661	0.0473	0.0324	0.0213	0.0135	0.0083	0.0050	0.0029
10	0.1194	0.1048	0.0859	0.0663	0.0486	0.0341	0.0230	0.0150	0.0095	0.0058
11	0.1194	0.1144	0.1015	0.0844	0.0663	0.0496	0.0355	0.0245	0.0164	0.0106
12	0.1094	0.1144	0.1099	0.0984	0.0829	0.0661	0.0504	0.0368	0.0259	0.0176
13	0.0926	0.1056	0.1099	0.1060	0.0956	0.0814	0.0658	0.0509	0.0378	0.0271
14	0.0728	0.0905	0.1021	0.1060	0.1024	0.0930	0.0800	0.0655	0.0514	0.0387
15	0.0534	0.0724	0.0885	0.0989	0.1024	0.0992	0.0906	0.0786	0.0650	0.0516
16	0.0367	0.0543	0.0719	0.0866	0.0960	0.0992	0.0963	0.0884	0.0772	0.0646
17	0.0237	0.0383	0.0550	0.0713	0.0847	0.0934	0.0963	0.0936	0.0863	0.0760
18	0.0145	0.0256	0.0397	0.0554	0.0706	0.0830	0.0909	0.0936	0.0911	0.0844
19	0.0084	0.0161	0.0272	0.0409	0.0557	0.0699	0.0814	0.0887	0.0911	0.0888
20	0.0046	0.0097	0.0177	0.0286	0.0418	0.0559	0.0692	0.0798	0.0866	0.0888
21	0.0024	0.0055	0.0109	0.0191	0.0299	0.0426	0.0560	0.0684	0.0783	0.0846
22	0.0012	0.0030	0.0065	0.0121	0.0204	0.0310	0.0433	0.0560	0.0676	0.0769
23	0.0006	0.0016	0.0037	0.0074	0.0133	0.0216	0.0320	0.0438	0.0559	0.0669
24	0.0003	0.0008	0.0020	0.0043	0.0083	0.0144	0.0226	0.0328	0.0442	0.0557
25	0.0001	0.0004	0.0010	0.0024	0.0050	0.0092	0.0154	0.0237	0.0336	0.0446
26	0.0000	0.0002	0.0005	0.0013	0.0029	0.0057	0.0101	0.0164	0.0246	0.0343
27	0.0000	0.0001	0.0002	0.0007	0.0016	0.0034	0.0063	0.0109	0.0173	0.0254
28	0.0000	0.0000	0.0001	0.0003	0.0009	0.0019	0.0038	0.0070	0.0117	0.0181
29	0.0000	0.0000	0.0001	0.0002	0.0004	0.0011	0.0023	0.0044	0.0077	0.0125
30	0.0000	0.0000	0.0000	0.0001	0.0002	0.0006	0.0013	0.0026	0.0049	0.0083
31	0.0000	0.0000	0.0000	0.0000	0.0001	0.0003	0.0007	0.0015	0.0030	0.0054
32	0.0000	0.0000	0.0000	0.0000	0.0001	0.0001	0.0004	0.0009	0.0018	0.0034
33	0.0000	0.0000	0.0000	0.0000	0.0000	0.0001	0.0002	0.0005	0.0010	0.0020
34	0.0000	0.0000	0.0000	0.0000	0.0000	0.0000	0.0001	0.0002	0.0006	0.0012
35	0.0000	0.0000	0.0000	0.0000	0.0000	0.0000	0.0000	0.0001	0.0003	0.0007
36	0.0000	0.0000	0.0000	0.0000	0.0000	0.0000	0.0000	0.0001	0.0002	0.0004
37	0.0000	0.0000	0.0000	0.0000	0.0000	0.0000	0.0000	0.0000	0.0001	0.0002
38	0.0000	0.0000	0.0000	0.0000	0.0000	0.0000	0.0000	0.0000	0.0000	0.0001
39	0.0000	0.0000	0.0000	0.0000	0.0000	0.0000	0.0000	0.0000	0.0000	0.0001

APPENDIX C: AREAS FOR THE STANDARD NORMAL DISTRIBUTION

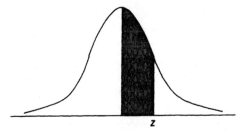

Entries in the table give the area under the curve between the mean and z standard deviations above the mean. For example, for z = 1.25 the area under the curve between the mean and z is 0.3944.

z	0.00	0.01	0.02	0.03	0.04	0.05	0.06	0.07	0.08	0.09
0.0	.0000	.0040	.0080	.0120	.0160	.0199	.0239	.0279	.0319	.0359
0.1	.0398	.0438	.0478	.0517	.0557	.0596	.0636	.0675	.0714	.0753
0.2	.0793	.0832	.0871	.0910	.0948	.0987	.1026	.1064	.1103	.1141
0.3	.1179	.1217	.1255	.1293	.1331	.1368	.1406	.1443	.1480	.1517
0.4	.1554	.1591	.1628	.1664	.1700	.1736	.1772	.1808	.1844	.1879
0.5	.1915	.1950	.1985	.2019	.2054	.2088	.2123	.2157	.2190	.2224
0.6	.2257	.2291	.2324	.2357	.2389	.2422	.2454	.2486	.2517	.2549
0.7	.2580	.2611	.2642	.2673	.2704	.2734	.2764	.2794	.2823	.2852
0.8	.2881	.2910	.2939	.2967	.2995	.3023	.3051	.3078	.3106	.3133
0.9	.3159	.3186	.3212	.3238	.3264	.3289	.3315	.3340	.3365	.3389
1.0	.3413	.3438	.3461	.3485	.3508	.3531	.3554	.3577	.3599	.3621
1.1	.3643	.3665	.3686	.3708	.3729	.3749	.3770	.3790	.3810	.3830
1.2	.3849	.3869	.3888	.3907	.3925	.3944	.3962	.3980	.3997	.4015
1.3	.4032	.4049	.4066	.4082	.4099	.4115	.4131	.4147	.4162	.4177
1.4	.4192	.4207	.4222	.4236	.4251	.4265	.4279	.4292	.4306	.4319
1.5	.4332	.4345	.4357	.4370	.4382	.4394	.4406	.4418	.4429	.4441
1.6	.4452	.4463	.4474	.4484	.4495	.4505	.4515	.4525	.4535	.4545
1.7	.4554	.4564	.4573	.4582	.4591	.4599	.4608	.4616	.4625	.4633
1.8	.4641	.4649	.4656	.4664	.4671	.4678	.4686	.4693	.4699	.4706
1.9	.4713	.4719	.4726	.4732	.4738	.4744	.4750	.4756	.4761	.4767
2.0	.4772	.4778	.4783	.4788	.4793	.4798	.4803	.4808	.4812	.4817
2.1	.4821	.4826	.4830	.4834	.4838	.4842	.4846	.4850	.4854	.4857
2.2	.4861	.4864	.4868	.4871	.4875	.4878	.4881	.4884	.4887	.4890
2.3	.4893	.4896	.4898	.4901	.4904	.4906	.4909	.4911	.4913	.4916
2.4	.4918	.4920	.4922	.4925	.4927	.4929	.4931	.4932	.4934	.4936
2.5	.4938	.4940	.4941	.4943	.4945	.4946	.4948	.4949	.4951	.4952
2.6	.4953	.4955	.4956	.4957	.4959	.4960	.4961	.4962	.4963	.4964
2.7	.4965	.4966	.4967	.4968	.4969	.4970	.4971	.4972	.4973	.4974
2.8	.4974	.4975	.4976	.4977	.4977	.4978	.4979	.4979	.4980	.4981
2.9	.4981	.4982	.4982	.4983	.4984	.4984	.4985	.4985	.4986	.4986
3.0	.4987	.4987	.4987	.4988	.4988	.4989	.4989	.4989	.4990	.4990

APPENDIX D: VALUES OF $e^{-\lambda}$

λ	$e^{-\lambda}$	λ	$e^{-\lambda}$	λ	$e^{-\lambda}$
0.05	0.9512	2.05	0.1287	4.05	0.0174
0.10	0.9048	2.10	0.1225	4.10	0.0166
0.15	0.8607	2.15	0.1165	4.15	0.0158
0.20	0.8187	2.20	0.1108	4.20	0.0150
0.25	0.7788	2.25	0.1054	4.25	0.0143
0.30	0.7408	2.30	0.1003	4.30	0.0136
0.35	0.7047	2.35	0.0954	4.35	0.0129
0.40	0.6703	2.40	0.0907	4.40	0.0123
0.45	0.6376	2.45	0.0863	4.45	0.0117
0.50	0.6065	2.50	0.0821	4.50	0.0111
0.55	0.5769	2.55	0.0781	4.55	0.0106
0.60	0.5488	2.60	0.0743	4.60	0.0101
0.65	0.5220	2.65	0.0707	4.65	0.0096
0.70	0.4966	2.70	0.0672	4.70	0.0091
0.75	0.4724	2.75	0.0639	4.75	0.0087
0.80	0.4493	2.80	0.0608	4.80	0.0082
0.85	0.4274	2.85	0.0578	4.85	0.0078
0.90	0.4066	2.90	0.0550	4.90	0.0074
0.95	0.3867	2.95	0.0523	4.95	0.0071
1.00	0.3679	3.00	0.0498	5.00	0.0067
1.05	0.3499	3.05	0.0474	5.05	0.0064
1.10	0.3329	3.10	0.0450	5.10	0.0061
1.15	0.3166	3.15	0.0429	5.15	0.0058
1.20	0.3012	3.20	0.0408	5.20	0.0055
1.25	0.2865	3.25	0.0388	5.25	0.0052
1.30	0.2725	3.30	0.0369	5.30	0.0050
1.35	0.2592	3.35	0.0351	5.35	0.0047
1.40	0.2466	3.40	0.0334	5.40	0.0045
1.45	0.2346	3.45	0.0317	5.45	0.0043
1.50	0.2231	3.50	0.0302	5.50	0.0041
1.55	0.2122	3.55	0.0287	5.55	0.0039
1.60	0.2019	3.60	0.0273	5.60	0.0037
1.65	0.1920	3.65	0.0260	5.65	0.0035
1.70	0.1827	3.70	0.0247	5.70	0.0033
1.75	0.1738	3.75	0.0235	5.75	0.0032
1.80	0.1653	3.80	0.0224	5.80	0.0030
1.85	0.1572	3.85	0.0213	5.85	0.0029
1.90	0.1496	3.90	0.0202	5.90	0.0027
1.95	0.1423	3.95	0.0193	5.95	0.0026
2.00	0.1353	4.00	0.0183	6.00	0.0025
				7.00	0.0009
				8.00	0.000335
				9.00	0.000123
				10.00	0.000045

APPENDIX E: RANDOM DIGITS

63271	59986	71744	51102	15141	80714	58683	93108	13554	79945
88547	09896	95436	79115	08303	01041	20030	63754	08459	28364
55957	57243	83865	09911	19761	66535	40102	26646	60147	15702
46276	87453	44790	67122	45573	84358	21625	16999	13385	22782
55363	07449	34835	15290	76616	67191	12777	21861	68689	03263
69393	92785	49902	58447	42048	30378	87618	26933	40640	16281
13186	29431	88190	04588	38733	81290	89541	70290	40113	08243
17726	28652	56836	78351	47327	18518	92222	55201	27340	10493
36520	64465	05550	30157	82242	29520	69753	72602	23756	54935
81628	36100	39254	56835	37636	02421	98063	89641	64953	99337
84649	38968	75215	75498	49539	74240	03466	49292	36401	45525
63291	11618	12613	75055	43915	26499	41116	64531	56827	30825
70502	53225	03655	05915	37140	57051	48393	91322	25653	06543
06426	24771	59935	49801	11082	66762	94477	02494	88215	27191
20711	55609	29430	70165	45406	78484	31639	52009	18873	96927
41990	70538	77191	25860	55204	73417	83920	69468	74972	38712
72452	36618	76298	26678	89334	33938	95567	29380	75906	91807
37042	40318	57099	10528	09925	89773	41335	96244	29002	46453
53766	52875	15987	46962	67342	77592	57651	95508	80033	69828
90585	58955	53122	16025	84299	53310	67380	84249	25348	04332
32201	96293	37203	64516	51530	37069	40261	61374	05815	06714
62606	64324	46354	72157	67248	20135	49804	09226	64419	29457
10078	28073	85389	50324	14500	15562	64165	06125	71353	77669
91561	46145	24177	15294	10061	98124	75732	00815	83452	97355
13091	98112	53959	79607	52244	63303	10413	63839	74762	50289
73864	83014	72457	22682	03033	61714	88173	90835	00634	85169
66668	25467	48894	51043	02365	91726	09365	63167	95264	45643
84745	41042	29493	01836	09044	51926	43630	63470	76508	14194
48068	26805	94595	47907	13357	38412	33318	26098	82782	42851
54310	96175	97594	88616	42035	38093	36745	56702	40644	83514
14877	33095	100924	58013	61439	21882	42059	24177	58739	60170
78295	23179	02771	43464	59061	71411	05697	67194	30495	21157
67524	02865	39593	54278	04237	92441	26602	63835	38032	94770
58268	57219	68124	73455	83236	08710	04284	55005	84171	42596
97158	28672	50685	01181	24262	19427	52106	34308	73685	74246
04230	16831	69085	30802	65559	09205	71829	06489	85650	38707
94879	56606	30401	02602	57658	70091	54986	41394	60437	03195
71446	15232	66715	26385	91518	70566	02888	79941	39684	54315
32886	05644	79316	09819	00813	88407	17461	73925	53037	91904
62048	33711	25290	21526	02223	75947	66466	06232	10913	75336

This table is reproduced with permission from The Rand Corporation, *A Million Random Digits*, The Free Press, New York, 1955 and 1983.